Digital Communications 2

To all our families,
far and wide

Digital Communications 2

Directed and Practical Work

Safwan El Assad
Dominique Barba

iSTE

WILEY

First published 2020 in Great Britain and the United States by ISTE Ltd and John Wiley & Sons, Inc.

ISTE Ltd
27-37 St George's Road
London SW19 4EU
UK

www.iste.co.uk

John Wiley & Sons, Inc.
111 River Street
Hoboken, NJ 07030
USA

www.wiley.com

Library of Congress Control Number: 2020940940

British Library Cataloguing-in-Publication Data
A CIP record for this book is available from the British Library
ISBN 978-1-78630-542-8

Contents

Foreword

We have written this training book on digital communications in the spirit of presenting – in an integrated form – the basic knowledge on which modern digital communication systems are based and, above all, the way in which they are technically implemented, both in principle, and in practice. This book is the product of a long experience of training in this field in engineering school (Polytech Nantes, France).

The training is comprehensive: courses, tutorials presenting many standard problems targeted with detailed solutions, practical work concretely illustrating various aspects of the techniques of implementation.

As we have mentioned, although our experience is primarily that of training engineers, we have, through adaptations of the content, wished to address broader audiences: first in initial training, engineers, Master 2, specialized telecommunications licenses or other related specialties. But also to the trainers by providing them, through tutorials and practices (Lab Works), content that can be very useful in the construction of the training they provide. In continuing education, this book is also addressed to telecommunication technicians or for an additional year of specialization (specific years complementary to training in IUT).

This book, which is composed of two associated volumes, is presented in its first aspect, as a very concise and complete synthesis of the foundations and techniques of digital communications (Volume 1). It is broken down into two parts. The first part concerns the theory of information itself, which deals with both sources of information and communication channels, in terms of the errors they introduce in the transmission of information, as well as ways to protect the latter by using appropriate coding methods. The second part deals with the technical aspects of transmission, we first present the baseband transmission with the important concept of equalization and its implementations. The performance evaluation, in terms of probability of errors, is systematically developed and detailed as well as the on-line codes used. We then present the transmissions with digital modulation of carriers used in transmission (radio transmissions but also on electric cables).

A second important aspect, teaching knowledge and skills, composes this book (first part of Volume 2). It concerns the tutorial aspect of a course. This is an ordered set of about 30 standard problems with detailed solutions covering the different parts of the course. The set should allow a learner to gradually and deeply understand the essentials of this field and acquire the necessary skills to practice them in the industrial world.

Finally, the last aspect concerns practices in the proper sense of the term, an indispensable complement to training progressing to know-how (second part of Volume 2). We propose here a set of five lab works. The interest of these is that they go from the basic measurements on the transmission cables, to the design in software simulation of modems and cyclic coders, through the use of blocks of electronic modules carrying out basic functions useful in digital communications.

For every book sold, we will provide the buyer with two practical pieces of software from MATLAB-Simulink: “Modem QPSK” and “Cyclic encoder-decoder”, free of charge. We will provide necessary explanations and endeavor to help with the set-up of the two pieces of practical material.

PART 1

Tutorials

1

Theory of Information: Problems 1 to 15

1.1. Problem 1 – Entropy

We consider the information transmission channel of memoryless binary symmetrical type of Figure 1.1.

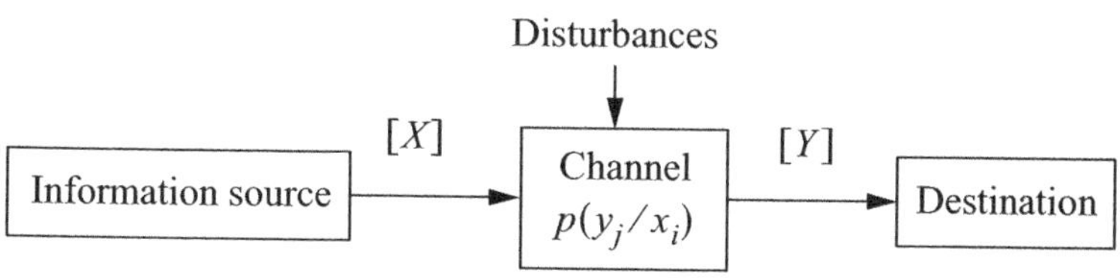

Figure 1.1. *Basic diagram of a digital communication*

It is assumed that the signal-to-noise ratio leads to the following values of conditional probabilities of errors:

$$p(y_j = 1/x_i = 0) = p(y_j = 0/x_i = 1) = p$$

$$p(y_i/x_i) = 1 - p$$

The source of binary information is considered to emit independent information with the following probabilities:

$$p(x_1) = p_1 \quad \text{and} \quad p(x_2) = p_2 = 1 - p_1$$

1) Calculate the source entropy $H(X)$.

2) Calculate the entropy $H(Y)$ at the receiver end.

3) Calculate the conditional entropy $H(Y/X)$ (entropy of transmission error).

4) Calculate the loss of information in the transmission channel $H(X/Y)$.

5) Deduce the average amount of information received by the recipient for each binary symbol sent $I(X,Y)$ (mutual information).

6) Determine the channel capacity C and show that it is obtained when $p_1 = 0.5$.

Solution of problem 1

1) By definition, we have:

$$H(X) = -\sum_{i=1}^{2} p(x_i) \log_2 p(x_i)$$

then:

$$H(X) = -\{p_1 \log_2 p_1 + (1 - p_1) \log_2(1 - p_1)\} = H(p_1)$$

2) By definition, we have:

$$H(Y) = -\sum_{j=1}^{2} p(y_j) \log_2 p(y_j)$$

and:

$$p(y_j) = \sum_{i=1}^{2} p(x_i) \times p(y_j/x_i)$$

hence:

$$H(Y) = -\{[p_1(1-p) + (1-p_1)p] \times \log_2[p_1(1-p) + (1-p_1)p]$$
$$+[p_1 p + (1-p_1)(1-p)]$$
$$\times \log_2[p_1 p + (1-p_1)(1-p)]\}$$

3) In the same way, we have:

$$H(Y/X = x_i) = -\sum_{j=1}^{2} p(y_j/x_i) \times \log_2 p(y_j/x_i)$$

and:

$$H(Y/X) = \sum_{i=1}^{2} p(x_i)\, H(Y/X = x_i)$$

Since we are dealing with a binary symmetric communication channel, it turns out that:

$$H(Y/X) = H(Y/X = x_i) = -\{(1-p)\log_2(1-p) + p\log_2 p\} = H(p)$$

4) We have:

$$H(X/Y) = -\sum_{i=1}^{2}\sum_{j=1}^{2} p(x_i) \times p(y_j/x_i)\log_2\left[\frac{p(x_i) \times p(y_j/x_i)}{p(y_j)}\right]$$

That is:

$$H(X/Y) = -\left\{p_1(1-p)\log_2\left[\frac{p_1(1-p)}{p_1(1-p) + (1-p_1)p}\right]\right.$$

$$+ p_1 p\log_2\left[\frac{p_1 p}{p_1 p + (1-p_1)(1-p)}\right]$$

$$+(1-p_1)p\log_2\left[\frac{(1-p_1)p}{p_1(1-p) + (1-p_1)p}\right]$$

$$\left.+(1-p_1)(1-p)\log_2\left[\frac{(1-p_1)(1-p)}{p_1 p + (1-p_1)(1-p)}\right]\right\}$$

5) By definition, we have:

$$I(X,Y) = H(Y) - H(Y/X)$$

6) By definition, we have:

$$C = \underset{\{p(x_i)\}}{Max}\, I(X,Y) = \underset{\{p(x_i)\}}{Max}\, H(Y) - H(Y/X)$$

$\underset{\{p_1\}}{Max}\, H(Y)$ is got for p_1 such that $\frac{\partial H(Y)}{\partial p_1} = 0$

$$\frac{\partial H(Y)}{\partial p_1} = -\left\{(1-2p)\log_2\left[\frac{(1-p)p_1 + p(1-p_1)}{pp_1 + (1-p)(1-p_1)}\right]\right\} = 0$$

You need to have the numerator of the log function equal to the denominator, hence:

$$2p_1(1-2p) = 1-2p \text{ ; hence } p_1 = 1/2$$

Thus, the maximum defines the capacity C of the communication channel and is obtained for:

$$p_1 = 1/2 \text{ , hence } Max\ H(Y) = 1 \text{ and therefore: } C = 1 - H(p)$$

1.2. Problem 2 – K-order extension of a transmission channel

A memoryless binary symmetric transmission channel is considered: whatever the binary information to be transmitted, the probability of the transmission error is constant, equal to p.

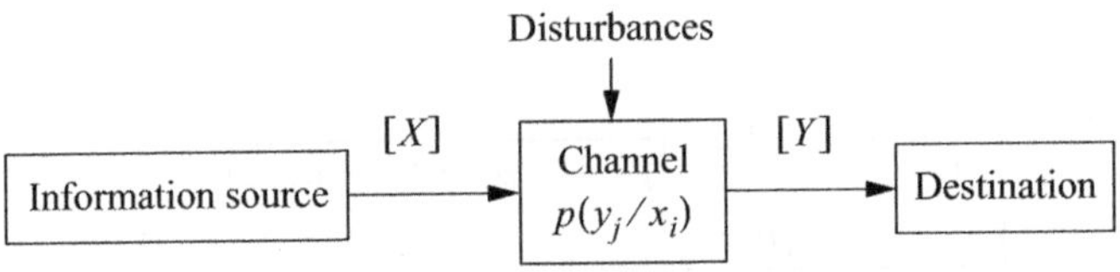

Figure 1.2. *Basic block diagram of a digital communication of a memoryless information source*

A. *K-order extension of a memoryless binary symmetric channel of error probability p*

The k-order extension channel has an input alphabet of 2^k binary words of length k and an output alphabet identical to that of the input alphabet. This channel is thus represented by a square matrix P_k of dimension $[2^k, 2^k]$ whose element p_{ij} corresponds to the probability of receiving y_j conditionally to have x_i transmitted $p(y_j/x_i)$.

1) If d is the Hamming distance between the two binary words of length k corresponding for one to the symbol x_i, and for the other to the symbol y_j, express the probability p_{ij} according to the three parameters: p, k, d.

B. *Second-order extension of a memoryless binary symmetric channel*

2) Write completely in literal form as a function of p the matrix P_2 representative of the second order extension of the binary symmetric channel.

3) The information source is considered to be transmitting equiprobable quaternary symbols x_i in the channel. Calculate the probability $p(y_j)$ to receive a symbol y_j.

4) Deduce the relationship which exists between the elements p_{ij} of the matrix P_2 representative of the second order extension of the binary symmetric channel and the probability $p(x_i/y_j)$ that the symbol x_i was emitted conditionally having received y_j.

5) Calculate the average amount of information $H(X/Y)$ lost in the channel due to transmission errors. You will express $H(X/Y)$ as a function of:

$$H(p) = -\{(1-p)\log_2(1-p) + p\log_2 p\}$$

C. *Fourth-order extension of a memoryless binary symmetric channel*

The size of the input alphabet of the source is then 16. The output alphabet is the same as that of the input alphabet.

The source is considered to emit equiprobable symbols x_i.

6) We extrapolate the result obtained in B-5 by considering that we have:

$$H(X/Y) = kH(p)$$

In the case $p = 0.03$, calculate the statistical mean of the information amount $H(X/Y)$ lost per symbol sent.

7) What is the entropy $H(X)$ of the source?

8) What is the maximum number of possible errors on a symbol received?

Solution of problem 2

A. K-order extension

1) The symbol x_i is made up of k bits. It is the same for the symbol y_j, so:

$$p_{ij} = p(y_j/x_i) = p(y_{j,1}, y_{j,2}, \cdots, y_{j,k}/x_{i,1}, x_{i,2}, \cdots, x_{i,k})$$

The communication channel is memoryless, so the probability of obtaining a given bit at the output depends only on the bit transmitted at the input (in addition to the intrinsic properties of the transmission channel itself), hence:

$$p(y_j/x_i) = p(y_{j,1}/x_{i,1}) \times p(y_{j,2}/x_{i,2}) \times \cdots \times p(y_{j,k}/x_{i,k})$$
$$= \prod_{n=1}^{k} p(y_{j,n}/x_{i,n})$$

because of the independence between the source of information and the communication channel.

The Hamming distance $d = d_H(y_j, x_i)$ is the number of bits of the same rank that are different between the symbol y_j and symbol x_i.

Then:

$$p(y_j/x_i) = p^d(1-p)^{k-d}$$

This law is close to the Binomial law because if p is the probability of a wrong decision on bit b, then $(1-p)$ is the probability of a right decision on bit b.

B. Second-order extension of the channel

2) We have:

$$k = 2 \rightarrow p(y_j/x_i) = p^d(1-p)^{2-d} \rightarrow \text{the matrix } P_2 \text{ (see Table 1.1)}$$

$$\sum_{j=1}^{4} p(y_j/x_i) = 1 = \sum_{i=1}^{4} p(y_j/x_i)$$

because of the symmetry.

	j	1	2	3	4
	y_j	0 0	0 1	1 0	1 1
i	x_i				
1	0 0	$(1-p)^2$	$p(1-p)$	$p(1-p)$	p^2
2	0 1	$p(1-p)$	$(1-p)^2$	p^2	$p(1-p)$
3	1 0	$p(1-p)$	p^2	$(1-p)^2$	$p(1-p)$
4	1 1	p^2	$p(1-p)$	$p(1-p)$	$(1-p)^2$

Table 1.1. *Matrix P_2 representative of second-order extension of a binary symmetric channel*

3) We have:

$$p(x_i, y_j) = p(x_i) \times p(y_j/x_i) = p(y_j) \times p(x_i/y_j)$$

$$p(y_j) = \sum_{i=1}^{4} p(x_i, y_j) = \sum_{i=1}^{4} p(x_i) \times p(y_j/x_i)$$

Yet, the symbols are equiprobable:

$$p(x_i) = \frac{1}{4} \ \forall\, i = 1, \cdots, 4$$

Then, the symbols y_j are also equiprobable:

$$p(y_j) = \frac{1}{4}[(1-p)^2 + 2p(1-p) + p^2] = \frac{1}{4} \ \forall\, j = 1, \cdots, 4$$

4) We have:

$$p(x_i/y_j) = \frac{p(x_i) \times p(y_j/x_i)}{p(y_j)} = p_{ij}$$

because:

$$p(x_i) = p(y_j) = 1/4$$

5) Average amount of bit of information $H(X/Y)$ lost in the transmission channel.

We have:

$$H(X/Y = y_j) = -\sum_{i=1}^{4} p(x_i/y_j) \log_2 p(x_i/y_j)$$

$$H(X/Y) = E\{H(X/Y = y_j)\} = \sum_{j=1}^{4} p(y_j) H(X/Y = y_j)$$

$$H(X/Y) = -\frac{1}{4} \sum_{j=1}^{4} \sum_{i=1}^{4} p(y_j/x_i) \log_2 p(y_j/x_i)$$

because here we have:

$$p(x_i/y_j) = p(y_j/x_i)$$

$$H(X/Y) = -\frac{1}{4}[(1-p)^2\log_2(1-p)^2$$
$$+2p(1-p)\log_2 p(1-p) + p^2\log_2 p^2] \times 4$$

$$H(X/Y) = -2\{(1-p)^2\log_2(1-p) + p(1-p)[\log_2 p + \log_2(1-p)]$$
$$+p^2\log_2 p\}$$

$$H(X/Y) = -2\{(1-p)[(1-p)\log_2(1-p) + p\log_2 p]$$
$$+p[(1-p)\log_2(1-p) + p\log_2 p]\}$$

$$H(X/Y) = 2[(1-p)H(p) + pH(p)] = 2H(p) = kH(p)$$

C. Fourth-order extension of the transmission channel

6) $p = 0.03$ and $H(X/Y) = 4H(p)$.

Average amount of information (in bit of information) lost per binary symbol sent?

We have:

$$H(X/Y) = -4[0.97 \times \log_2(0.97) + 0.03 \times \log_2(0.03)]$$
$$= 0.7777 \text{ bit of information/symbol}$$

7) Entropy of the source?

$$H(X) = H(S^4) = 4H(S)$$

and:

$$H(S) = -\sum_{i=1}^{2} p(b_i)\log_2 p(b_i) = 1 \text{ because } p(b_1) = p(b_2) = \frac{1}{2}$$

hence:

$$H(X) = 4 \text{ bits of information/symbol}$$

8) Maximum number of possible errors?

$$d_{max} = 4$$

1.3. Problem 3 – Compressed speech digital transmission and Huffman coding

In the context of the transmission of the highly compressed speech signal over the telephone channel entirely in digital form, let us look at the problem of statistical source coding.

An information source S delivering elementary symbols s belonging to a symbol dictionary of size 6 is considered. The probabilities of transmission of this simple source of information are given in Table 1.2.

s_i	s_1	s_2	s_3	s_4	s_5	s_6
$Pr\{s_i\}$	0.05	0.20	0.22	0.33	0.15	0.05

Table 1.2. *Probabilities of emitting symbols s by the information source*

The symbols are delivered by the source S every $T = 10^{-3}$ s.

1) Determine the entropy $H(S)$ of the source. Deduce the entropy bitrate D_S.

2) Construct the statistical Huffman coding, called code C_1, which generates a binary code associated with each symbol s_i.

3) Deduce the average length $\overline{l_1}$ of code C_1 and the bitrate D_1 per second.

4) What are the efficiency η_1 and redundancy ρ_1 of code C_1?

5) If we chose a fixed-length code (code C_2), what would be its efficiency η_2? What do you conclude?

6) Would it be possible to transmit this source of information over a transmission channel having a bitrate capacity of 2,400 bit/second?

Solution of problem 3

1) The entropy of the source is:

$$H(S) = -\sum_{i=1}^{6} p(s_i) \log_2 p(s_i)$$

Recall:

$$\log_2(Z) = \frac{\log_e(Z)}{\log_e(2)} \quad \text{and} \quad \frac{1}{\log_e(2)} \cong 1.44$$

$$H(S) \cong -1.44[0.05\log_e(0.05) + 0.2\log_e(0.2) + 0.22\log_e(0.22)$$
$$+0.33\log_e(0.33) + 0.15\log_e(0.15) + 0.05\log_e(0.05)]$$
$$\cong 2.31 \text{ bits of information/symbol}$$

The entropy bitrate of the source is:

$$D_s = \frac{H(S)}{T} = 2.31 \times 10^3 = 2.31 \text{ Kbits of information/s}$$

2) Construction of the Huffman code.

Symbol s_i	$p(s_i)_0$	$p(s_i)_1$	$p(s_i)_2$	$p(s_i)_3$	$p(s_i)_4$	Code C_1
s_4	0.33 →	0.33 →	0.33	0.42	0.58 0	0 0
s_3	0.22 →	0.22	0.25	0.33 0	0.42 1	1 0
s_2	0.20 →	0.20	0.22 0	0.25 1		1 1
s_5	0.15 →	0.15 0	0.20 1			0 1 0
s_1	0.05 0	0.10 1				0 1 1 0
s_6	0.05 1					0 1 1 1

Table 1.3. *Construction of the Huffman code* C_1

3) Average length of codewords:

$$\overline{l_1} = \sum_{i=1}^{6} p(s_i) \times l_i = 0.05 \times 4 + 0.20 \times 2 + 0.22 \times 2 + 0.33 \times 2$$
$$+ 0.15 \times 3 + 0.05 \times 4 = 2.35 \text{ bit/symbol}$$

Bitrate per second:

$$D_1 = \frac{\overline{l_1}}{T} = 2.35 \text{ Kbit/second}$$

4) Efficiency and redundancy of the Huffman code:

$$\eta_1 = \frac{H(S)}{\overline{l_1}} = \frac{2.31}{2.35} \cong 98.3\ \%$$

$$\rho_1 = 1 - \eta_1 = 0.017$$

5) Fixed-length code C_2.

Since we have 6 messages, we need 3 bits as: $2^2 < 6 < 2^3$, then:

$$\eta_2 = \frac{H(S)}{3} = \frac{2.31}{3} = 77\ \%$$

The fixed-length code C_2 is less efficient than the Huffman code C_1.

The bitrate per second with code C_2 is: $3 \times 1{,}000 = 3\ \text{Kbit/s}$.

6) The capacity of the channel is 2.4 Kbit/s, so we can transmit the code C_1 but not the code C_2 because the bitrate of C_2 is more important than the capacity of the channel.

1.4. Problem 4 – Coding without and with information compression

We consider a digital communication system, designed for the transmission of a signal $s(t)$ in digital form on a 34 Mbit/s transmission channel. Subsequently, we are only interested in a part of the transmitter, composed of a device for digitization and serialization (sampling, linear quantization on 8 bits, parallel to serial bytes transformation) represented in Figure 1.3. The sampling frequency is 10 MHz.

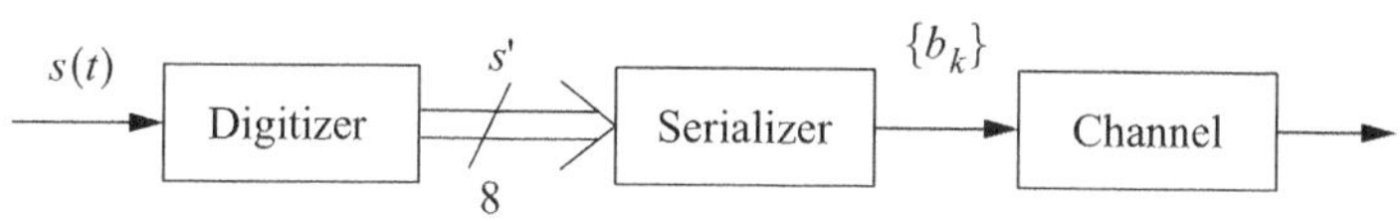

Figure 1.3. *Block diagram of a digital transmission system for analog signal*

1) With the system in Figure 1.3, is it possible to transmit this signal on the channel?

The bitrate D_1 is important, so we try to reduce it. For this purpose, a coding system with information compression of DPCM type (Differential Pulse Code

Modulation) is interposed between the digitization and serialization blocks. The DPCM coding system transforms the 256-level representation $s'(t_k)$ into a 9-level representation $s(t_k)$. The symbol s (corresponding to the encoded amplitude of the sample $s(t_k)$), is represented according to a natural binary code.

2) What is the bitrate D_2 at the output of the serialization unit?

To further reduce the bitrate, a block coding C which groups two consecutive symbols to form bijectively a single code symbol $S(t_k)$ is inserted after the encoding system DPCM (thus it has a frequency half that of s) : $\{s(t_{2k}), s(t_{2k+1})\} \leftrightarrow S(t_k)$.

3) The coding C does not using any statistical properties of s, what is its bitrate D_3?

To further reduce the bitrate, a Huffman code C_4 is used as the code C but without grouping by two the symbols s. The probabilities of realization of s are the following:

$$Pr(s = s_1) = Pr(s = s_3) = Pr(s = s_4) = 0.0625$$

$$Pr(s = s_2) = Pr(s = s_5) = 0.125$$

$$Pr(s = s_6) = Pr(s = s_9) = 0.03125$$

$$Pr(s = s_7) = Pr(s = s_8) = 0.25$$

4) Construct the Huffman code C_4. You will explicitly determine the codewords associated with each of the possible realizations of s.

5) Determine the average length $\overline{l_4}$ of the codewords of C_4 and the entropy $H(s)$.

6) What is the bitrate D_4 of the code C_4? What is its efficiency η_4? Can the signal be transmitted on the transmission channel?

7) We want to protect the binary information transmitted against transmission errors. The block encoding technique is used. This technique adds 15 bits of protection (packet error detection code) to a packet of 240 useful bits. What is the new average bitrate D_5 and is it compatible with the transmission channel capacity?

Solution of problem 4

1) We have:

$$D_1 = 8 \times 10^7 = 80 \text{ Mbit/s}$$

The rate D_1 is greater than the channel capacity, thus we cannot transmit the signal on this channel.

2) It is a 9-level encoding, therefore it takes 4 bits per sample (fixed-length code), hence:

$$D_2 = 4 \times 10^7 = 40 \text{ Mbit/s}$$

3) We have:

$$\{s(t_{2k}), s(t_{2k+1})\} \quad \overset{Code\ C}{\leftrightarrow} \quad S(t_k)$$

The pair $\{s(t_{2k}), s(t_{2k+1})\}$ has $9 \times 9 = 81$ different configurations possible, and since: $2^6 < 81 < 2^7$, it takes 7 bits to encode a pair of samples, hence:

$$D_3 = 7 \times \frac{1}{2} \times 10^7 = 35 \text{ Mbit/s}$$

4) Huffman coding.

s_i	$p(s_i)_0$	$p(s_i)_1$	$p(s_i)_2$	$p(s_i)_3$	$p(s_i)_4$	$p(s_i)_5$	$p(s_i)_6$	$p(s_i)_7$	C_4
s_8	0.25	0.25	0.25	0.25	0.25	0.25	0.5	0.5 0	10
s_7	0.25	0.25	0.25	0.25	0.25	0.25	0.25 0	0.5 1	11
s_5	0.125	0.125	0.125	0.125	0.25	0.25 0	0.25 1		010
s_2	0.125	0.125	0.125	0.125	0.125 0	0.25 1			011
s_1	0.0625	0.0625	0.125	0.125 0	0.125 1				0010
s_3	0.0625	0.0625	0.625 0	0.125 1					0011
s_4	0.0625	0.0625 0	0.625 1						0000
s_6	0.03125 0	0.0625 1							00010
s_9	0.03125 1								00011

Table 1.4. *Construction of the Huffman code C_4*

5) By definition, we have for the average length:

$$\overline{l_4} = \sum_{i=1}^{9} p(s_i) \times l_i$$

$$\overline{l_4} = 2 \times \left(\frac{1}{4} \times 2\right) + 2 \times \left(\frac{1}{8} \times 3\right) + 3 \times \left(\frac{1}{16} \times 4\right) + 2 \times \left(\frac{1}{32} \times 5\right)$$
$$= 2.8125 \text{ bit/codeword}$$

and for the entropy :

$$H(s) = -\sum_{i=1}^{9} p(s_i) \log_2 p(s_i)$$

Thus, by replacing:

$$H(s) = -\{2 \times 2^{-2} \log_2 2^{-2} + 2 \times 2^{-3} \log_2 2^{-3} + 3 \times 2^{-4} \log_2 2^{-4} + 2 \times 2^{-5} \log_2 2^{-5}\}$$

$$H(s) = 4 \times \frac{1}{4} + 6 \times \frac{1}{8} + 12 \times \frac{1}{16} + 10 \times \frac{1}{32}$$
$$= 2.8125 \text{ bits of information/codeword}$$

6) Bitrate D_4 of the code C_4 and its efficiency η_4 for this source:

$$D_4 = \overline{l_4} \times 10^7 = 28.125 \text{ Mbit/s}$$

$$\eta_4 = \frac{H(s)}{\overline{l_4}} = 1$$

The code C_4 is optimal absolute, because the probabilities are of the form: $p_i = 2^{-l_i}$.

Since the bitrate D_4 is smaller than the capacity of the channel, it turns out that the signal can be transmitted on the channel.

7) Block coding for protection against transmission errors:

$$D_5 = D_4 \times \frac{255}{240} = 29.883 \text{ Mbit/s}$$

In the same way, since the bitrate D_5 is smaller than the capacity of the channel, it turns out this signal can also be transmitted in a protected manner on the communication channel.

1.5. Problem 5 – Digital transmission of a TV signal (luminance component only) with information compression and Huffman coding

An information encoding and transmitting system for transmitting a monochrome television signal $s(t)$ in digital form is considered. The general scheme of the preliminary part of this system is given in Figure 1.4.

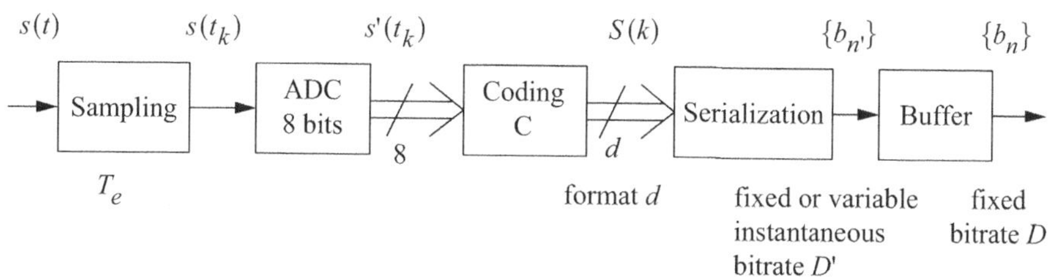

Figure 1.4. *General scheme of a digital transmission of a TV signal with information compression*

The analog signal (luminance component) is sampled with a sampling period $T_e = 100$ ns. In an analog/digital converter, each sample is then quantized linearly and converted to an integer $s'$of 8 bits (natural binary code). A coding block C converts this number of 8 bits into another binary codeword S of fixed or variable length d depending on the cases that we will examine. The codeword S of format d is then serialized and thus generates a bit stream with a fixed or variable bitrate D', depending on the case selected. A buffer is used to output a fixed bitrate D sequence such that D can be considered equal to $E[D']$ (E is the expected value). The transmission channel has a capacity of 34 Mbit/s of which only 32 Mbit/s can be used for the transmission of the video signal itself.

1) We first consider a very simplified version where the coding block C does not exist: the word $S(k)$ is strictly identical to the binary representation $s'(t_k)$ of the sampled signal $s(t_k)$.

What is the bitrate D' (in bit/s) at the output of the serialization block and the fixed bitrate D at the output of the buffer?

The bitrate D being considered too significant one seeks to reduce it. A Huffman encoding C_2 is used, constructed from the knowledge (by estimation) of the amplitude probability law represented by the discrete random variable associated with s' is used. The entropy $H(s')$ is equal to 6.65 bit of information per amplitude and the efficiency η_2 of the code C_2 is 0.95.

2) What is the average length $\bar{l_2} = E(d)$ of the codewords S? Deduce the fixed bitrate D_2.

Since the bitrate is still too big, a differential pulse code modulation (DPCM) coding system with information compression type, shown in Figure 1.5, is used.

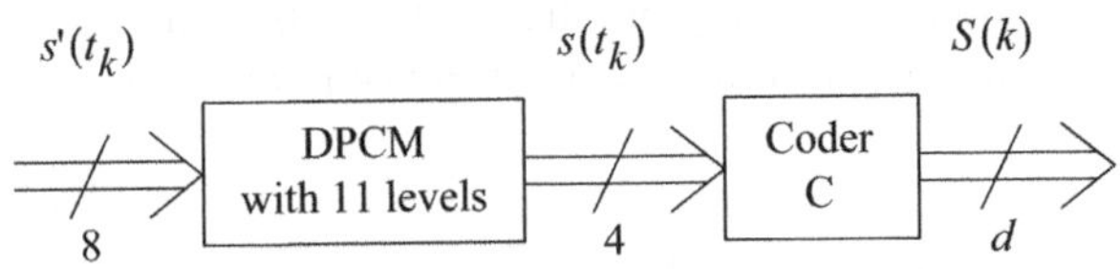

Figure 1.5. *Information compression using a DPCM system and a Huffman code C_4*

From a 256-level representation, the DPCM system generates a representation of $s(t_k)$ with 11 levels. The number s is represented according to a natural binary code.

3) We first consider in Figure 1.5 that the coding C does not exist. What are the bitrate D_3' at the output of the serialization block and the fixed bitrate D_3? Is it too large?

An alternative is now considered to further reduce the bitrate D_3. The coding C (called coding C_3) groups two consecutive symbols s to form bijectively a single codeword S (thus this one has a frequency half of that of s) : $\{s(t_{2k}), s(t_{2k+1})\} \leftrightarrow S(t_k)$.

4) Since the code C_3 does not use any statistical properties of s, show that the minimum length l_3 of the codeword s is 7 bits. What is the fixed bitrate D_{31}? Are we able to transmit the image on the transmission channel?

To further reduce the bitrate, a Huffman code C_4 is used as code C but without grouping the s symbols by two. The probabilities of realization of s are as follows:

$$Pr(s = s_1) = Pr(s = s_2) = Pr(s = s_{10}) = Pr(s = s_{11}) = 0.03125$$

$$Pr(s = s_3) = Pr(s = s_4) = Pr(s = s_8) = Pr(s = s_9) = 0.0625$$

$$Pr(s = s_5) = Pr(s = s_7) = 0.125$$

$$Pr(s = s_6) = 0.375$$

5) Design the Huffman code C_4. You will determine explicitly the codewords associated with each of the possible realizations of s.

6) What is the average length $\overline{l_4}$ of the codewords S. What is the efficiency η_4 of code C_4 since the entropy $H(s)$ is 2.905 bit/amplitude?

7) What is the fixed bitrate D_4? Are we able to transmit the image on the channel?

We want to protect the binary information transmitted against transmission errors. A coding block is used which adds a 16-bit protection to a useful 256-bit packet (packet error detector code).

8) What is the new average bitrate D_{41} and is it compatible with the capacity of the transmission channel?

Solution of problem 5

1) Bitrate at the output?

$$T_e = 100 \text{ ns} \rightarrow f_e = 10 \text{ MHz}$$

$$D' \text{ is fixed} \rightarrow D = D' = 8 \times 10^7 = 80 \text{ Mbit/s}$$

2) Average length of codewords and fixed bitrate?

$$H(s') = -\sum_{i=1}^{256} p(s_i') \log_2 p(s_i') = 6.65 \text{ bits of information/amplitude}$$

$$\text{Code } C_2$$

$$s \quad \rightarrow \quad S$$

$$\eta_2 = \frac{H(s')}{\overline{l_2}} \quad \rightarrow \quad \overline{l_2} = \frac{H(s')}{\eta_2} = \frac{6.65}{0.95} = 7 \text{ bit/amplitude}$$

$$D_2 = \overline{l_2} \times f_e = 7 \times 10^7 = 70 \text{ Mbit/s}$$

3) 11-level DPCM coding, thus 4 bits per sample are needed because $2^3 < 11 < 2^4$.

The bitrate D'_3 is fixed, hence:

$$D'_3 = D_3 = 4 \times 10^7 = 40 \text{ Mbit/s}$$

Yes, the bitrate D_3 is too large because it is greater than the capacity of the channel.

4) Coding C_3:

$$\{s_{2k}, s_{2k+1}\} \quad \overset{Code\ C_3}{\leftrightarrow} \quad S_k$$

The pair $\{s_{2k}, s_{2k+1}\}$ has $11 \times 11 = 121$ different configurations possible and since: $2^6 < 121 < 2^7$, 7 bits are necessary to encode a pair of samples, so:

$$l_3 = 7 \text{ bit/pair of samples.}$$

D'_{31} is fixed, hence:

$$D_{31} = D'_{31} = 7 \times \frac{f_e}{2} = 7 \times \frac{1}{2} \times 10^7 = 35 \text{ Mbit/s}$$

It is not possible to transmit the image on the channel because the bitrate D_{31} is greater than the capacity of the channel.

5) Huffman coding C_4.

6) Average length $\overline{l_4}$ and efficiency η_4:

$$\overline{l_4} = \sum_{i=1}^{11} p(s_i) \times l_i$$

$$\overline{l_4} = 0.375 \times 1 + 0.125 \times 3 + 0.125 \times 4 + 4 \times 0.0625 \times 4 + 4 \times 0.03125 \times 6 = 3 \text{ bit/codeword}$$

$$\eta_4 = \frac{H(s)}{\overline{l_4}} = 96.83\ \%$$

7) Fixed bitrate: $D_4 = \overline{l_4} \times f_e = 3 \times 10^7 = 30$ Mbit/s.

The bitrate D_4 is lower than the capacity of the channel, therefore we can transmit it on this communication channel.

8) Block coding protection:

$$D_{41} = D_4 \times \frac{272}{256} = 31.875 \text{ Mbit/s}$$

The D_{41} bitrate is less than the channel capacity, so it is compatible with the channel transmission.

s_i	$p(s_i)_0$	$p(s_i)_1$	$p(s_i)_2$	$p(s_i)_3$	$p(s_i)_4$	$p(s_i)_5$	$p(s_i)_6$	$p(s_i)_7$	$p(s_i)_8$	$p(s_i)_9$
s_6	0.375	0.375	0.375	0.375	0.375	0.375	0.375	0.375	0.375	0.625 0
s_5	0.125	0.125	0.125	0.125	0.125	0.125	0.25	0.25	0.375 0	0.375 1
s_7	0.125	0.125	0.125	0.125	0.125	0.125	0.125	0.25 0	0.25 1	
s_3	0.0625	0.0625	0.0625	0.125	0.125	0.125	0.125 0	0.125 1		
s_4	0.0625	0.0625	0.0625	0.0625	0.125	0.125 0	0.125 1			
s_8	0.0625	0.0625	0.0625	0.0625	0.0625 0	0.125 1				
s_9	0.0625	0.0625	0.0625	0.0625 0	0.0625 1					
s_1	0.03125	0.0625	0.0625 0	0.0625 1						
s_2	0.03125	0.03125 0	0.0625 1							
s_{10}	0.03125 0	0.03125 1								
s_{11}	0.03125 1									

s_i	s_6	s_5	s_7	s_3	s_4	s_8
Code C_4	1	001	0000	0110	0111	0100
s_i	s_9	s_1	s_2	s_{10}	s_{11}	
Code C_4	0101	000110	000111	000100	000101	

Table 1.5. *Construction of Huffman code C_4*

1.6. Problem 6 – Information, entropy, codes (1)

A color image coding system is considered for both storage and efficient transmission over a transmission channel. A bank of still images considered as an $S1$ source of information, are in VGA format (video graphics array) 640 × 480 pixels with only 16 color levels per pixel (luminance and chrominance jointly).

The statistics on this bank of images show that out of the 16 colors:

– 4 are used 60% of the time, with equal frequency;

– 4 others are used 30% of the time, with equal frequency;

– the others are used 10% of the time, also with equal frequency.

1) What is the amount Q_1 of binary information required to store an image with a fixed format binary code (code C_1)?

We want to reduce this amount by using a variable length code like a Huffman code.

2) Construct the code associated with this type of information (code C_2). For that, you can use a simple technique of grouping words to encode a class of words (important gain of time).

Deduce the average length $\overline{l_2}$, the amount Q_2 of binary information needed to store an image and the compression rate τ given by this code.

3) What is the entropy H of this source of information (per pixel)?

Deduce the efficiency η_2 of code C_2.

One wants to transmit the coded images with code C_2 to a recipient through a memoryless binary symmetric channel (BSC) having a fixed bitrate D. Let $S2$ be the binary information source that is at the serial output of the Huffman coding.

4) What are for $S2$ the probability p_0 to issue $x_i = 0$ and the probability p_1 to issue $x_i = 1$?

The transmission channel is a memoryless binary symmetric channel (BSC). It introduces transmission errors with an error probability p (the numerical application will be $p = 10^{-4}$).

5) Determine the entropies $H(X)$, $H(Y)$ and $H(Y/X)$.

6) Determine the amount of information received by the recipient for each binary symbol sent $I(X, Y)$, as well as the entropy $H(X/Y)$ (called ambiguity).

7) What is the average loss of information per image transmitted?

8) Determine the average number of received pixels per image, whose value is wrong.

9) Would it be possible to add a protection code after the coding C_2?

What do you suggest and justify your proposal?

Does it work at a codeword level or a block-code level?

Solution of problem 6

1) VGA image: $640 \times 480 = 307\ 200$ pixels, 16 colors per pixel.

4 bit/pixel (because $16 = 2^4$) are necessary. Thus it needs:

$$Q_1 = 307{,}200 \times 4 = 1{,}228{,}800 \text{ bits} = 153{,}600 \text{ bytes}$$

2) The 16 colors are divided into 3 groups:

$$g_1 \leftrightarrow (c_0, \cdots, c_3)$$

$$g_2 \leftrightarrow (c_4, \cdots, c_7)$$

$$g_3 \leftrightarrow (c_8, \cdots, c_{15})$$

Construction of Huffman's code on groups: code C_2.

Group g	$p(g)_0$	$p(g)_1$	Code
g_1	0.6	0.6 → 0	0
g_2	0.3 → 0	0.4 → 1	1 0
g_3	0.1 → 1		1 1

Group g_1	Code
c_0	0 0 0
c_1	0 0 1
c_2	0 1 0
c_3	0 1 1

Group g_2	Code
c_4	1 0 0 0
c_5	1 0 0 1
c_6	1 0 1 0
c_7	1 0 1 1

Group g_3	Code
c_8	1 1 0 0 0
c_9	1 1 0 0 1
c_{10}	1 1 0 1 0
c_{11}	1 1 0 1 1
c_{12}	1 1 1 0 0
c_{13}	1 1 1 0 1
c_{14}	1 1 1 1 0
c_{15}	1 1 1 1 1

Table 1.6. *Construction of Huffman code C_2. For a color version of this table, see www.iste.co.uk/assad/digital2.zip*

So, the average length of this coding is:

$$\overline{l_2} = \sum_{i=0}^{15} p_i \times l_i = 4 \times \left(\frac{0.6}{4} \times 3\right) + 4 \times \left(\frac{0.3}{4} \times 4\right) + 8 \times \left(\frac{0.1}{8} \times 5\right)$$

$$= 3.5 \text{ bit/color } = 3.5 \text{ bit/pixel}$$

NOTE.– This average length would actually be equal to 3.45 bit/color (or pixel) for direct Huffman coding on colors c_0 to c_{15}.

Thus, with this code, one needs:

$$Q_2 = 307{,}200 \times \overline{l_2} = 307{,}200 \times 3.5 = 1{,}075{,}200 \text{ bits} = 134{,}400 \text{ bytes}$$

The compression rate τ is given by:

$$\tau = \frac{Q_1}{Q_2} = \frac{4}{3.5} = 1.142857$$

3) The entropy of this code is:

$$H(c) = -\sum_{i=1}^{15} p(c_i) \log_2 p(c_i)$$

$$H(c) = -\left\{4 \times \frac{0.6}{4} \log_2\left(\frac{0.6}{4}\right) + 4 \times \frac{0.3}{4} \log_2\left(\frac{0.3}{4}\right) + 8 \times \frac{0.1}{8} \log_2\left(\frac{0.1}{8}\right)\right\}$$

$$= 3.395462 \text{ bits of information/color}$$

$$= 3.395462 \text{ bits of information/pixel}$$

Its efficiency is:

$$\eta_2 = \frac{H(c)}{\overline{l_2}} = \frac{3.395462}{3.5} \cong 97\ \%$$

4) In the group g_1, there are 8 bits at 0 out of 12 bits.

In the group g_2, there are 8 bits at 0 out of 16 bits.

In the group g_3, there are 12 bits at 0 out of 40 bits.

So the probability of having a bit at 0 is:

$$p_0 = Pr\{x_i = 0\} = p(x_1) = 0.6 \times \frac{8}{12} + 0.3 \times \frac{8}{16} + 0.1 \times \frac{12}{40} = 0.58$$

and a bit at 1 is:

$$p_1 = Pr\{x_i = 1\} = p(x_2) = 1 - p_0 = 0.42$$

5) Recall that the source of information considered here is the binary source $S2$.

The three entropies are given successively by:

$$H(X) = -\sum_{i=1}^{2} p(x_i) \log_2 p(x_i) = -\{p_0 \log_2 p_0 + p_1 \log_2 p_1\}$$

$$H(X) \cong -1.44\{0.58 \times \log_e 0.58 + 0.42 \times \log_e 0.42\}$$
$$= 0.981454 \text{ bits of information/binary symbol}$$

$$H(Y) = -\sum_{j=1}^{2} p(y_j) \log_2 p(y_j)$$

with:

$$p(y_j) = \sum_{i=1}^{2} p(x_i) \times p(y_j/x_i)$$

$$p(y_1) = Pr\{y_j = 0\} = 0.58 \times (1-p) + 0.42 \times p = 0.58 - 0.16p$$
$$= 0.579984$$

$$p(y_2) = Pr\{y_j = 1\} = 1 - p(y_1) = 0.420016$$

$$H(Y) \cong -1.44\{p(y_1) \times \log_e p(y_1) + p(y_2) \times \log_e p(y_2)\}$$
$$= 0.981461 \text{ bits of information/binary symbol}$$

$$H(Y/X = x_i) = -\sum_{j=1}^{2} p(y_j/x_i) \times \log_2 p(y_j/x_i)$$

and:

$$H(Y/X) = \sum_{i=1}^{2} p(x_i)\, H(Y/X = x_i)$$

Since we are dealing with a binary symmetric communication channel, we get:

$$H(Y/X) \cong -1.44\{(1-p)\log_e(1-p) + p\log_e p\} = H(p)$$

$$H(Y/X) \cong 1.4730335 \times 10^{-3} \text{ bits of information/binary symbol}$$

6) The amount of information transmitted is:

$$I(X,Y) = H(Y) - H(Y/X) = 0.9799879 \text{ bits of information/binary symbol}$$

$$H(X/Y) = H(X) - I(X,Y) = 1.46661 \times 10^{-3} \text{ bits of information/binary symbol}$$

7) The average loss of information per image is:

$$H(X/Y) \times Q_2 = 1{,}576.35 \text{ bits of information/image}$$

8) Average number of wrong pixels received:

– if transmission of group g_1: coding on 3 bits, $p(g_1) = 0.6$, then the probability of error-free transmission of the group's codewords g_1 is: $(1-p)^3$;

– if transmission of group g_2: coding on 4 bits, $p(g_2) = 0.3$, then the probability of error-free transmission of the group's codewords g_2 is: $(1-p)^4$;

– if transmission of group g_3: coding on 5 bits, $p(g_3) = 0.1$, then the probability of error-free transmission of the group's codewords g_3 is: $(1-p)^5$.

Then the probability of an error-free transmission of a pixel is:

$$Pr\{error - free\ pixel\} = 0.6 \times (1-p)^3 + 0.3 \times (1-p)^4 + 0.1 \times (1-p)^5$$

And if $p \ll 1 \rightarrow (1-p)^n \cong 1 - np$, hence:

$$Pr\{error - free\ pixel\} \cong 0.6 \times (1-3p) + 0.3 \times (1-4p) + 0.1 \times (1-5p)$$

$$\cong 1 - 3.5p$$

The probability of a transmission error of a pixel is then:

$$Pr\{\text{error of a pixel}\} \cong 3.5\ p$$

This result is quite logical since the average length of a codeword is:

$$\overline{l_2} = 3.5 \text{ bit/pixel}$$

The average number of erroneous pixels received per image is then:

$$307{,}200 \times 3.5\ p \cong 108 \text{ pixels}$$

9) The codewords ∈ to the code C_2 are of variable lengths (3 or 4 or 5 bits), but the protection codes studied in this course are dependent on the length of the codewords, so the error correction will be difficult at the level of each codeword. This is why the protection (error correction) will be built at the level of blocks of bits instead of at the level of codewords.

1.7. Problem 7 – Information, entropy, codes (2)

Let us take a facsimile-type digitized image coding system, images with black parts on a white background (handwritten or printed text, diagram, graphic, etc.), for storage and efficient transmission on a communication channel. The scanned images are in 1,600 x 2,400 pixels format with 2 grey levels per pixel. Pixels here are considered to be independent in terms of random variables (it is a great simplification).

The statistics made on the facsimile images show that the 0 label pixels associated with white color are observed with a frequency equal to 0.9 and that 1 label pixels associated with black color are therefore observed with a frequency equal to 0.1.

1) What is the quantity Q_1 of binary information needed to store an image with a fixed format binary code (code C_1)? Can a Huffman coding of this 2-symbol information source reduce this quantity Q_1 and why?

NOTE.– In the Huffman codes that will be constructed later, the suffix 1 will always be used for the lowest probability element and the suffix 0 for the highest probability.

2) What is the entropy $H(S_1)$ of the source of information per pixel? Deduce the efficiency η_1 of the code C_1.

We want to increase efficiency by using a code associated not with each pixel but associated with each group of 2 pixels (second-order extension code).

3) Construct the Huffman code associated with this new source S_2 of information (code C_2). Deduce the average length $\overline{l_2}$, the quantity Q_2 of coding bits necessary for the storage of an image, the compression ratio τ_2 obtained by this code C_2 (with respect to the code C_1) and its efficiency η_2.

It is still necessary to increase the efficiency of the coding by using a code associated with each group of 3 pixels (code with an extension of order 3).

4) Construct the Huffman code associated with this new source S_3 of information (code C_3). Deduce the average length $\overline{l_3}$, the quantity Q_3 of binary information necessary for storing an image, the compression ratio τ_3 obtained by this code C_3 (still with respect to the code C_1) and its efficiency η_3.

One could thus go on increasing the number of grouped pixels to increase the efficiency of the coding.

5) What would be the compression ratio τ obtained by an almost infinite order extension code (very large in practice) with respect to the code C_1 and its efficiency η?

We want to transmit the coded images with the code C_3 to a recipient on a transmission line having a fixed bitrate D. Let S_3 be the source of binary information that we have at the serial output of the Huffman code C_3.

6) What is for code C_3, the probability p_0 to issue $x_i = 0$ and the probability p_1 to issue $x_i = 1$?

The transmission channel is a memoryless binary symmetric channel. It introduces transmission errors with a probability p (the numerical application will be: $p = 10^{-6}$).

7) Determine the entropies $H(X)$, $H(Y)$ and $H(Y/X)$.

8) Determine the amount of information received by the recipient for each binary symbol sent $I(X,Y)$, as well as the entropy $H(X/Y)$.

9) What is the average loss of information per image transmitted?

10) Determine the average number of pixels received per image, whose value is wrong.

Solution of problem 7

1) Image size: $N = 1{,}600 \times 2{,}400 = 3{,}840{,}000$ pixels.

In order to memorize an image with code C_1: 1 bit/pixel, is needed:

$$Q_1 = 1 \times N = 3{,}840{,}000 \text{ bits} = 480{,}000 \text{ bytes}$$

The use of a Huffman coding of the source S_1 with 2 symbols $\{s_1 = 0, s_2 = 1\}$, gives an average length of $\overline{l_1} = 1$ bit/symbol, so there is no compression at all.

2) The entropy is given by:

$$H(S_1) = -\sum_{i=1}^{2} p(s_i) \log_2 p(s_i) \cong -1.44\{0.9 \times \log_e 0.9 + 0.1 \times \log_e 0.1\}$$

$$\cong 0.46812 \text{ bits of information/binary symbol}$$

and its efficiency is:

$$\eta_1 = \frac{H(S_1)}{\overline{l_1}} = 46.81\ \%$$

3) Second-order extension code: grouping 2 pixels together (symbol $s_{ij} = s_i s_j$) hence there are four possible events.

Huffman coding: code C_2

Symbols s_{ij}	$p(s_{ij})_0$	$p(s_{ij})_1$	$p(s_{ij})_2$	Code C_2
$s_{11} = s_1 s_1$	0.81	0.81	0.81 0	0
$s_{12} = s_1 s_2$	0.09	0.1 0	0.19 1	1 1
$s_{21} = s_2 s_1$	0.09 0	0.09 1		1 0 0
$s_{22} = s_2 s_2$	0.01 1			1 0 1

Table 1.7. *Construction of Huffman code C_2*

The average length of the codewords is:

$$\overline{l_2} = \sum_{i=1}^{4} p_i \times l_i = 0.81 \times 1 + 0.09 \times 2 + 0.09 \times 3 + 0.01 \times 3$$

$$= 1.29 \text{ bits/symbol}$$

The amount of coding bits needed to store an image is:

$$Q_2 = \frac{N}{2} \times \overline{l_2} = \frac{3{,}840{,}000}{2} \times 1.29 = 2{,}476{,}800 \text{ bits}$$

The compression rate τ_2 relative to code C_1 and its efficiency η_2 are respectively:

$$\tau_2 = \frac{Q_1}{Q_2} = \frac{\overline{l_1}}{\overline{l_2}/2} \cong 1.550387579$$

$$\eta_2 = \frac{H(S_2)}{\overline{l_2}}$$

and:

$$H(S_2) = H(S_1^2) = 2H(S_1) = 0.93624 \text{ bit of information/symbol}$$

hence:

$$\eta_2 = \frac{H(S_2)}{\overline{l_2}} = \frac{0.93624}{1.29} \cong 72.57\,\%$$

4) Third-order extension code C_3: grouping of 3 pixels together (symbol $s_{ijk} = s_i s_j s_k$), so there are eight possible events.

The average length of the codewords is:

$$\overline{l_3} = \sum_{i=1}^{8} p_i \times l_i = 0.729 \times 1 + 3 \times 0.081 \times 3 + 3 \times 0.009 \times 5 + 0.001 \times 5$$

$$= 1.598 \text{ bits/symbol}$$

$SI = s_{ijk}$	$p(SI)_0$	$p(SI)_1$	$p(SI)_2$	$p(SI)_3$	$p(SI)_4$	$p(SI)_5$	$p(SI)_6$	Code C_3
$S1 = s_{000}$	0.729	0.729	0.729	0.729	0.729	0.729	0.729 0	0
$S2 = s_{001}$	0.081	0.081	0.081	0.081	0.109	0.162 0	0.271 1	1 0 0
$S3 = s_{010}$	0.081	0.081	0.081	0.081	0.081 0	0.109 1		1 0 1
$S4 = s_{100}$	0.081	0.081	0.081	0.081 0	0.081 1			1 1 0
$S5 = s_{110}$	0.009	0.01	0.081 0	0.028 1				11100
$S6 = s_{101}$	0.009	0.009 0	0.01 1					11101
$S7 = s_{011}$	0.009 0	0.009 1						11110
$S8 = s_{111}$	0.001 1							11111

Table 1.8. *Construction of Huffman code C_3*

The amount of coding bits needed to store an image is:

$$Q_3 = \frac{N}{3} \times \overline{l_3} = \frac{3{,}840{,}000}{3} \times 1.598 = 2{,}045{,}440 \text{ bits}$$

The relative compression rate τ_3 and the efficiency η_3 compared to the code C_1 are respectively:

$$\tau_3 = \frac{Q_1}{Q_3} = \frac{\overline{l_1}}{\overline{l_3}/3} = \frac{1}{1.598/3} = 1.877346683$$

$$\eta_3 = \frac{H(S_3)}{\overline{l_3}} = \frac{3H(S_1)}{1.598} = 87.88\ \%$$

5) Quasi infinite-order extension code: $n \rightarrow \infty \rightarrow \overline{l} = \overline{l_{min}} = H(S_1)$ hence the compression rate and efficiency, respectively:

$$\tau = \frac{\overline{l_1}}{H(S_1)} = \frac{1}{0.46812} = 2.136204392$$

$$\eta = 1$$

6) Probability p_0 to issue 0 and p_1 to issue 1 for the code C_3:

$$p_0 = Pr\{x_i = 0\} = p(x_1)$$

$$= 1 \times 0.729 \times + \frac{2}{3} \times 0.081 + 2 \times \frac{1}{3} \times 0.081 + \frac{2}{5} \times 0.009 + 2 \times \frac{1}{5} \times 0.009$$

$$= 0.8442$$

$$p_1 = Pr\{x_i = 1\} = p(x_2) = 1 - p_0 = 0.1558$$

7) The three entropies are successively the following:

$$H(X) = -\sum_{i=1}^{2} p(x_i) \log_2 p(x_i) = -\{p_0 \log_2 p_0 + p_1 \log_2 p_1\}$$

$$\cong -1.44\{0.8442 \times \log_e 0.8442 + 0.1558 \times \log_e 0.1558\}$$

$$\cong 0.6230004 \text{ bits of information/binary symbol}$$

$$H(Y) = -\sum_{j=1}^{2} p(y_j) \log_2 p(y_j)$$

with:

$$p(y_j) = \sum_{i=1}^{2} p(x_i) \times p(y_j/x_i)$$

$$p(y_1) = Pr\{y_j = 0\} = 0.8442 \times (1 - p) + 0.1558p = 0.8442 - 0.6884p$$

$$= 0.8441993$$

$$p(y_2) = Pr\{y_j = 1\} = 1 - p(y_1) = 0.1558006$$

$$H(Y) \cong -1.44\{p(y_1) \times \log_e p(y_1) + p(y_2) \times \log_e p(y_2)\}$$

$$= 0.623002 \text{ bits of information/binary symbol}$$

$$H(Y/X = x_i) = -\sum_{j=1}^{2} p(y_j/x_i) \times \log_2 p(y_j/x_i)$$

and:

$$H(Y/X) = \sum_{i=1}^{2} p(x_i)\, H(Y/X = x_i)$$

Since we are dealing with a binary symmetric communication channel, we get:

$$H(Y/X) \cong -1.44\{(1-p)\log_e(1-p) + p\log_e p\} = H(p)$$

$$H(Y/X) \cong 2.1334334 \times 10^{-5} \text{ bits of information/binary symbol}$$

8) Amount of information received by the recipient and entropy (ambiguity):

$$I(X,Y) = H(Y) - H(Y/X) = 0.6229806 \text{ bits of information/binary symbol}$$

$$H(X/Y) = H(X) - I(X,Y) = 1.98 \times 10^{-5} \text{ bits of information/binary symbol}$$

9) The average loss of information per image is:

$$H(X/Y) \times Q_3 = 1.98 \times 10^{-5} \times 2{,}045{,}440$$
$$= 40.499712 \text{ bits of information/image}$$

10) Group g_1: transmission of symbol $S1$; coding on 1 bit, $p(g_1) = 0.729$.

Then, the probability of error-free transmission of the codeword $S1$ of group g_1 is: $(1-p)$.

Group g_2: transmission of symbols $S2$ or $S3$ or $S4$; coding on 3 bits, $p(g_2) = 3 \times 0.081 = 0.243$.

Then, the probability of error-free transmission of the codewords of group g_2 is: $(1-p)^3$.

Group g_3: transmission of symbols $S5$ or $S6$ or $S7$ or $S8$; coding on 5 bits, $p(g_3) = 3 \times 0.009 + 1 \times 0.001 = 0.028$.

Then, the probability of error-free transmission of the codewords of group g_3 is: $(1-p)^5$.

So the probability of error-free transmission of a 3-pixel packet is:

$$Pr\{3 \text{ error} - \text{free pixels}\} = 0.729 \times (1-p) + 0.243 \times (1-p)^3 + 0.028 \times (1-p)^5$$

And, if $p \ll 1 \rightarrow (1-p)^n \cong 1 - np$, hence:

$$Pr\{3 \text{ error} - \text{free pixels}\} \cong 0.729 \times (1-p) + 0.243 \times (1-3p) + 0.028 \times (1-5p) \cong 1 - 1.598\,p$$

The probability of error in the transmission of a packet of 3 pixels is then:

$$Pr\{3 \text{ wrong pixels}\} \cong 1.598\, p$$

This result is quite logical since the average length of the codewords is:

$$\overline{l_3} = 1.598 \text{ bit/3 pixels}$$

The average number of erroneous pixels received per image is then:

$$\frac{N}{3} \times 1.598\, p = \frac{3{,}840{,}000}{3} \times 1.598 \times 10^{-6} = 2.045 \text{ pixels} \cong 3 \text{ pixels}$$

1.8. Problem 8 – Coding and transmission of a television-type information source

Let us take a coding system of analog television signal. The analog color TV signals are digitized and encoded in (4:2:2) format. However, to simplify the problem discussed here, only the luminance component will be considered. This leads to having per frame:

– 576 useful lines, with 720 pixels of luminance per line (rectangular sampling structure);

– 25 frames per second (50 fields per second);

– with 256 gray levels per monochrome pixel (initial binary coding on 8 bits).

To simplify, we consider that the pixel levels are independent (in terms of random variables), denoted by: "U".

However, the probability law $Pr(U)$ of grey levels U is absolutely non-uniform.

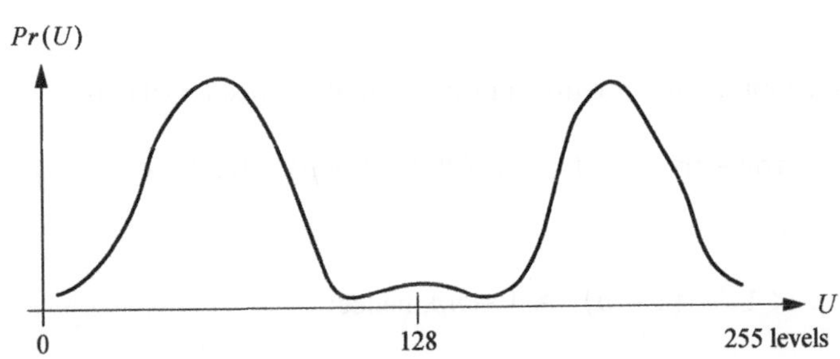

Figure 1.6. *Probability law $Pr(U)$ of TV frames grey levels U*

1) What is the amount Q_0 of binary symbols needed to store a second of monochrome TV frames with the initial fixed format binary code (code C_0)? We assume that the entropy $H(U) = 6\text{ bits of information/pixel}$. Deduce the efficiency η_0 of the code C_0.

2) Can a Huffman coding (called code C_1) of this source S_0 of information reduce this amount and why?

If we consider that the Huffman coding C_1 performs the absolute optimal coding, deduce the quantity Q_1 of binary symbols necessary to store a second of digital monochrome TV frames.

We are looking at increasing the efficiency by using an information compression of the source S_0. For this, an adaptive (and thus non-linear) re-quantization of the 256 grey levels U of each pixel is carried out on 8 levels, now denoted "Z". The probability law $Pr(Z)$ resulting from this new source of information (denoted S_2) and its binary code C_2 are shown in Table 1.9.

3) What is the amount Q_2 of binary symbols needed to store one second of digital monochrome TV frames (code C_2)?

Grey levels of *Z*	**0**	**1**	**2**	**3**	**4**	**5**	**6**	**7**
$\boldsymbol{Pr(Z)}$	0.0625	0.0625	0.15	0.21	0.14	0.0625	0.25	0.0625
Code $\boldsymbol{C_2}$	000	001	010	011	100	101	110	111

Table 1.9. *Probability law $Pr(Z)$ of S_2 and binary code C_2*

4) Can a Huffman coding (called code C_3) of this source of information reduce this amount? What is the entropy per pixel and the total entropy of one second of digital TV frames?

5) Design the Huffman code associated with this new source S_2 (code C_3).

NOTE.– In the Huffman code that will be designed, the suffix 1 will always be used for the element with the lowest probability and therefore the suffix 0 for the element with the highest probability.

6) From the Huffman code C_3, deduce:

– the average number of coding bits per pixel;

– the amount Q_3 of binary symbols needed to store one second of digital monochrome TV frames;

– the compression rate τ_3 obtained by this code C_3 (with respect to the code C_2);

– its efficiency η_3 and redundancy ρ_3.

It is desired to transmit the coded frames with the code C_3 to a recipient over a digital transmission line, having a given capacity, denoted "*Cap*". Let S_3 be the source of binary information X that we have at the serial output of the Huffman coding.

7) What is the probability p_0 to issue $x_i = 0$ and the probability p_1 to issue $x_i = 1$ for S_3? Deduce its entropy $H(X)$.

The transmission channel is a memoryless binary symmetric channel. It introduces transmission errors with a probability p (the numerical application will be $p = 10^{-2}$). The output of the binary transmission channel is called Y when its input is X (the binary output of the source S_3).

8) What is the entropy of Y. Deduce the amount of information $I(X,Y)$ received by the recipient for each binary symbol sent by S_3.

9) What is the average loss of information in the channel per binary symbol sent $H(X/Y)$ and the average loss of information per second of transmitted TV frames?

10) Determine the average number of received pixels per second of TV frames whose value is wrong.

11) What is the capacity Cap of the binary transmission channel and the capacity Cap_s per second of TV frames?

We model $Pr(U)$ by a weighted sum (factors λ_a and λ_b respectively, with $\lambda_a = 0.6225$ and so $\lambda_b = 1 - \lambda_a$) of two discrete Gaussian probability laws G_a and G_b, with: $G_a(U) = \text{Gauss}(64, 8)$ and $G_b(U) = \text{Gauss}(160, 4)$ where, in $\text{Gauss}(m, \sigma)$, m is the mean value and σ is the standard deviation of the Gaussian probability law.

12) What is (with justification) among the eight following values: 8; 7; 6; 5; 4; 3; 2 and 1 bit/pixel the order of magnitude of the entropy H_0 of the source information S_0 per pixel?

Solution of problem 8

1) The image size is: $N = 576 \times 720 = 414{,}720$ pixels.

One second of digital TV frames has:

$$Ns = N \times 25 = 10{,}368{,}000 \text{ pixel/s}$$

Monochrome TV: $\overline{l_0} = 8$ bits/pixel

Code C_0: $Q_0 = Ns \times \overline{l_0} = 82{,}944{,}000$ bits/s

Efficiency: $\eta_0 = \frac{H(U)}{\overline{l_0}} = \frac{6}{8} = 75\ \%$

2) The source of information generating U is non-uniform on [0, 255], so:

$$H(U) < 8 \text{ bits of information/pixel}$$

Therefore, the entropy coding is quite interesting.

If the Huffman code C_1 performs an absolute optimal coding, then $\overline{l_1} = H(U)$, hence:

$$Q_1 = Ns \times \overline{l_1} = 62{,}208{,}000 \text{ bits/s}$$

3) The amount of bits per second for the code C_2 is:

$$Q_2 = Ns \times \overline{l_2} = Ns \times 3 = 31{,}104{,}000 \text{ bit/s}$$

4) Since Z has a non-uniform probability law, then the Huffman coding is efficient.

The entropy per pixel is:

$$H(Z) = -\sum_{i=0}^{7} p_i \log_2 p_i$$

$$H(Z) \cong -1.44 \begin{Bmatrix} 4 \times 0.0625 \times \log_2(0.0625) + 0.15 \times \log_2(0.15) \\ +0.21 \times \log_2(0.21) + 0.14 \times \log_2(0.14) + 0.25 \times \log_2(0.25) \end{Bmatrix}$$

$\cong 2.7804781$ bits of information/pixel

The total entropy of one second of monochrome TV frames is:

$$H(TVs) = Ns \times H(Z) = 28{,}827{,}997 \text{ bits of information/s}$$

5) Huffman coding: code C_3.

Level Z	$p(Z)_0$	$p(Z)_1$	$p(Z)_2$	$p(Z)_3$	$p(Z)_4$	$p(Z)_5$	$p(Z)_6$	Code C_3
6	0.25	0.25	0.25	0.25	0.29	0.46	0.54 0	01
3	0.21	0.21	0.21	0.25	0.25	0.29 0	0.46 1	11
2	0.15	0.15	0.15	0.21	0.25 0	0.25 1		000
4	0.14	0.14	0.14	0.15 0	0.21 1			001
0	0.0625	0.125	0.125 0	0.14 1				1010
1	0.0625	0.0625 0	0.125 1					1011
5	0.0625 0	0.0625 1						1000
7	0.0625 1							1001

Table 1.10. *Construction of Huffman code C_3*

6) The average codewords length is:

$$\overline{l_3} = \sum_{i=1}^{8} p_i \times l_i = 0.25 \times 2 + 0.21 \times 2 + 0.15 \times 3 + 0.14 \times 3 + 4 \times 0.0625 \times 4$$

$$= 2.79 \text{ bit/grey level} = 2.79 \text{ bits/pixel}$$

The number of bits per second is:

$$Q_3 = Ns \times \overline{l_3} = 28{,}926{,}720 \text{ bits/s}$$

The compression ratio, efficiency and redundancy are respectively:

$$\tau_3 = \frac{Q_2}{Q_3} = \frac{\overline{l_2}}{\overline{l_3}} = \frac{3}{2.79} = 1.07527$$

$$\eta_3 = \frac{H(Z)}{\overline{l_3}} = \frac{2.7804781}{2.79} \cong 99.66\ \%$$

$$\rho_3 = 1 - \eta_3 = 0.0034128$$

7) Probability p_0 to issue $x_i = 0$ and probability p_1 to issue $x_i = 1$ for S_3?

The probability of sending a bit at zero is given by:

$$p_0 = Pr\{x_i = 0\} = p(x_1) = 0.25 \times \frac{1}{2} + 0.21 \times \frac{0}{2} + 0.15 \times \frac{3}{3} + 0.14 \times \frac{2}{3} + 0.0625 \times \frac{2}{4} + 0.0625 \times \frac{1}{4} + 0.0625 \times \frac{3}{4} + 0.0625 \times \frac{2}{4} = 0.4933$$

and that of sending a bit at 1 is therefore:

$$p_1 = Pr\{x_i = 1\} = p(x_2) = 1 - p_0 = 0.5067$$

The entropy is given by:

$$H(X) = -\sum_{i=1}^{2} p(x_i) \log_2 p(x_i) = -\{p_0 \log_2 p_0 + p_1 \log_2 p_1\}$$

$$\cong -1.44\{0.4933 \times \log_e 0.4933 + 0.5067 \times \log_e 0.5067\}$$

$$= 0.99987 \text{ bits of information/binary symbol}$$

8) The entropy of Y at the output of the communication channel is:

$$H(Y) = -\sum_{j=1}^{2} p(y_j) \log_2 p(y_j)$$

with:

$$p(y_j) = \sum_{i=1}^{2} p(x_i) \times p(y_j/x_i)$$

$$p(y_1) = Pr\{y_j = 0\} = 0.4933 \times (1 - p) + 0.5067 \times p = 0.493434$$

$$p(y_2) = Pr\{y_j = 1\} = 1 - p(y_1) = 0.506566$$

$$H(Y) \cong -1.44\{p(y_1) \times \log_e p(y_1) + p(y_2) \times \log_e p(y_2)\}$$

$$= 0.99800773 \text{ bits of information/binary symbol}$$

The amount of information transmitted through the channel is:

$$I(X,Y) = H(Y) - H(Y/X) = H(X) - H(X/Y)$$

since we are dealing with a binary symmetric communication channel, we have:

$$H(Y/X) = H(Y/X = x_i) = -\sum_{j=1}^{2} p(y_j/x_i) \times \log_2 p(y_j/x_i)$$

$$H(Y/X) \cong -1.44\{(1-p)\log_e(1-p) + p\log_e p\} = H(p)$$

$$H(Y/X) \cong 0.080642209 \text{ bits of information/binary symbol}$$

hence:

$$I(X,Y) = H(Y) - H(Y/X) = 0.917365521 \text{ bits of information/binary symbol}$$

9) The average loss of information in the channel per binary symbol sent is given by:

$$H(X/Y) = H(X) - I(X,Y) = 0.082504479 \text{ bits of information/binary symbol}$$

The average loss of information per second of transmitted TV frames is:

$$Ns \times \overline{l_3} \times H(X/Y) = 2{,}386{,}583.963 \text{ bits of information/s}$$

10) We have to consider each of the codeword lengths:

– the group g_1 corresponds to the set of levels $\{0, 1, 5, 7\}$ each of which is coded on 4 bits and of probability equal to 0.0625, hence:

$$p(g_1) = p(0) + p(1) + p(5) + p(7) = 0.25$$

– the group g_2 corresponds to the set of levels $\{2, 4\}$ each of which is coded on 3 bits and of probability equal to 0.15 and 0.14 respectively, hence:

$$p(g_2) = p(2) + p(4) = 0.15 + 0.14 = 0.29$$

– the group g_3 corresponds to the set of levels $\{3, 6\}$ each of which is coded on 2 bits and of probability equal to 0.21 and 0.25 respectively, hence:

$$p(g_3) = p(3) + p(6) = 0.21 + 0.25 = 0{,}46$$

The probability of error-free transmission of:

g_1 is $(1-p)^4$; g_2 is $(1-p)^3$; g_3 is $(1-p)^2$

Thus, the probability of having no error with the code C_3 is:

$$Pr\{\text{no error}\} = 0.25 \times (1-p)^4 + 0.29 \times (1-p)^3 + 0.46 \times (1-p)^2$$

But if $p \ll 1 \rightarrow (1-p)^n \cong 1 - np$, then:

$$Pr\{\text{no error}\} \cong 0.25 \times (1 - 4\,p) + 0.29 \times (1 - 3\,p) + 0.46 \times (1 - 2\,p) \cong 1 - 2.79\,p$$

The probability of having at least one error with the code C_3 is then:

$$Pr\{\text{error}\} = 1 - Pr\{\text{no error}\} = 2.79\,p = 0.0279$$

The average number of wrong pixels received per second of frames is:

$$Ns \times 0.0279 = 289{,}267 \text{ pixels}$$

11) Since the transmission channel is binary symmetric:

$$Cap = Max\, I(X, Y) = 1 - H(p) = 0.91935779 \text{ bits of information/binary symbol}$$

and:

$$Cap_s = Cap \times H(TVs) = 26{,}503{,}243 \text{ bits of information/s}$$

12) We have:

$$Pr(U) = \lambda_a G_a(U) + (1 - \lambda_a) G_b(U)$$

If we consider that a Gaussian law has a practical range of $\pm 3\,\sigma$ around its mean value m, then:

$$H(U) = \lambda_a H(U_1) + (1 - \lambda_a) H(U_2) + H(\lambda_a)$$

with U_1, the random variable associated with the Gaussian law is (64, 8) and U_2, the random variable associated with the Gaussian law is (160, 4). In addition, we have:

$H(U_1) < H(U_1')$: entropy of a uniform law on $[64 - (3 \times 8), 64$
$+(3 \times 8)] = [40, 88]$

$H(U_2) < H(U_2')$: entropy of a uniform law on $[160 - (3 \times 4), 160$
$+(3 \times 4)] = [148, 172]$

$$H(U_1') = \log_2(88 - 40) = \log_2(48) = 5.585$$

$$H(U_2') = \log_2(172 - 148) = \log_2(24) = 4.585$$

Consequently, we have:

$$H(U) < \lambda_a \times 5.585 + (1 - \lambda_a) \times 4.585 + H(\lambda_a)$$

with:

$$H(\lambda_a) = 0.9544$$

hence:

$$H(U) < 6.1619 \text{ bits of information/pixel} \rightarrow H(U) \leq 6 \text{ bits of information/pixel}$$

So, the order of magnitude of the entropy is:

$H(U) \cong 6$ bits of information/pixel.

1.9. Problem 9 – Entropy and motion information encoding of multimedia source

The context is that of the transmission of coded video. Several categories of information are represented and coded in a compressed form. One of these categories is motion information. Each frame I_t of a sequence SI of frames:

$$SI\{\cdots, I_{t-1}, I_t, I_{t+1}, \cdots\}$$

is divided into K macro-blocks MB_k of size 16 x 16 pixels (we have: $k = 1, \cdots, K$).

The sequence SI consists of L frames per second (typically in Europe L = 25). Each macro-block MB is associated with a displacement vector $\vec{D}$ which makes it possible to predict its content from previous frame(s). $\vec{D}$ is a vector with two components: dx and dy, taking their values on integers and half integers. For simplicity, it is assumed that in practice only seven values for dx and dy, respectively, are of significant probabilities. Each value of dx and dy is associated with a symbol s. These values are given in Table 1.11 with their probabilities (for the sake of simplicity, it has been assumed that the components dx and dy of the displacement vector $\vec{D}$ have the same statistics. This is not really the case).

Value	-1.5	-1	-0.5	0	0.5	1	1.5
Symbol	s_1	s_2	s_3	s_4	s_5	s_6	s_7
Probability	0.014	0.024	0.117	0.701	0.101	0.027	0.016

Table 1.11. *Probabilities of a component d of motion vector $\vec{D}$*

1) Determine the entropy $H(d)$ of a component dx or dy of the displacement vector $\vec{D}$. Deduce the entropy $H(\vec{D})$ of the displacement vector $\vec{D}$ for a separate coding of dx and of dy. What would be the efficiency η_1 of a fixed length code C_1 (length L_1) coding $\vec{D}$ and the bitrate DB_1 per second for a number $K = 396$ macro-blocks per frame and $L = 25$ frame/second for coding the vectors $\vec{D}$?

2) Taking code C_1 the natural binary coding in the ascending order of the symbols s_i, determine the probability p_0 of having a bit at zero in the bitstream encoding the displacement vector $\vec{D}$. Deduce the probability p_1 of having a bit at one.

3) Construct the Huffman code C_2 giving the codeword S_i associated with each of the symbols s_i of a vector component $\vec{D}$.

NOTE.– In the construction of the code C_2, the coding suffix associated with the element of lowest probability will be systematically set to 0.

Deduce from this: the average length $\overline{l_2}$ of the codewords of C_2, the average length $\overline{L_2}$ of the codewords encoding the vector $\vec{D}$ (again with a separate coding of dx and dy), its efficiency η_2 and the average bitrate per second DB_2 for coding the vectors $\vec{D}$.

4) It is considered that the source S delivers the following SS time sequence of symbols s:

$$\cdots\cdots s_4 \quad s_2 \quad s_4 \quad s_4 \quad s_3 \quad s_5 \quad s_4 \quad s_6 \quad s_4 \quad s_7 \cdots\cdots \rightarrow \text{time}$$

Deduce the corresponding sequence SB of bits obtained at the output of the Huffman coding C_2. Sequence SB is of the form $\{\cdots, b_{k-1}, b_k, b_{k+1}, \cdots\}$.

What do you observe?

Taking code C_2, determine the probability p_0 to have a bit at zero in the bit stream encoding the displacement vector $\vec{D}$. Deduce the probability p_1 of having a bit at one.

Solution of problem 9

1) The entropy of a component of the motion vector $\vec{D}$ is given by:

$$H(d) = -\sum_{i=1}^{7} p(s_i) \times \log_2 p(s_i)$$

$$H(d) \cong -1.44 \times \{0.014 \times \log_e(0.014) + 0.024 \times \log_e(0.024) + 0.117 \times \log_e(0.117) + 0.701 \times \log_e(0.701) + 0.101 \times \log_e(0.101) + 0.027 \times \log_e(0.027) + 0.016 \times \log_e(0.016)\}$$
$$\cong 1.499 \text{ bits of information/component}$$

The components dx and dy have the same statistics and are coded separately, hence:

$$H(D) = 2 \times H(d) = 2.998 \text{ bits of information/vector } \vec{D}$$

There are seven values possible per dx or dy component, so for fixed length coding, it takes 3 bits to encode dx and 3 bits to encode dy. So, a total of:

$$L_1 = 6 \text{ bits}$$

to encode each displacement vector $\vec{D}$.

The efficiency of this simple encoding technique is then:

$$\eta_1 = \frac{H(D)}{L_1} = \frac{2.998}{6} \cong 50\,\%$$

and the bitrate for coding one second of $\vec{D}$ is:

$$DB_1 = L_1 \times K \times L = 6 \times 396 \times 25 = 59{,}400 \text{ bits}$$

2) Since dx and dy have the same statistics, it is sufficient to consider only one component to determine the probability p_0 of having a bit at zero.

s_i	s_1	s_2	s_3	s_4	s_5	s_6	s_7
Code C_1	000	001	010	011	100	101	110
Number of 0	3	2	2	1	2	1	1

Table 1.12. *Code C_1 and number of 0 in each codeword*

$$p_0 = \sum_{i=1}^{7} p(s_i) \times \frac{\text{Number of zeros in } s_i}{l_i}$$

$$p_0 = \left\{0.014 \times \frac{3}{3} + 0.024 \times \frac{2}{3} + 0.117 \times \frac{2}{3} + 0.701 \times \frac{1}{3} + 0.101 \times \frac{2}{3} \right.$$
$$\left. +0.027 \times \frac{1}{3} + 0.016 \times \frac{1}{3}\right\} = 0.4233$$

The probability p_1 of having a bit at 1 is then:

$$p_1 = 1 - p_0 = 0.5767$$

3) Huffman coding:

s_i	$p(s_i)_0$	$p(s_i)_1$	$p(s_i)_2$	$p(s_i)_3$	$p(s_i)_4$	$p(s_i)_5$	C_2
s_4	0.701	0.701	0.701	0.701	0.701	0.701 1	1
s_3	0.117	0.117	0.117	0.117	0.182 1	0.299 0	0 0
s_5	0.101	0.101	0.101	0.101 1	0.117 0		0 1 1
s_6	0.027	0.03	0.051 1	0.081 0			0 1 0 1 1
s_2	0.024	0.027 1	0.03 0				0 1 0 1 0
s_7	0.016 1	0.024 0					0 1 0 0 1
s_1	0.014 0						0 1 0 0 0

Table 1.13. *Construction of Huffman code C_2 of a component of vector $\vec{D}$*

The average length of the codeword (one component d of the motion vector only) is:

$$\overline{l_2} = \sum_{i=1}^{7} p(s_i) \times l_2(i)$$

$$\overline{l_2} = \{0.014 \times 5 + 0.024 \times 5 + 0.117 \times 2 + 0.701 \times 1 + 0.101 \times 3$$
$$+0.027 \times 5 + 0.016 \times 5\} = 1.643 \text{ bits/codeword}$$

So, per a full motion vector:

$$\overline{L_2} = 2 \times \overline{l_2} = 3.286 \text{ bits/vector } \vec{D}$$

Its efficiency is:

$$\eta_2 = \frac{H(D)}{\overline{L_2}} = \frac{2.998}{3.286} = 91.23\,\%$$

And per second of TV frames, an average rate of:

$$DB_2 = \overline{L_2} \times K \times L = 3.286 \times 396 \times 25 = 32{,}531.2 \text{ bits/s} \cong 32{,}531 \text{ bits/s}$$

4) From the coded sequence (Table 1.14), there are rapid changes in the length of the codewords from one motion vector to the next.

SS	s_4	s_2	s_4	s_4	s_3	s_5	s_4	s_6	s_4	s_7
SB	1	01010	1	1	00	011	1	01011	1	01001

Table 1.14. *Coding of a sequence of a motion vector component*

The probability of having a bit at 0 is:

$$p_0 = \sum_{i=1}^{7} p(s_i) \times \frac{\text{Numbre of zeros in } s_i}{l_i}$$

$$p_0 = \Big\{0.014 \times \frac{4}{5} + 0.024 \times \frac{3}{5} + 0.117 \times \frac{2}{2} + 0.701 \times \frac{0}{1} + 0.101 \times \frac{1}{3}$$

$$\left.+0.027 \times \frac{2}{5} + 0.016 \times \frac{3}{5}\right\} = 0.1966$$

The probability of having a bit at 1 is then:

$$p_1 = 1 - p_0 = 0.8034$$

1.10. Problem 10 – Hamming coding

The problem of coding binary words for protection against transmission errors is tackled. The code C considered here is a single error correcting Hamming code. We successively call:

– i^t: binary word of information to be transmitted, of length m:

$$i^t = [i_1, i_2, \cdots, i_m]$$

– u^t: binary codeword resulting from the coding of i^t, of length n:

$$u^t = [u_1, u_2, \cdots, u_n]$$

– v^t: binary word obtained at the output of the transmission channel associated with u^t transmitted in the channel, also of length n:

$$v^t = [v_1, v_2, \cdots, v_n]$$

The construction of the codeword u^t from the information word i^t is done by using the generator matrix $[G]$ of the code C:

$$u^t = i^t \times [G]$$

The decoding of the word v^t received is carried out in two phases:

a) the detection of a possible transmission error and correction of the error;

b) the decoding by itself.

In the first phase, we calculate the syndrome ${s'}^t$ on the word received:

$${s'}^t = v^t \times [H']^t$$

Where $[H']$ is the parity matrix of the code C associated with the matrix $[G]$ *via* the matrix $[H]$. The latter is obtained from $[H']$ by permutation of columns to satisfy the form of $[H]$.

The syndrome makes it possible to detect the presence of a transmission error and to locate its position in the word received.

The characteristics imposed on the code C are as follows:

– correction of single errors;

– it is a systematic code;

– it is a Hamming code, with $m = 4$.

1) What is the minimum distance of this code?

2) Show that the length n of the codewords is 7.

3) Deduce that the generator matrix of the code is of the form:

$$[G] = [I_{4,4} \mid P_{4,3}]$$

where $I_{4,4}$ is the identity matrix.

4) Show that the parity matrix of the code is of the form:

$$[H] = [P_{4,3}^t \mid I_{3,3}]$$

5) Show that the presence of a transmission error on the jth bit of u^t generates a syndrome s'^t equal to the jth line h_j' of $[H']^t$ (h_j' is the natural binary representation of the number j).

6) Determine the matrices $[H']^t$ and $[H]$.

7) Determine the generator matrix $[G]$ of the code.

8) Construct the codewords u^t corresponding to the three information words:

$$i^t = \begin{bmatrix} 0 & 0 & 1 & 0 \\ 0 & 1 & 1 & 0 \\ 1 & 1 & 0 & 1 \end{bmatrix}$$

We receive the three following words:

$$v^t = \begin{bmatrix} 0 & 1 & 1 & 0 & 1 & 1 & 0 \\ 0 & 1 & 1 & 1 & 0 & 1 & 1 \\ 1 & 1 & 0 & 1 & 0 & 0 & 1 \end{bmatrix}$$

9) Check each of the words for a membership or non-membership to the code.

10) Make a block diagram of the encoder and the decoder.

Solution of problem 10

1) The minimum distance of the code is given by:

$$d_{min} = 2r + 1; \quad (r = 1)\text{: single errors, 1 wrong bit} \rightarrow d_{min} = 3$$

2) Length of Hamming code (corrector of one erroneous bit):

$$2^k \geq 1 + n\,; \text{ with } n = m + k \rightarrow 2^k \geq 1 + m + k$$

$$\rightarrow 2^k \geq 5 + k \rightarrow k = 3 \text{ and } n = 4 + 3 = 7$$

with:

– n: number of bits of a codeword;

– m: number of bits of an information word;

– k: number of bits of a control word.

3) It is a systematic code:

$$u^t = i^t \times [G] = [i^t, i^t \times [P]\,] \rightarrow [G]_{4,7} = \left[I_{4,4} \mid P_{4,3}\right]$$

4) We should have:

$$[G]_{4,7} \times [H]_{3,7}^t = [0]_{4,3}\,;\text{ and also : } [G]_{4,7} \times [H']_{3,7}^t = [0]_{4,3}$$

$$\rightarrow \left[I_{4,4} \mid P_{4,3}\right] \times [H]_{3,7}^t = [0]_{4,3}$$

Note that: $[M]_{3,7}^t$ means $\left[[M]_{3,7}\right]^t$.

If:

$$[H]_{3,7}^t = \begin{bmatrix} P_{4,3} \\ -- \\ I_{3,3} \end{bmatrix}$$

then:

$$\left[I_{4,4} \mid P_{4,3}\right] \times \begin{bmatrix} P_{4,3} \\ -- \\ I_{3,3} \end{bmatrix} = [P]_{4,3} \oplus [P]_{4,3} = [0]_{4,3}$$

$$\rightarrow [H] = \left[P_{4,3}^t \mid I_{3,3}\right]$$

5) If we receive:

$$v^t = u^t \oplus \varepsilon^t \text{ with } \varepsilon^t = [0, 0, \cdots, 1, 0, \cdots, 0] \quad j^{th} \text{ position}$$

then:

$$v_{7,1}^t \times [H']_{3,7}^t = s_{3,1}'^t = [u^t \oplus \varepsilon^t] \times [H']^t = u^t \times [H']^t \oplus \varepsilon^t \times [H']^t$$
$$= \varepsilon_{7,1}^t \times [H']_{3,7}^t$$

$$\rightarrow s'^t = \varepsilon^t \times [H']^t = j^{th} \text{ row of } [H']^t$$

6) If s'^t gives the position of the error coded in a natural binary code, the form of the matrix $[H']^t$ is then:

$$[H']_{3,7}^t = \begin{bmatrix} 0 & 0 & 1 \\ 0 & 1 & 0 \\ 0 & 1 & 1 \\ 1 & 0 & 0 \\ 1 & 0 & 1 \\ 1 & 1 & 0 \\ 1 & 1 & 1 \end{bmatrix}$$

Hence:

$$[H']_{3,7} = \begin{bmatrix} 0 & 0 & 0 & 1 & 1 & 1 & 1 \\ 0 & 1 & 1 & 0 & 0 & 1 & 1 \\ 1 & 0 & 1 & 0 & 1 & 0 & 1 \end{bmatrix} \rightarrow [H]_{3,7} = \begin{bmatrix} 0 & 1 & 1 & 1 & 1 & 0 & 0 \\ 1 & 0 & 1 & 1 & 0 & 1 & 0 \\ 1 & 1 & 0 & 1 & 0 & 0 & 1 \end{bmatrix}$$

We get $[H]$ verifying the systematic code from $[H']$.

7) The generator matrix $[G]$ of the code is such that:

$$[H]_{3,7}^t = \begin{bmatrix} 0 & 1 & 1 \\ 1 & 0 & 1 \\ 1 & 1 & 0 \\ 1 & 1 & 1 \\ - & - & - \\ 1 & 0 & 0 \\ 0 & 1 & 0 \\ 0 & 0 & 1 \end{bmatrix} = \begin{bmatrix} P_{4,3} \\ -- \\ I_{3,3} \end{bmatrix} \rightarrow [G]_{4,7} = [I_{4,4} \mid P_{4,3}]$$

$$= \begin{bmatrix} 1 & 0 & 0 & 0 & | & 0 & 1 & 1 \\ 0 & 1 & 0 & 0 & | & 1 & 0 & 1 \\ 0 & 0 & 1 & 0 & | & 1 & 1 & 0 \\ 0 & 0 & 0 & 1 & | & 1 & 1 & 1 \end{bmatrix}$$

8) Construction of the codeword u^t corresponding to the information word:

$$i^t_{4,1} \times [G]_{4,7} = u^t_{7,1}$$

$$[u_1 \quad u_2 \quad u_3 \quad u_4] \times [G] = [u_1 \quad u_2 \quad u_3 \quad u_4 \quad u_5 \quad u_6 \quad u_7]$$

$$\begin{bmatrix} 0 & 0 & 1 & 0 \\ 0 & 1 & 1 & 0 \\ 1 & 1 & 0 & 1 \end{bmatrix} \times \begin{bmatrix} 1 & 0 & 0 & 0 & 0 & 1 & 1 \\ 0 & 1 & 0 & 0 & 1 & 0 & 1 \\ 0 & 0 & 1 & 0 & 1 & 1 & 0 \\ 0 & 0 & 0 & 1 & 1 & 1 & 1 \end{bmatrix} = \begin{bmatrix} 0 & 0 & 1 & 0 & 1 & 1 & 0 \\ 0 & 1 & 1 & 0 & 0 & 1 & 1 \\ 1 & 1 & 0 & 1 & 0 & 0 & 1 \end{bmatrix}$$

Here:

– $[u_1 \quad u_2 \quad u_3 \quad u_4] = [i_1 \quad i_2 \quad i_3 \quad i_4]$ is the information word;

– $[u_5 \quad u_6 \quad u_7]$ is the control word concatenated to the information word.

9) Checking of code membership or code non-membership:

$$[v_1 \quad v_2 \quad v_3 \quad v_4 \quad v_5 \quad v_6 \quad v_7] \times [H']^t = [s'_3 \quad s'_2 \quad s'_1]$$

$$\begin{bmatrix} 0 & 1 & 1 & 0 & 1 & 1 & 0 \\ 0 & 1 & 1 & 1 & 0 & 1 & 1 \\ 1 & 1 & 0 & 1 & 0 & 0 & 1 \end{bmatrix} \times \begin{bmatrix} 0 & 0 & 1 \\ 0 & 1 & 0 \\ 0 & 1 & 1 \\ 1 & 0 & 0 \\ 1 & 0 & 1 \\ 1 & 1 & 0 \\ 1 & 1 & 1 \end{bmatrix} = \begin{bmatrix} 0 & 1 & 0 \\ 1 & 0 & 0 \\ 0 & 0 & 0 \end{bmatrix}$$

The first word received is not a member of the code: error on the 2nd bit.

The second word received is not a member of the code: error on the 4th bit.

The 3rd word received is a member of the code: no error detected.

10) The relationship:

$$i^t \times [G] = u^t$$

makes it possible to determine the control bits as a function of the information bits:

$$[u_5 \quad u_6 \quad u_7] = f[u_1 \quad u_2 \quad u_3 \quad u_4]$$

and, from 8), we get:

$$u_5 = u_2 \oplus u_3 \oplus u_4$$

$$u_6 = u_1 \oplus u_3 \oplus u_4$$

$$u_7 = u_1 \oplus u_2 \oplus u_4$$

These same equations can be obtained from the following relationships:

$$u_{7,1}^t \times [H']_{3,7}^t = [0]_{1,3} \quad \text{or again : } \quad u_{7,1}^t \times [H]_{3,7}^t = [0]_{1,3}$$

Hamming coder

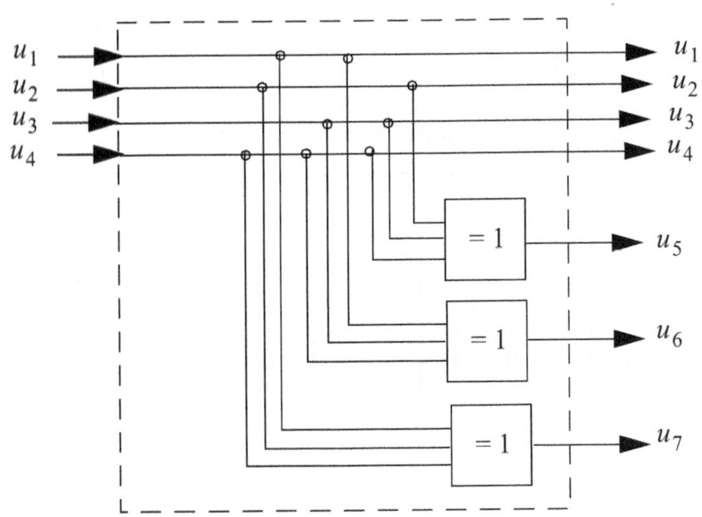

Figure 1.7. *Block diagram of Hamming coder* $C(7,4)$

Hamming decoder

The decoder is based on:

1) The calculation of the syndrome given by the relation: $s_{3,1}^{\prime t} = v_{7,1}^t \times [H']_{3,7}^t$:

$$[s_3' \quad s_2' \quad s_1'] = [v_1 \quad v_2 \quad v_3 \quad v_4 \quad v_5 \quad v_6 \quad v_7] \times \begin{bmatrix} 0 & 0 & 1 \\ 0 & 1 & 0 \\ 0 & 1 & 1 \\ 1 & 0 & 0 \\ 1 & 0 & 1 \\ 1 & 1 & 0 \\ 1 & 1 & 1 \end{bmatrix}$$

hence:

$$s_3' = v_4 \oplus v_5 \oplus v_6 \oplus v_7$$

$$s_2' = v_2 \oplus v_3 \oplus v_6 \oplus v_7$$

$$s_1' = v_1 \oplus v_3 \oplus v_5 \oplus v_7$$

2) The identification of the position of the error and its possible correction:

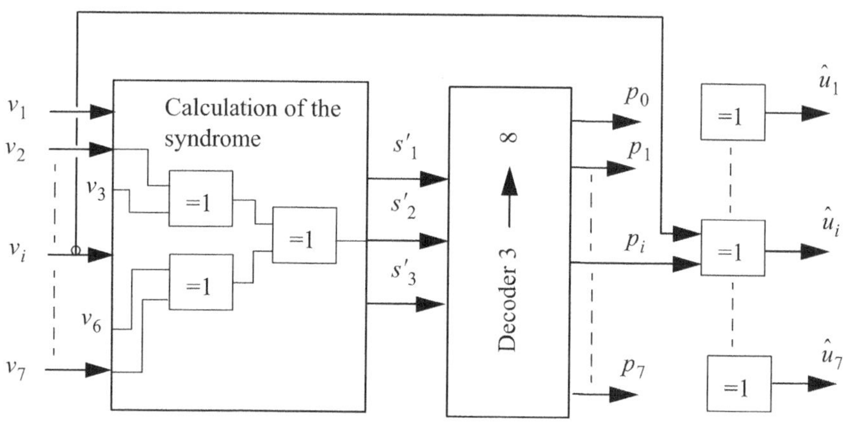

Figure 1.8. *Block diagram of Hamming decoder* $C(7,4)$

1.11. Problem 11 – Cyclic coding (1)

The problem of coding the information to be transmitted in order to protect it against transmission errors is tackled. For that, we propose to use a cyclic code C defined by its generator polynomial $g(x)$ of degree $k = 3$:

$$g(x) = x^3 + x^2 + 1$$

with $n = 7$, the length of the codes generated by $g(x)$.

1) What is the necessary and sufficient condition for the code generated by $g(x)$ to be a cyclic code?

2) Give explicitly the generator matrix $[G]$ of the code C.

3) Determine the polynomial $h(x)$, then the corresponding matrix $[H]$.

4) Determine the expressions of the control bits according to the information bits, based on:

a) the matrix $[H]$;

b) the generator polynomial $g(x)$.

Let $i(x) = x^3 + 1$ be the polynomial information (information word) to encode.

5) Determine the polynomials $c(x)$ and $u(x)$ corresponding to the control word and to the codeword, respectively.

6) Give the implementation scheme of the encoder (based on D flip-flops) providing a systematic code after n clock cycles.

7) Give the implementation scheme of the decoder associated with the coder from question 6.

8) Give the implementation scheme of the encoder based on LFSR register (linear feedback shift register) providing a systematic code after n clock cycles.

9) Give the implementation scheme of the decoder associated with the coder from question 8.

10) Does the generated cyclic code detect single, double or triple errors? Justify your answers.

11) Determine the length-percentage pairs of detectable error packets by this code.

12) Give the implementation scheme of the pseudo-random number generator based on the generator polynomial $g(x)$. Starting from the initial state $[Q]^t = [Q_0 = 0 \quad Q_1 = 0 \quad Q_2 = 1]$, give the state of the register at each clock cycle and until the register returns to its initial state.

What is the length of the cycle produced at the output of this pseudo-random number generator?

Solution of problem 11

1) The necessary and sufficient condition for $g(x)$ to generate a cyclic code is that $g(x)$ divides $(x^n + 1)$ but does not divide $(x^{n_1} + 1)$, with $n_1 < n = 7$:

$g(x)$ divides $(x^7 + 1)$

because:

$$(x^7 + 1) = (x^3 + x^2 + 1) \times (x^4 + x^3 + x^2 + 1)$$

but does not divide $(x^{n_1} + 1)$, with $n_1 < n = 7$.

2) We have: $n = m + k$, with $n = 7$ and $k = 3 \rightarrow m = 4$.

Generator matrix of the cyclic code:

$$[G]_{m,n} = [G]_{4,7} = \begin{bmatrix} 1 & 1 & 0 & 1 & 0 & 0 & 0 \\ 0 & 1 & 1 & 0 & 1 & 0 & 0 \\ 0 & 0 & 1 & 1 & 0 & 1 & 0 \\ 0 & 0 & 0 & 1 & 1 & 0 & 1 \end{bmatrix} \leftarrow g(x)$$

3) Polynomial $h(x)$ and matrix $[H]$:

$$h(x) = \frac{x^7 + 1}{g(x)} = \frac{x^7 + 1}{x^3 + x^2 + 1} = x^4 + x^3 + x^2 + 1; \quad d^\circ h(x) = m = 4$$

$$[H]_{k,n} = [H]_{3,7} = \begin{bmatrix} 1 & 0 & 1 & 1 & 1 & 0 & 0 \\ 0 & 1 & 0 & 1 & 1 & 1 & 0 \\ 0 & 0 & 1 & 0 & 1 & 1 & 1 \end{bmatrix} \quad h(x) \rightarrow \text{(first row)}$$

4) a) Expression of the control bits from the matrix $[H]$.

We have:

$$[H]_{k,n} \times [u]_{n,1} = [0]_{k,1} \rightarrow [H]_{3,7} \times [u]_{7,1} = [0]_{3,1}$$

$$\begin{bmatrix} 1 & 0 & 1 & 1 & 1 & 0 & 0 \\ 0 & 1 & 0 & 1 & 1 & 1 & 0 \\ 0 & 0 & 1 & 0 & 1 & 1 & 1 \end{bmatrix} \times \begin{bmatrix} u_6 \\ u_5 \\ u_4 \\ u_3 \\ u_2 \\ u_1 \\ u_0 \end{bmatrix} = \begin{bmatrix} 0 \\ 0 \\ 0 \end{bmatrix}$$

$$\rightarrow \begin{cases} u_6 \oplus u_4 \oplus u_3 \oplus u_2 = 0 \\ u_5 \oplus u_3 \oplus u_2 \oplus u_1 = 0 \\ u_4 \oplus u_2 \oplus u_1 \oplus u_0 = 0 \end{cases} \rightarrow \begin{cases} u_2 = u_3 \oplus u_4 \oplus u_6 \\ u_1 = u_5 \oplus u_3 \oplus u_3 \oplus u_4 \oplus u_6 = u_4 \oplus u_5 \oplus u_6 \\ u_0 = u_4 \oplus u_2 \oplus u_1 = u_3 \oplus u_4 \oplus u_5 \end{cases}$$

4) b) Expression of the control bits from the generator polynomial $g(x)$:

$$c(x) = \text{Remainder}\left\{\frac{x^k i(x)}{g(x)}\right\} = \text{Remainder}\left\{\frac{x^3[u_6x^3 + u_5x^2 + u_4x + u_3]}{x^3 + x^2 + 1}\right\}$$

$$\begin{array}{rl} & x^3 + x^2 + 1 \\ u_6x^6 + u_5x^5 + u_4x^4 + u_3x^3 & \overline{} \\ \oplus \quad u_6x^6 + u_6x^5 + u_6x^3 & u_6x^3 + (u_5 + u_6)x^2 + (u_4 + u_5 + u_6)x \\ \hline & +(u_3 + u_4 + u_5) \end{array}$$

$$\begin{array}{c} (u_5 + u_6)x^5 + u_4x^4 + (u_3 + u_6)x^3 \\ (u_5 + u_6)x^5 + (u_5 + u_6)x^4 + (u_5 + u_6)x^2 \\ \hline (u_4 + u_5 + u_6)x^4 + (u_3 + u_6)x^3 + (u_5 + u_6)x^2 \\ (u_4 + u_5 + u_6)x^4 + (u_4 + u_5 + u_6)x^3 + (u_4 + u_5 + u_6)x \\ \hline (u_3 + u_4 + u_5)x^3 + (u_5 + u_6)x^2 + (u_4 + u_5 + u_6)x \\ (u_3 + u_4 + u_5)x^3 + (u_3 + u_4 + u_5)x^2 + (u_3 + u_4 + u_5) \\ \hline \underbrace{(u_3 + u_4 + u_6)}_{u_2}x^2 + \underbrace{(u_4 + u_5 + u_6)}_{u_1}x + \underbrace{(u_3 + u_4 + u_5)}_{u_0} \end{array}$$

$$\rightarrow c(x) = u_2x^2 + u_1x + u_0$$

5) Control word and codeword generation:

$$i(x) = x^3 + 1 \rightarrow x^k i(x) = x^3(x^3 + 1) = x^6 + x^3$$

$$\rightarrow \begin{cases} u_6 = u_3 = 1 \\ u_4 = u_5 = 0 \end{cases}$$

$$\rightarrow \begin{cases} u_2 = 0 \\ u_1 = u_0 = 1 \end{cases}$$

$$\rightarrow c(x) = x + 1 \text{ and } u(x) = x^k i(x) + c(x) = x^6 + x^3 + x + 1$$

6) Design of the encoder implementation scheme:

The multiplexers (Muxs) $c1$ and $c2$ are in position 1 during $m = 4$ clock cycles. At the clock cycles $m + 1, m + 2, \cdots, n$ that is from 5 to 7, the multiplexers are in position 2.

7) Design of the decoder implementation scheme:

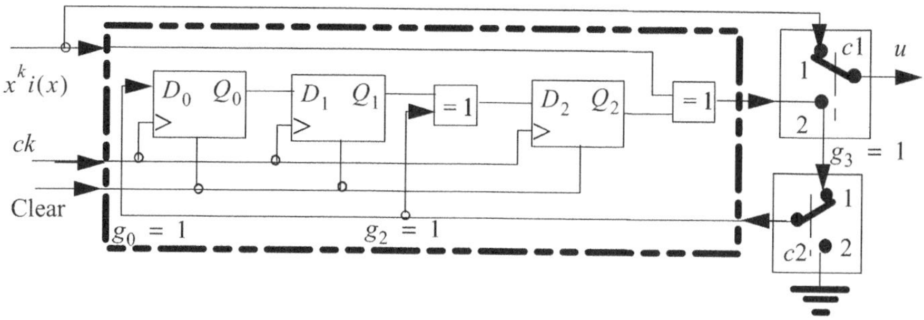

Figure 1.9. *Implementation scheme of the encoder*

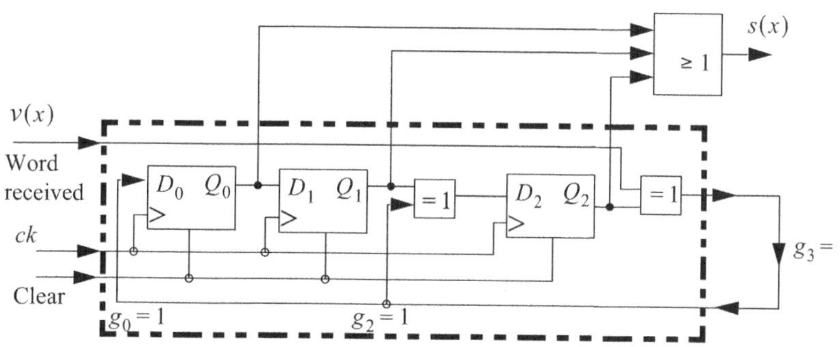

Figure 1.10. *Implementation scheme of the decoder*

After $n = 7$ clock cycles, we look at the value of the syndrome $s(x)$:

$$\rightarrow \text{if} \begin{cases} s(x) = 0 \rightarrow \text{no transmission error detected} \\ s(x) = 1 \rightarrow \text{detection of transmission error} \end{cases}$$

8) Coder based on a linear feedback shift register (LFSR).

The multiplexer (Mux) c is in position 1 during $m = 4$ clock cycles, then in position 2 for the next clock cycles $m + 1, m + 2, \cdots, n$, that is from 5 to 7.

9) Decoder based on a linear feedback shift register (LFSR).

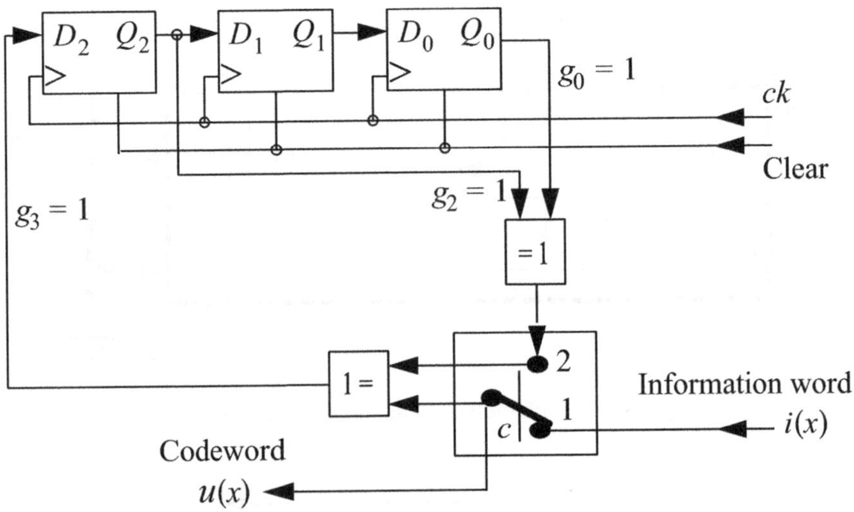

Figure 1.11. *Implementation scheme of the coder based on a linear feedback shift register*

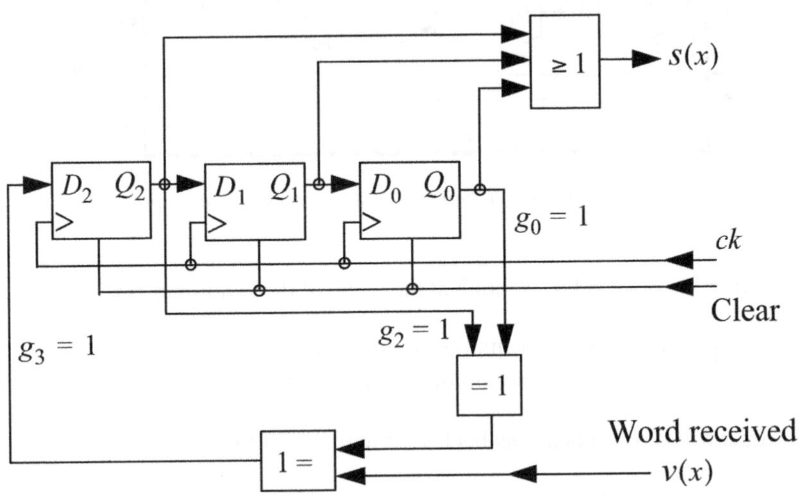

Figure 1.12. *Implementation scheme of the decoder based on a linear feedback shift register*

10) The received word is: $v(x) = u(x) + \varepsilon(x)$

The syndrome is given by:

$$s(x) = \text{Remainder}\left[\frac{v(x)}{g(x)}\right] = \text{Remainder}\left[\frac{\varepsilon(x)}{g(x)}\right]$$

Error detection is possible if $v(x)$ does not belong to the code and if $g(x)$ does not divide $\varepsilon(x)$.

– *Single errors*: in this case, $\varepsilon(x)$ is of the form $\varepsilon(x) = x^i$ which is not divisible by $g(x)$ of the form $g(x) = 1 + \cdots$, consequently, detection of all the single errors.

– *Triple errors*: if $g(x) \neq (1 + x)p(x)$, then no detection of all the triple errors (see question 11).

– *Double errors*: in this case, $\varepsilon(x)$ is of the form $\varepsilon(x) = x^i + x^j = x^i(x^{j-i} + 1)$. Since $g(x)$ does not divide x^i, it then suffices that $g(x)$ does not divide $(x^{j-i} + 1)$ either. The generator polynomial $g(x)$ divides $x^n + 1$ but does not divide $x^{n_1} + 1$, with $n_1 < n$, then $g(x)$ is said to be of order n. If $n = 2^k - 1$, then $g(x)$ is a primitive polynomial. Here, $n = 7$, $k = 3$, and $7 = 2^3 - 1$, thus, this code is able to detect all the double errors because $(j - i) < n$.

11) A packet of errors that starts in position j and is of length l is written:

$$\varepsilon(x) = x^j + \varepsilon_{j+1}x^{j+1} + \cdots + x^{j+l-1}$$

where the first and the last coefficients are at 1 and the others can be 0 or 1:

$$\varepsilon(x) = x^j \times \left[1 + \varepsilon_{j+1}x + \cdots + x^{l-1}\right] = x^j \times \varepsilon 1(x)$$

Several cases are to be considered:

– $l - 1 < k : k = 3 \rightarrow l = 3$, then detection of 100% of the error packets with $l \leq k$;

– $l - 1 = k \rightarrow l = 4$, and the proportion of detectable error packets is then: $1 - 2^{-(k-1)} = 1 - 2^{-2} = 0.75$, i.e. 75% of the error packets;

– $l - 1 > k \rightarrow l > 4$, and the proportion of detectable error packets is then: $1 - 2^{-k} = 1 - 2^{-3} = 0.875$, i.e. 87.5% of the error packets.

12) Pseudo-random number generator.

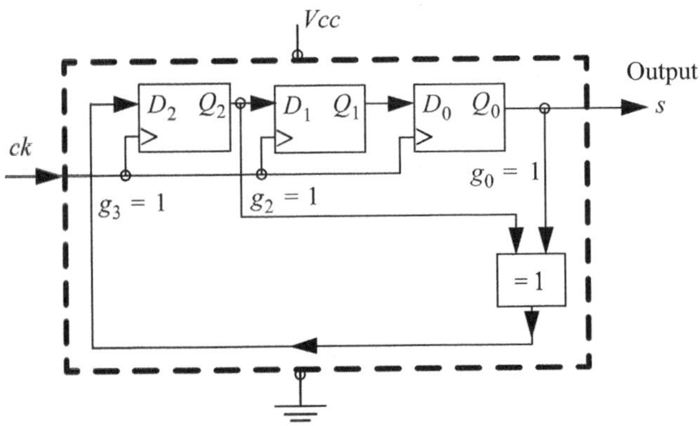

Output: $s = Q_0$

Initial state

	Q_2	Q_1	Q_0
N° ck	1	0	0
1°	1	1	0
2°	1	1	1
3°	0	1	1
4°	1	0	1
5°	0	1	0
6°	0	0	1
7°	1	0	0

Return to initial state

Figure 1.13. *Pseudo-random number generator and register states*

The cycle length is: $l = 2^k - 1 = 2^3 - 1 = 7 \rightarrow \begin{cases} 4 \text{ bits at } 1 \\ 3 \text{ bits at } 0 \end{cases}$: a quasi-balanced sequence.

1.12. Problem 12 – Cyclic coding (2)

The problem of coding the information to be transmitted in order to protect it against transmission errors is considered. For that, we use a cyclic code C defined by its generator polynomial $g(x)$ of degree k and the polynomial $h(x)$ of degree m, orthogonal to $g(x)$ modulo $(x^n + 1)$.

We set $n = 15$ and the associated generator polynomial is:

$$g(x) = x^5 + x^4 + x^2 + 1$$

1) Does the cyclic code C detect double errors? Justify your answer.

We impose that the cyclic code be a systematic code, that will be denoted code C_1. In this case, a word to be encoded is represented by the polynomial $i(x)$, and from this the coded word represented by the polynomial $u(x)$ is obtained.

2) What is the structure of the polynomial $u(x)$: format of each of the two parts of $u(x)$?

3) From the construction mechanism of the codewords u by the code C_1, determine the implementation scheme of the coder associated with the code C_1 (using only the operators: D flip-flop; multiplexer 2 to 1; XOR).

Taking as an example the word to be coded i represented by the polynomial $i(x) = x^8 + x^6 + x^3 + x + 1$, describe the operation of the pre-multiplied encoder: internal state and values of the input and output at each clock cycle.

Deduce the polynomial code $u_1(x)$ associated with $i(x)$.

4) Determine the implementation scheme of the decoder associated with the code C_1 making it possible for the detection of errors and explain how it operates.

We no longer impose the cyclic code C to be systematic. Let C_2 be the code C such that $u(x)$ is obtained by multiplication of $i(x)$ by $g(x)$.

5) Determine the implementation scheme of the coder associated with the code C_2 (using only the operators: D flip-flop; XOR).

Taking as an example the word to be coded i from question 3, describe the operation of the coder. Deduce the polynomial code $u_2(x)$ associated with $i(x)$.

Solution of problem 12

1) We have $n = 15; k = 5$ and:

$$g(x) = x^5 + x^4 + x^2 + 1 = (x + 1)(x^4 + x + 1) = (x + 1) \times p(x)$$

$g(x)$ is not primitive, but $p(x)$ which is of degree 4, is primitive because: $n = 15 = 2^4 - 1$.

Two errors occurring in position i and j of a codeword are characterized by a polynomial error of type:

$$\varepsilon(x) = x^i + x^j = x^i(x^{j-i} + 1) \quad \text{with} \quad n > j > i$$

The polynomial $p(x)$ being primitive, thus $p(x)$ does not divide any of the polynomials of the form $(x^{n_1} + 1)$ with $n_1 < n$. Then $(j - i)$ is at most equal to $(n - 1)$. In addition, $p(x)$ does not divide x^i, hence this cyclic code detects all the double errors.

It should also be noted that the polynomial $(x + 1)$ detects all the single and triple errors.

2) Structure of the polynomial:

$$x^k i(x) = g(x) \times q(x) + c(x) \rightarrow x^k i(x) + c(x) = g(x) \times q(x) = u(x)$$

with:

– $x^k i(x)$: polynomial information cyclically shifted from k positions to the left;

– $c(x)$: polynomial control.

3) We have:

$$x^k i(x) = g(x) \times q(x) + c(x)$$

hence:

$$x^5 \times (x^8 + x^6 + x^3 + x + 1) = (x^5 + x^4 + x^2 + 1) \times q(x) + c(x)$$

$$x^{13} + x^{11} + x^8 + x^6 + x^5 = (x^5 + x^4 + x^2 + 1)$$
$$\times (x^8 + x^7 + x^5 + x + 1) + x^4 + x^3 + x^2 + x + 1$$

So finally:

$$u_1(x) = x^k i(x) + c(x)$$

$$u_1(x) = x^{13} + x^{11} + x^8 + x^6 + x^5 + x^4 + x^3 + x^2 + x + 1$$

Diagram of implementation of the coder associated with code C_1 (block diagram of Table 1.15) and description of its operation.

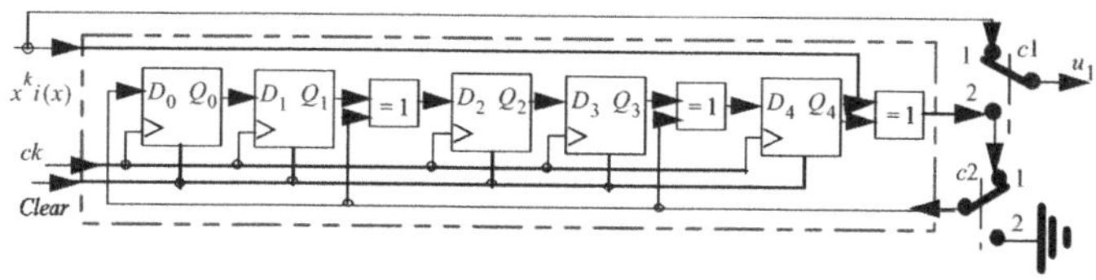

c1c2 1 1	ck	$i(x)$	D_0	Q_0 D_1	Q_1	D_2	Q_2 D_3	Q_3	D_4	Q_4	u_1
		1 x^8	1	**0**	**0**	1	**0**	**0**	1	**0**	1
	1°			1	0		1	0		1	
		0	1	1		1	1		1		0
	2°			1	1		1	1		1	
		1	0	1		1	1		1		1
	3°			0	1		1	1		1	
		0	1	0		0	1		0		0
	4°			1	0		0	1		0	
		0	0	1		0	0		1		0
	5°			0	1		0	0		1	
		1	0	0		1	0		0		1
	6°			0	0		1	0		0	
		0	0	0		0	1		0		0
	7°			0	0		0	1		0	
		1	1	0		1	0		0		1
	8°			1	0		1	0		0	
		1	1	1		1	1		1		1
	9°			1	1		1	1		1	
2 2		0	0	1		1	1		1		1
	10°			0	1		1	1		1	
		0	0	0		1	1		1		1
	11°			0	0		1	1		1	
		0	0	0		0	1		1		1
	12°			0	0		0	1		1	
		0	0	0		0	0		1		1
	13°			0	0		0	0		1	
		0	0	0		0	0		0		1
	14°			0	0		0	0		0	

Table 1.15. *Description of the operations of the premultiplied coder. For a color version of this table, see www.iste.co.uk/assad/digital2.zip*

4) The structure of the decoder for error detection is given in Figure 1.14.

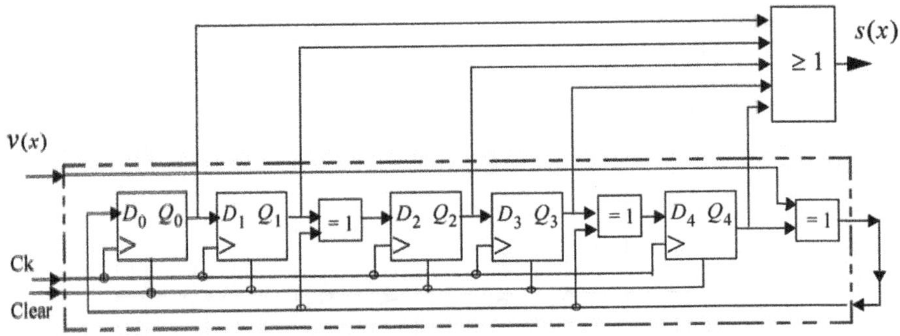

Figure 1.14. *Structure of the decoder for the detection of errors. For a color version of this figure, see www.iste.co.uk/assad/digital2.zip*

The detection process is as follows:

– initialization: reset the register by performing the action Clear;

– during n clock cycles, the received word $v(x)$ enters the divisor. The remainder of the division $x^k s(x)$ is stored in the register at the nth clock cycle, the output of the OR gate will then indicate whether there is an error or not.

5) We have:

$$u_2(x) = i(x) \times g(x) \mod (x^n + 1)$$

$$u_2(x) = \sum_{s=0}^{m-1} i_s x^s \sum_{j=0}^{k} g_j x^j = \sum_{s=0}^{m-1} \sum_{j=0}^{k} i_s \times g_j \times x^{s+j}$$

Let's set: $l = s + j$:

$$u_2(x) = \sum_{l=0}^{m+k-1} \left[\sum_{s=0}^{m-1} i_s \times g_{l-s} \right] x^l \text{ with } (l - s) \in [0, \cdots, k]$$

again:

$$u_2(x) = i_0 g_0 + (i_0 g_1 + i_1 g_0)x + (i_0 g_2 + i_1 g_1 + i_2 g_0)x^2 + \cdots$$
$$+(i_{m-2} g_k + i_{m-1} g_{k-1})x^{m+k-2} + i_{m-1} g_k x^{m+k-1}$$

A hardware implementation of this relation defines the coder associated with the code C_2 (see the block diagram of Table 1.16). The information word is entered in a shift register, least significant bit first, and the bits corresponding to the terms present in the register are added (modulo 2). The bits of the product come out, least significant bit first.

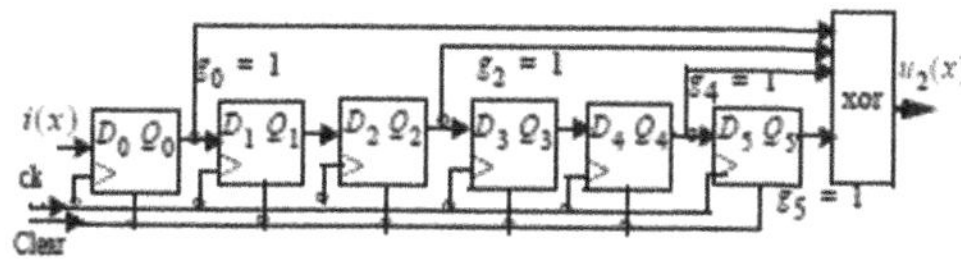

ck	$i(x)$	Q_0	Q_1	Q_2	Q_3	Q_4	Q_5	$u_2(x)$
	1	0	0	0	0	0	0	
1•		1	0	0	0	0	0	$1\,x^0$
	1							
2•		1	1	0	0	0	0	1
	0							
3•		0	1	1	0	0	0	1
	1							
4•		1	0	1	1	0	0	0
	0							
5•		0	1	0	1	1	0	1
	0							
6•		0	0	1	0	1	1	1
	1							
7•		1	0	0	1	0	1	0
	0							
8•		0	1	0	0	1	0	1
	1							
9•		1	0	1	0	0	1	1
	0							
10•		0	1	0	1	0	0	0
	0							
11•		0	0	1	0	1	0	0
	0							
12•		0	0	0	1	0	1	1
	0							
13•		0	0	0	0	1	0	1
	0							
14•		0	0	0	0	0	1	$1\,x^{13}$
	0							
15•		0	0	0	0	0	0	

Table 1.16. *Description of the operations of the encoder* C_2

Indeed:

$$u_2(x) = i(x) \times g(x) = (x^8 + x^6 + x^3 + x + 1) \times (x^5 + x^4 + x^2 + 1)$$
$$= x^{13} + x^{12} + x^{11} + x^8 + x^7 + x^5 + x^4 + x^2 + x + 1$$

1.13. Problem 13 – Cyclic coding and Hamming coding (1)

We consider a linear block code C of parameter $n = 7$ and of primitive generator polynomial: $g(x) = x^3 + x^2 + 1$.

1) Show that this code is cyclic. Deduce the second primitive generator polynomial $g_1(x)$.

2) Determine a matrix $[G]$ generating this code. Deduce the generator matrix $[G_s]$ from the systematic version of the code in question.

3) Determine the codeword $u(x)$ in systematic form which is associated with the information word: $i(x) = x^3 + 1$.

4) Design the premultiplied coder making it possible to generate the codeword $u(x)$ from the information word: $i(x) = x^3 + 1$.

5) Give the control matrix $[H]$ of the dual code to the code C.

6) Find, from the relation linking the control matrix $[H]$ and the codeword u, the control bits as a function of the information bits of question 3.

7) Make your comments about the code C and its dual.

Solution of problem 13

1) The code is cyclic if $g(x)$ divides $(x^n + 1)$ but does not divide $(x^{n_1} + 1)$ with $n_1 < n$.

Here $n = 7$ and:

$$(x^7 + 1) = (x^3 + x^2 + 1) \times (x^4 + x^3 + x^2 + 1)$$

$$(x^7 + 1) = (x^3 + x^2 + 1) \times (x^3 + x + 1) \times (x + 1)$$

So, $g(x)$ divides $(x^7 + 1)$, and the code C is a cyclic code.

The second primitive generator polynomial $g_1(x)$ is:

$$g_1(x) = (x^3 + x + 1)$$

2) The generator matrix $[G]$ of the code $C(7,4)$ is given from the generator polynomial $g(x)$ as follows:

$$[G]_{4,7} = \begin{bmatrix} 1 & 1 & 0 & 1 & 0 & 0 & 0 \\ 0 & 1 & 1 & 0 & 1 & 0 & 0 \\ 0 & 0 & 1 & 1 & 0 & 1 & 0 \\ 0 & 0 & 0 & 1 & 1 & 0 & 1 \end{bmatrix} \begin{matrix} x^3 g(x) \\ x^2 g(x) \\ x g(x) \\ g(x) \end{matrix}$$

To get a systematic code, the generating matrix $[G_s]$ must have the form $[G_s] = [I_{4,4} \mid P_{4,3}]$ obtained from the arithmetic operations on the rows of the matrix $[G]_{4,7}$. Indeed, from the form of the matrix $[G]$ we find that:

– the row 1 of the matrix $[G_s]$ is obtained by the sum of the rows: 1 + 2 + 3 of the matrix $[G]$;

– the row 2 of the matrix $[G_s]$ is obtained by the sum of the rows: 2 + 3 + 4 of the matrix $[G]$;

– the row 3 of the matrix $[G_s]$ is obtained by the sum of the rows: 3 + 4 of the matrix $[G]$;

– the row 4 of the matrix $[G_s]$ is identical to the row 4 of the matrix $[G]$, hence:

$$[G_s] = \begin{bmatrix} 1 & 0 & 0 & 0 & | & 1 & 1 & 0 \\ 0 & 1 & 0 & 0 & | & 0 & 1 & 1 \\ 0 & 0 & 1 & 0 & | & 1 & 1 & 1 \\ 0 & 0 & 0 & 1 & | & 1 & 0 & 1 \end{bmatrix}$$

3) We have:

$$x^k i(x) = g(x) \times q(x) + c(x) \rightarrow x^k i(x) + c(x) = g(x) \times q(x) = u(x)$$

with:

$$i(x) = x^3 + 1\,; k = 3 \;\rightarrow x^k i(x) = x^3 \times (x^3 + 1\,) = x^6 + x^3$$

hence:

$$\begin{array}{r|l} x^6 + x^3 & x^3 + x^2 + 1 \leftarrow g(x) \\ \oplus & \overline{x^3 + x^2 + x + 1 \leftarrow q(x)} \\ x^6 + x^5 + x^3 & \\ \hline \end{array}$$

$$
\begin{array}{c}
x^5 \\
x^5 + x^4 + x^2 \\
\hline
x^4 + x^2 \\
x^4 + x^3 + x \\
\hline
x^3 + x^2 + x \\
x^3 + x^2 + 1 \\
\hline
x + 1 \leftarrow c(x)
\end{array}
$$

$$\rightarrow u(x) = x^6 + x^3 + x + 1$$

$$u = [1 \quad 0 \quad 0 \quad 1 \quad 0 \quad 1 \quad 1]$$

4) Construction of the pre-multiplied coder.

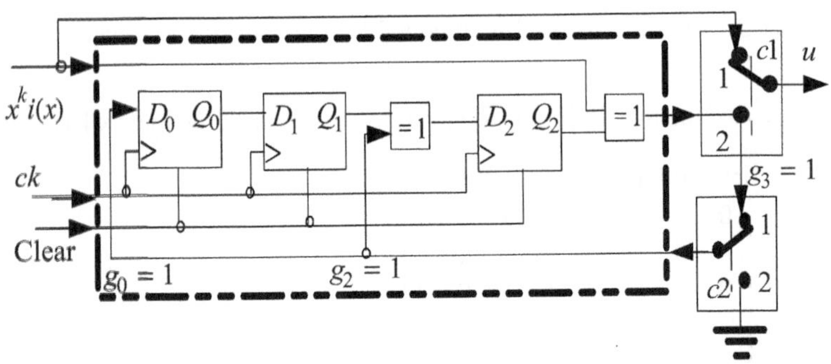

Figure 1.15. *Implementation scheme of the pre-multiplied coder*

5) Control matrix $[H]$ of the dual code to the code C.

It is such that we have:

$$[G_s] \times [H]^t = [0]\,; \quad [G_s] = [I_{4,4} \mid P_{4,3}] \rightarrow [H] = [P^t_{4,3} \mid I_{3,3}]$$

$$\rightarrow [H]_{3,7} = \begin{bmatrix} 1 & 0 & 1 & 1 & | & 1 & 0 & 0 \\ 1 & 1 & 1 & 0 & | & 0 & 1 & 0 \\ 0 & 1 & 1 & 1 & | & 0 & 0 & 1 \end{bmatrix}$$

6) Control bits according to the information bits of question 3:

$$[H]_{3,7} \times u_{7,1} = [0]_{3,1}$$

$$\begin{bmatrix} 1 & 0 & 1 & 1 & 1 & 0 & 0 \\ 1 & 1 & 1 & 0 & 0 & 1 & 0 \\ 0 & 1 & 1 & 1 & 0 & 0 & 1 \end{bmatrix} \times \begin{bmatrix} u_6 \\ u_5 \\ u_4 \\ u_3 \\ u_2 \\ u_1 \\ u_0 \end{bmatrix} = \begin{bmatrix} 0 \\ 0 \\ 0 \end{bmatrix}$$

$$\rightarrow \begin{cases} u_6 + u_4 + u_3 + u_2 = 0 \\ u_6 + u_5 + u_4 + u_1 = 0 \\ u_5 + u_4 + u_3 + u_0 = 0 \end{cases} \rightarrow \begin{cases} u_2 = u_3 + u_4 + u_6 \\ u_1 = u_4 + u_5 + u_6 \\ u_0 = u_3 + u_4 + u_5 \end{cases}$$

Thus, we have:

$$u^t = \Big[\underbrace{u_6\, u_5\, u_4\, u_3}_{\text{information bits}} \quad \underbrace{u_2\, u_1\, u_0}_{\text{control bits}}\Big] = [1\,0\,0\,1\,0\,1\,1]$$

7) The dual code of a maximum length cyclic code is the Hamming code.

1.14. Problem 14 – Cyclic coding and Hamming coding (2)

We consider a linear block code defined by its following generator matrix:

$$[G]_{m,n} = \begin{bmatrix} 1 & 1 & 1 & 0 & 1 & 0 & 0 \\ 0 & 1 & 1 & 1 & 0 & 1 & 0 \\ 0 & 0 & 1 & 1 & 1 & 0 & 1 \end{bmatrix}$$

1) Give the expression of the generator polynomial $g(x)$ associated with $[G]_{m,n}$.

2) Is the code generated by $g(x)$ cyclic? Justify your answer.

It is required that the cyclic code generated by $g(x)$ is systematic.

3) Determine the polynomial (codeword) $u(x)$ from the polynomial (information word): $i(x) = x^2 + 1$.

4) Give the implementation scheme of the pre-multiplied encoder making it possible to generate the codeword $u(x)$ from the information word $i(x) = x^2 + 1$ and describe its operation: internal state and input and output values for three clock cycles.

5) Does the generated code detect odd numbers of errors and double errors? Justify your answer.

6) Determine the proportion of error packets of length $l > 5$, detectable by the generated code.

7) Give explicitly the generating matrix $[G_s]$, from the matrix $[G]$ given above, which allows the generation of a systematic code C.

8) Determine explicitly the form of the control matrix $[H]$ which enables the generation of a code D dual to the code C.

9) Find, from the relation between the control matrix $[H]$ and the codeword u, the control bits as a function of the information bits.

10) Give the implementation scheme of the pseudo-random number generator (PRNG) based on $g(x)$.

Solution of problem 14

1) The last row of $[G]_{m,n} = [G]_{3,7}$ is the lower-level codeword that represents the generator polynomial $g(x)$:

$$\rightarrow g(x) = x^4 + x^3 + x^2 + 1 \quad \rightarrow k = 4$$

2) From $[G]_{3,7} \rightarrow m = 3$ and $n = m + k = 3 + 4 = 7$. The generated code is cyclic if $g(x)$ divides $(x^n + 1)$ but does not divide $(x^{n_1} + 1)$, with $n_1 < n = 7$:

$$
\begin{array}{l|l}
\quad x^7 + 1 & x^4 + x^3 + x^2 + 1 \leftarrow g(x) \\
\oplus & \text{-----------} \\
\quad x^7 + x^6 + x^5 + x^3 & x^3 + x^2 + 1 \leftarrow q(x) \\
\quad \text{--------} & \\
x^6 + x^5 + x^3 + 1 & \\
x^6 + x^5 + x^4 + x^2 & \\
\text{-------} & \\
x^4 + x^3 + x^2 + 1 & \\
x^4 + x^3 + x^2 + 1 & \\
\text{--------} &
\end{array}
$$

0 0 0 0

$$\rightarrow (x^7 + 1) = \underbrace{(x^4 + x^3 + x^2 + 1)}_{g(x)} \times \underbrace{(x^3 + x^2 + 1)}_{q(x)}$$

Therefore, since $g(x)$ divides $(x^7 + 1)$, but does not divide $(x^6 + 1)$, $(x^5 + 1)$, or $(x^4 + 1)$, then the code generated by $g(x)$ is a cyclic one.

3) Determination of the polynomial $u(x)$ associated to the polynomial: $i(x) = x^2 + 1$.

We have:

$$x^k i(x) = g(x) \times q(x) + c(x) \rightarrow x^k i(x) + c(x) = g(x) \times q(x) = u(x)$$

$$x^k i(x) = x^4 \times (x^2 + 1) = x^6 + x^4$$

$$\begin{array}{l|l} x^6 + x^4 & x^4 + x^3 + x^2 + 1 \leftarrow g(x) \\ \oplus & \text{-------------} \\ x^6 + x^5 + x^4 + x^2 & x^2 + x + 1 \leftarrow q(x) \\ \text{--------} & \\ x^5 + x^2 & \\ x^5 + x^4 + x^3 + x & \\ \text{--------} & \\ x^4 + x^3 + x^2 + x & \\ x^4 + x^3 + x^2 + 1 & \\ \text{---------} & \end{array}$$

$$c(x) = x + 1 \rightarrow u(x) = x^k i(x) + c(x) = x^6 + x^4 + x + 1$$

4) Block diagram of the pre-multiplied encoder generating the codeword $u(x)$ from the information word $i(x) = x^2 + 1$ and description of its operation (see the block diagram in Table 1.17).

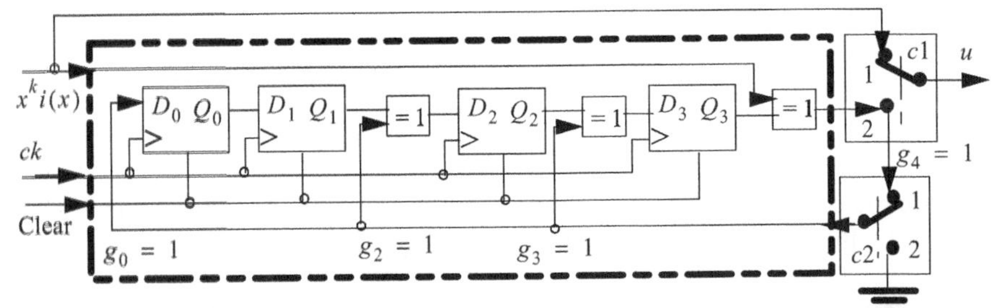

c1c2	ck	$x^k i(x)$	D_0	Q_0 D_1	Q_1	D_2	Q_2	D_3	Q_3	u_1
1 1		1 x^6	1	0	0	1	0	1	0	1
	1°			1	0		1		1	
		0 x^5	1	1		1		0		0
	2°			1	1		1		0	
		1 x^4	1	1		0		0		1
	3°			1	1		0		0	

Table 1.17. *Block diagram of the premultiplied encoder and encoder operation: internal state and input and output values for three clock cycles*

5) a) Detection of an odd number of errors.

If $g(x)$ can be set in the form $g(x) = (x+1)p(x)$, then $g(x)$ detects odd number of errors with $(x+1)$ (see Volume 1, Chapter 4).

$$
\begin{array}{l|l}
\; x^4 + x^3 + x^2 + 1 & \\
\oplus \qquad\qquad\qquad & \;\; x+1 \\
\quad x^4 + x^3 & \text{-----------} \\
\quad \text{-----} & x^3 + x + 1 \leftarrow q(x) \\
\quad x^2 + 1 & \\
\quad x^2 + x & \\
\quad \text{----} & \\
\quad x + 1 & \\
\quad x + 1 & \\
\quad \text{----} & \\
\quad 0 \;\; 0 &
\end{array}
$$

$$\rightarrow g(x) = (x+1) \times (x^3 + x + 1) = (x+1) \times p(x)$$

So, detection of an odd number of errors.

5) b) Detection of double errors.

The generator polynomial $g(x)$ contains the polynomial $p(x) = x^3 + x + 1$ that is primitive, because $2^3 - 1 = 7 = n$, so it makes it possible $p(x)$ to detect all the double errors.

6) Proportion of detectable error packets of length $l > 5$.

We have $k = 4$, so the proportion of detectable error packets of length $l > k + 1 \rightarrow l > 5$ is:

$$1 - 2^{-k} = 1 - 2^{-4} = 93.75\,\%$$

7) The matrix $[G_s]$ is taken from the matrix $[G]_{3,7}$ by shifting the positions of some columns verifying the expected form of $[G_s]_{m,n} = [G_s]_{3,7} = \left[I_{3,3} \mid P_{4,3}\right]$:

$$[G]_{3,7} = \begin{bmatrix} 1 & 1 & 1 & 0 & 1 & 0 & 0 \\ 0 & 1 & 1 & 1 & 0 & 1 & 0 \\ 0 & 0 & 1 & 1 & 1 & 0 & 1 \end{bmatrix} \rightarrow [G_s]_{3,7}$$

$$= \begin{bmatrix} 1 & 0 & 0 & | & 1 & 1 & 0 & 1 \\ 0 & 1 & 0 & | & 1 & 1 & 1 & 0 \\ 0 & 0 & 1 & | & 0 & 1 & 1 & 1 \end{bmatrix}$$

8) Form of the control matrix $[H]$:

$$[G_s]_{m,n} \times [H]_{k,n}^{t} = [0]_{m,k}$$

$$[G_s]_{m,n} = \left[I_{m,m} \mid P_{m,k}\right] \rightarrow [H]_{k,n} = \left[P_{m,k}^{t} \mid I_{k,k}\right]$$

$$[G_s]_{3,7} = \left[I_{3,3} \mid P_{3,4}\right] \rightarrow [H]_{4,7} = \left[P_{3,4}^{t} \mid I_{4,4}\right]$$

$$\rightarrow [H]_{4,7} = \begin{bmatrix} 1 & 1 & 0 & | & 1 & 0 & 0 & 0 \\ 1 & 1 & 1 & | & 0 & 1 & 0 & 0 \\ 0 & 1 & 1 & | & 0 & 0 & 1 & 0 \\ 1 & 0 & 1 & | & 0 & 0 & 0 & 1 \end{bmatrix}$$

9) Control bits as a function of information bits?

We have:

$$[H]_{4,7} \times [u]_{7,1} = [0]_{4,1}$$

$$\begin{bmatrix} 1 & 1 & 0 & 1 & 0 & 0 & 0 \\ 1 & 1 & 1 & 0 & 1 & 0 & 0 \\ 0 & 1 & 1 & 0 & 0 & 1 & 0 \\ 1 & 0 & 1 & 0 & 0 & 0 & 1 \end{bmatrix} \times \begin{bmatrix} u_1 \\ u_2 \\ u_3 \\ u_4 \\ u_5 \\ u_6 \\ u_7 \end{bmatrix} = \begin{bmatrix} 0 \\ 0 \\ 0 \\ 0 \end{bmatrix}$$

with:

u_1, u_2, u_3: information bits

u_4, u_5, u_6, u_7: control bits

$$\rightarrow \begin{cases} u_1 \oplus u_2 \oplus u_4 = 0 \\ u_1 \oplus u_2 \oplus u_3 \oplus u_5 = 0 \\ u_2 \oplus u_3 \oplus u_6 = 0 \\ u_1 \oplus u_3 \oplus u_7 = 0 \end{cases} \rightarrow \begin{cases} u_4 = u_1 \oplus u_2 \\ u_5 = u_1 \oplus u_2 \oplus u_3 \\ u_6 = u_2 \oplus u_3 \\ u_7 = u_1 \oplus u_3 \end{cases}$$

10) Implementation scheme of the pseudo-random number generator (PRNG) based on:

$$g(x) = x^4 + x^3 + x^2 + 1$$

The implementation scheme of the pseudo-random generator (PRNG) based on the polynomial $g(x)$ is given in Figure 1.16.

NOTE.– At start, the initial state $[Q_3, Q_2, Q_1, Q_0]$ of the register must be different from zero.

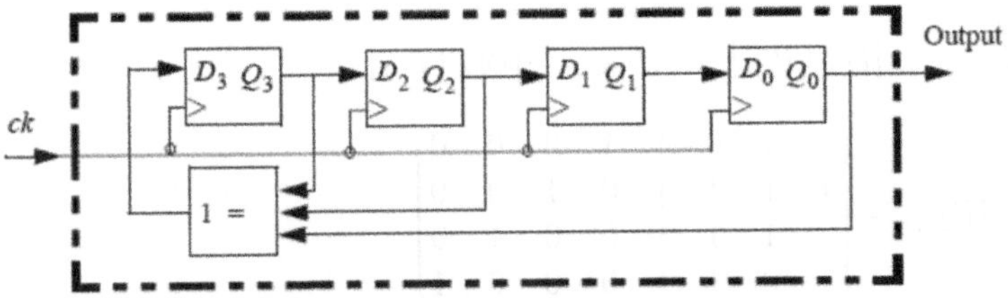

Figure 1.16. *Implementation scheme of the pseudo-random number generator (PRNG)*

1.15. Problem 15 – Cyclic code, M-sequences, and Gold sequences

We consider the problem of coding the information to be transmitted so as to protect it against transmission errors. For this purpose, a cyclic code C defined by its following generator polynomial: $g(x) = x^5 + x^2 + 1$, and $n = 31$ is used.

1) What is the necessary and sufficient condition for the proposed polynomial $g(x)$ to be primitive and generate a cyclic code?

It is desired to produce a systematic code C.

2) Give the expression of the codeword represented by the polynomial $u(x)$ corresponding to the information word represented by the polynomial: $i(x) = x^7 + x^4 + x + 1$.

3) Give the implementation scheme of the encoder based on a division circuit pre-multiplied by x^k, where k is the degree of the generator $g(x)$. Describe how it works.

4) Give the implementation scheme of the decoder associated with the code C allowing the detection of errors and explain how it works.

5) Does the generated cyclic code detect single, double or triple errors? Justify your answer in each case.

6) Determine the length-percentage pairs of error packets detectable by this code.

7) Give the wiring diagram of the pseudo-random number generator of maximum length (M-sequences), based on the primitive polynomial $g(x)$ defined above.

8) Give the expression of the generator polynomial $g_{rec}(x)$ reciprocal of the generator polynomial $g(x)$. What is the essential characteristic of the M-sequence generated by $g_{rec}(x)$ compared to that generated by $g(x)$?

9) Give the number of M-sequences generated by $g(x)$ and the ratio between the maximum of cross-correlation and that of the autocorrelation.

10) Show that the generator $g_1(x) = x^5 + x^4 + x^2 + x + 1$ associated with $g(x)$ forms a preferred pair.

11) Give the wiring diagram of the Gold generator based on $g(x)$ and $g_1(x)$, to generate all the Gold sequences.

12) Give the number of Gold sequences generated and the ratio between the maximum of cross-correlation and that of the autocorrelation.

Solution of problem 15

1) The necessary and sufficient condition that $g(x)$ is primitive is that:

$$2^k - 1 = n = 2^5 - 1$$

so $g(x)$ is primitive.

The polynomial $g(x)$ is generating a cyclic code, if it divides $x^{31} + 1$ but does not divide $(x^{n_1} + 1)$, with $n_1 < 31$.

The generator $g(x)$ divides $x^{31} - 1$, because, after division we get a null remainder:

$$x^{31} + 1 = (x^5 + x^2 + 1)$$
$$\times (x^{26} + x^{23} + x^{21} + x^{20} + x^{17} + x^{16} + x^{15} + x^{14} + x^{13}$$
$$+x^9 + x^8 + x^6 + x^5 + x^4 + x^2 + 1)$$

2) Expression of the codeword represented by the polynomial $u(x)$ corresponding to the information word represented by the polynomial $i(x)$:

We have:

$$x^k i(x) = g(x) \times q(x) + c(x) \rightarrow x^k i(x) + c(x) = g(x) \times q(x) = u(x)$$

or again:

$$x^k i(x) = x^5 \times (x^7 + x^4 + x + 1) = x^{12} + x^9 + x^6 + x^5$$

$$\begin{array}{l|l} x^{12} + x^9 + x^6 + x^5 & x^5 + x^2 + 1 \leftarrow g(x) \\ \oplus & \text{-----------} \\ x^{12} + x^9 + x^7 & x^7 + x^2 + x + 1 \leftarrow q(x) \\ \text{-------} & \end{array}$$

$$x^7 + x^6 + x^5$$

$$x^7 + x^4 + x^2$$

$$\text{-------}$$

$$x^6 + x^5 + x^4 + x^2$$

$$x^6 + x^3 + x$$

$x^5 + x^4 + x^3 + x^2 + x$

$x^5 + x^2 + 1$

$c(x) = x^4 + x^3 + x + 1$

$$\rightarrow u(x) = \underbrace{x^{12} + x^9 + x^6 + x^5}_{x^k i(x)} + \underbrace{x^4 + x^3 + x + 1}_{c(x)}$$

3) Implementation scheme of the coder based on a division circuit premultiplied by x^k.

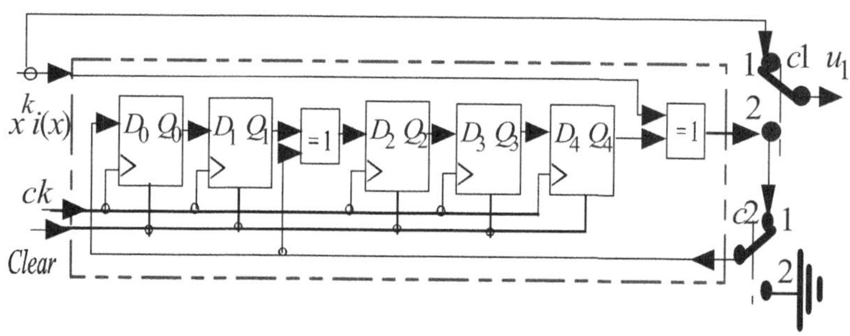

Figure 1.17. *Implementation scheme of the coder. For a color version of this figure, see www.iste.co.uk/assad/digital2.zip*

The operation of the encoder is as follows:

– resetting the D flip-flops;

– during $m = 8$ clock cycles, the multiplexers (Muxs) 1 and 2 are in position 1. The information bits are applied simultaneously to the divider and to the output. The k control bits ($k = 5$) are in the k flip-flops of the register;

– during k clock cycles, multiplexers (Muxs) 1 and 2 are in position 2; zeros enter the register and the control bits go out. The encoder uses $(m + k) = (8 + 5 = 13)$ clock cycles and the transmission channel is used throughout the operation. At the 13th clock cycle, the register flip-flops are zero and the encoder is ready to receive another information word to code. The encoder has a good efficiency.

4) Implementation scheme of the decoder associated with the code C.

The received word $v(x)$ is written:

$$v(x) = u(x) + \varepsilon(x)$$

with $\varepsilon(x)$ as a possible error word.

The syndrome is defined by:

$$s(x) = \text{Remainder}\left[\frac{v(x)}{g(x)}\right] = \text{Remainder}\left[\frac{u(x)}{g(x)}\right] + \text{Remainder}\left[\frac{\varepsilon(x)}{g(x)}\right]$$
$$= \text{Remainder}\left[\frac{\varepsilon(x)}{g(x)}\right]$$

So: if $\varepsilon(x)$ is non null, and if $v(x) \notin C(n,m)$, then $s(x) \neq 0$, hence the decoder implementation scheme.

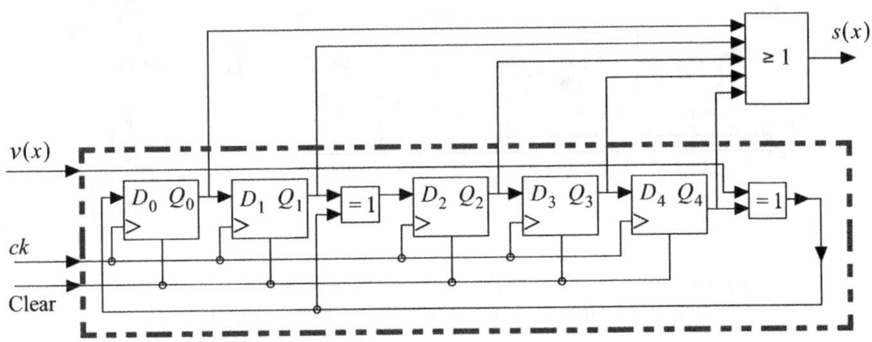

Figure 1.18. *Implementation scheme of the decoder. For a color version of this figure, see www.iste.co.uk/assad/digital2.zip*

The received word $v(x)$ is divided by $g(x)$ during $n = m + k = 8 + 5 = 13$ clock cycles. Then, the contents of the register are verified by a simple OR logic gate. If the content of the register is zero, then the received word is decided to be correct. Otherwise (the content of the register is not zero), the received word is decided to be erroneous.

5) Cyclic code capability to detect single, double or triple errors?

We know that:

$$s(x) = \text{Remainder}\left[\frac{\varepsilon(x)}{g(x)}\right]$$

Thus, error detection is possible if $g(x)$ does not divide $\varepsilon(x)$.

– *Single errors*: in this case an error in position i, represented by $\varepsilon(x) = x^i$ is not divisible by $g(x) = 1 + \cdots$, thus detection of all the single errors.

– *Triple errors*: in this case, $\varepsilon(x) = x^i + x^j + x^l$, and as $g(x) \neq (1 + x)p(x)$ then in principle, no detection of triple errors (see Volume 1, Chapter 4).

– *Double errors*: in this case, $\varepsilon(x)$ is of the form $\varepsilon(x) = x^i + x^j = x^i(x^{j-i} + 1)$ with $i < j < n$. Since $g(x)$ does not divide x^i, it suffices then that $g(x)$ does not divide either $(x^{j-i} + 1)$. The generator $g(x)$ divides $x^n + 1$ but does not divide $x^{n_1} + 1$, with $n_1 < n$, so $g(x)$ is said to be of order n. The primitive polynomials are irreducible. They detect all double errors because $(j - i) < n$.

6) Determination of the length-percentage pairs of detectable error packets.

An error packet that starts in position j and has a length l is written:

$$\varepsilon(x) = x^j + \varepsilon_{j+1}x^{j+1} + \cdots + x^{j+l-1}$$

where the first and the last coefficients of $\varepsilon(x)$ are at 1, the other coefficients can be at 1 or 0:

$$\varepsilon(x) = x^j \times \left[1 + \varepsilon_{j+1}x + \cdots + x^{l-1}\right] = x^j \times \varepsilon 1(x)$$

Three cases are encountered:

– $l - 1 < k$ $(k = 5)$ $\rightarrow l = 5$, hence detection at 100% of all the error packets of length $l \leq k$;

– $l - 1 = k$ $\rightarrow l = k + 1 = 6$, the proportion of error packets detectable is then: $1 - 2^{-(k-1)} = 1 - 2^{-4} = 0.9375$, i.e. 93.75% of the packets;

– $l - 1 > k$ $\rightarrow l > 6$, the proportion of error packets detectable is then: $1 - 2^{-k} = 1 - 2^{-5} = 0.9687$, i.e. 96.87%.

7) The implementation scheme of the pseudo-random number generator (PRNG) based on $g(x)$ is given in Figure 1.19.

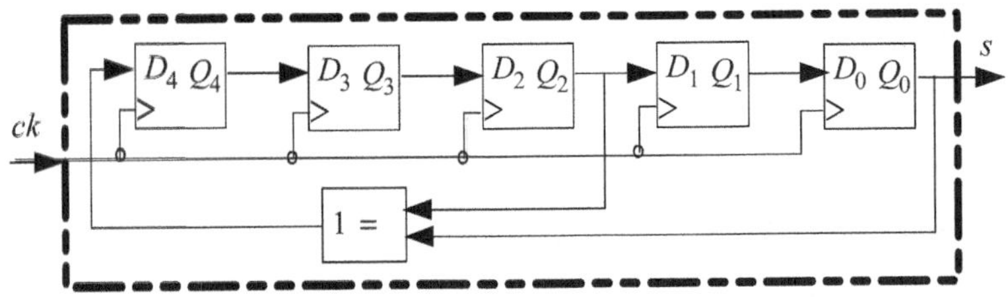

Figure 1.19. *Implementation scheme of the pseudo-random number generator based on* $g(x)$

NOTE.– At the start, the initial state of the D flip-flops $[Q_4, Q_3, Q_2, Q_1, Q_0]$ of the register should be different of zero.

8) Expression of the generator polynomial $g_{rec}(x)$ reciprocal of the generator $g(x)$.

We have:

$$g_{rec}(x) = x^k \times g(1/x) = x^5 \times (x^{-5} + x^{-2} + 1) = x^5 + x^3 + 1$$

The M-sequence generated by $g_{rec}(x)$ corresponds to the one generated by $g(x)$ but in a reverse sense.

9) Number of M-sequences generated by $g(x)$.

$k = 5$, so the number of M-sequences generated by $g(x)$ is 6 (see volume 1, chapter 4). The ratio $R_{sq_{Max}}/R_{ss}(0) = 0.35$ (see Volume 1, Chapter 4).

10) Does the generator polynomial $g_1(x) = x^5 + x^4 + x^2 + x + 1$, form with the polynomial $g(x)$ as a preferred pair?

Let α be a root of: $g(x) = x^5 + x^2 + 1$.

The polynomial $g_1(x) = x^5 + x^4 + x^2 + x + 1$, forms a preferred pair with $g(x)$ because:

$$\text{if } \begin{cases} 1)\ k \text{ is odd, since } k = 5, \text{conditions in 1) are satisfied} \\ 2)\ g_1(x) \text{ is such that } \alpha^{2^{\left[\frac{k-1}{2}\right]}+1} = \alpha^5 \text{ is a root of } g_1(x) \end{cases}$$

It means that $g(\alpha)$ divides $g_1(\alpha^5)$. This condition is also satisfied, because:

$$g_1(\alpha^5) = \alpha^{25} + \alpha^{20} + \alpha^{10} + \alpha^5 + 1$$

$$g_1(\alpha^5) = g(\alpha) \times (\alpha^{20} + \alpha^{17} + \alpha^{14} + \alpha^{12} + \alpha^{11} + \alpha^8 + \alpha^7 + \alpha^6 + \alpha^4 + \alpha^2 + 1)$$

with: $g(\alpha) = (\alpha^5 + \alpha^2 + 1)$

11) Implementation scheme of the Gold generator based on $g(x)$ and $g_1(x)$.

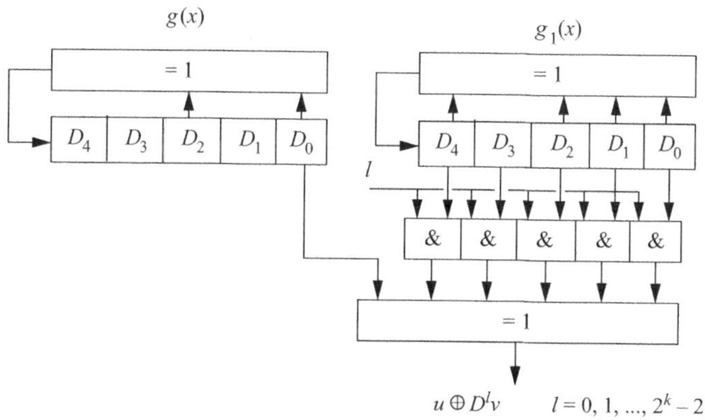

Figure 1.20. *Implementation scheme of the Gold generator*

12) The number of Gold sequences is $G(u, v) = \{u, v, u \oplus D^l v\}$, a set of $n + 2$ sequences. The ratio is $I(k)/R_{ss}(0) = 0.29$ (see Volume 1, Chapter 4).

2

Baseband Digital Transmission: Problems 16 to 26

2.1. Problem 16 – Entropy and information to signal source coding

We consider the problem of long-distance transmission over low cost electric cable of a compressed information source S from a video signal compression system. The compression system used makes sure that the source S delivers words s taken in a dictionary with only five words: $[s_1, s_2, s_3, s_4, s_5]$. The probabilities of issuing symbols are given in Table 2.1.

s	s_1	s_2	s_3	s_4	s_5
$Pr(s)$	0.11	0.19	0.40	0.21	0.09

Table 2.1. *Probability of emission of source S*

The symbols are delivered by the source with a rate of 13.5×10^6 symbols per second.

1) Determine the entropy $H(S)$ of the source S. Deduce the entropy rate per second. What would be the efficiency η_1 of a fixed-length code C_1, its length L_1, and the bitrate per second, denoted D_1?

2) Construct the Huffman code C_2 generating the codeword S_i associated with each of the symbols s_i.

NOTE.– In the design of the code C_2, the coding suffix associated with the element of lower probability will be systematically set to 1.

Deduce the average length L_2 of the codeword C_2, its efficiency η_2 and the bitrate per second D_2.

3) It is considered here that the source S delivers the following time sequence SS of symbols s:

$$\cdots\cdots s_2 \quad s_5 \quad s_3 \quad s_4 \quad s_3 \quad s_3 \quad s_2 \quad s_3 \quad s_5 \quad s_4 \cdots\cdots \longrightarrow \text{time}$$

Deduce the corresponding sequence SB of bits obtained at the output of the Huffman coding C_2. SB is of the form $\{\cdots\cdots b_{k-1}, b_k, b_{k+1} \cdots\cdots\}$. What do you observe?

4) The transmitter constructs a baseband signal supporting the transmitted information bits.

a) It first uses a bipolar encoder (called CODBip) of RZ type, with amplitude V and duration T_b. Draw a graph of the signal portion associated with SB transmitted by this CODBip encoder. What are the problems encountered in reception?

NOTE.– Both here and also in question b), we will consider, at the start of the sequence, that the parity flip-flop of the number of "1" is equal to 1.

b) It then uses an encoder (called CODHDB) of HDB-2 type. Draw a graph of the signal portion associated with SB transmitted by this CODHDB encoder. Are some problems solved now and why?

c) What is the approximate bandwidth of the signal emitted by the bipolar or HDB-2 code to encode the S source? (We can rely on the properties of the power spectral density $\Gamma(f)$ of the signal transmitted). To transmit this source of information on this type of cable, is there a good fit?

5) We want to reduce the bandwidth of the transmitted signal. Thus, it is desired to use an information-to-signal coder of partial response linear encoder type. This encoder will be very simple, of the form $1 - D^2$ (here D is the delay operator of T_b, time slice allocated to the transmission of a binary symbol). The signal $x(t)$ which carries the symbols c_k is of NRZ type, amplitude $V/2$ and duration T_b.

The following SBB binary sequence will be used for the rest of this problem:

0 0 1 0 0 0 1 1 0 1 1 1 0 0 1 1 0 0 0 1 0 1

a) This type of encoder needs to be preceded by an appropriate precoder. Why?

b) Describe the relationship between the pre-encoder output (giving the symbols b'_k) and its input b_k.

c) Describe the relationship between the encoder output (producing the symbols c_k) and its input a_k.

d) Represent graphically, the signal portion associated with SBB transmitted by this whole partial response coder (for a pulse amplitude modulation of duration T_b). It will be necessary to first determine the sequence obtained at the output of the precoder (it will be assumed that the two symbols b'_k not known at the beginning of the sequence are zero).

e) What is the approximate bandwidth of the signal emitted by this partial response linear code to encode the source S? Have we gained any bandwidth reduction?

6) For this partial response linear code, how does decoding produce the symbols b_k from c_k? Justify your answer. What happens to the reconstructed binary information $\hat{b}_k$ if a transmission error occurs for one of the symbols $\hat{c}_k$ reconstructed on reception?

Solution of problem 16

1) Entropy $H(S)$:

$$H(S) = -\sum_{i=1}^{5} p(s_i) \times \log_2 p(s_i)$$

$$H(S) = -\begin{Bmatrix} 0.11 \times \log_2(0.11) + 0.19 \times \log_2(0.19) \\ +0.4 \times \log_2(0.4) + 0.21 \times \log_2(0.21) + 0.09 \times \log_2(0.09) \end{Bmatrix}$$

$\cong 2.12$ bits of information/symbol

Entropy bitrate:

$$D(S) = H(S) \times 13.5 \times 10^6 = 28.26 \text{ Mbits of information/s}$$

Fixed-length code C_1 (length L_1): we have five symbols to encode, hence: $L_1 = 3$ bits.

Efficiency:

$$\eta_1 = \frac{H(S)}{L_1} = \frac{2.12}{3} = 70.67\ \%$$

Bitrate of code C_1:

$$D_1 = 3 \times 13.5 \times 10^6 = 40.5 \text{ Mbits/s}$$

2) Huffman code C_2.

s_i	$p(s_i)_0$	$p(s_i)_1$	$p(s_i)_2$	$p(s_i)_3$	C_2
s_3	0.4	0.4	0.4	0.6 0	1
s_4	0.21	0.21	0.39 0	0.4 1	0 1
s_2	0.19	0.20 0	0.21 1		0 0 1
s_1	0.11 0	0.19 1			0 0 0 0
s_5	0.09 1				0 0 0 1

Table 2.2. *Construction of Huffman code C_2*

Average length of the codewords:

$$L_2 = \sum_{i=1}^{5} p(s_i) \times l_i = 0.11 \times 4 + 0.19 \times 3 + 0.4 \times 1 + 0.21 \times 2 + 0.09 \times 4$$

$$= 2.19 \text{ bits/symbol}$$

Efficiency of code C_2:

$$\eta_2 = \frac{H(S)}{L_2} = \frac{2.12}{2.19} = 96.8\ \%$$

Bitrate of code C_2:

$$D_2 = 2.19 \times 13.5 \times 10^6 = 29.565 \text{ Mbit/s}$$

3) Sequence SB of bits obtained at the output of the Huffman coding.

SS	s_2	s_5	s_3	s_4	s_3	s_3	s_2	s_3	s_5	s_4
SB	001	0001	1	01	1	1	001	1	0001	01

Table 2.3. *Construction of binary sequence SB*

More bits are observed at zero than at one and, in addition, we have sequences of three consecutive zeros from time to time.

4) Information for baseband signal encoding.

a) Bipolar CODBip encoder of RZ type (see the graph in Table 2.4). The problems encountered in reception are related to the difficulties of getting a correct clock recovery in some cases, here three consecutive zeros, because the encoder produces no pulse for duration $3T$.

b) CODHDB coder of HDB-2 type: look at the graph in Table 2.4.

Sequences of three consecutive zeros are replaced by sequences of type "0 0 V" or "B 0 V" and thus, we can have a maximum duration of $2T$ without impulse.

c) The power spectral density of the RZ bipolar code is given by:

$$\Gamma_{RZ}(f) = \frac{V^2 T_b}{4}\left[\frac{\sin(\pi f\, T_b/2)}{\pi f\, T_b/2}\right]^2 \times [\sin(\pi f T_b)]^2$$

The zeros of $\Gamma_{RZ}(f)$ occur every $1/T_b$, so its bandwidth is $1/T_b$. It is substantially the same for the HDB-n code (here HDB-2).

Furthermore, in the vicinity of the frequency $f = 0$, the power is zero, this code can therefore be used for long-distance cable transmission. However, the presence of long sequences of zeros is detrimental to the clock recovery, therefore we use the code HDB-n.

5) Partial response code (PRC).

a) Yes, it is necessary to use a pre-encoder so that on reception, the decoding is instantaneous (without recursion) and therefore, if a decoding error occurs, it does not propagate recursively.

b) Pre-encoder output (symbol b'_k) as a function of its input b_k.

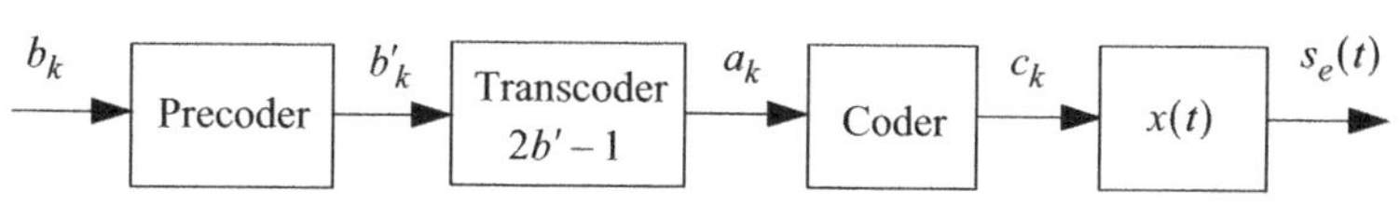

Figure 2.1. *Partial response linear coding scheme*

$$H(D) = 1 - D^2 \Leftrightarrow H(z) = 1 - z^{-2} \Rightarrow \frac{B'_k(z)}{B_k(z)} = \frac{1}{H(z)} = \frac{1}{1 - z^{-2}}$$

$$\Leftrightarrow B'_k(z) = B_k(z) + B'_k(z) \times z^{-2} \Rightarrow b'_k = b_k \oplus b'_{k-2}$$

c) Relationship between the encoder output (symbol c_k) and its input a_k:

$$H(z) = 1 - z^{-2} = \frac{C(z)}{A(z)} \Leftrightarrow C(z) = A(z) - A(z) \times z^{-2} \Rightarrow c_k = a_k - a_{k-2}$$

d) Graphical representation of the signal portion associated with SBB transmitted by this whole partial response coder: look at the graph in Table 2.4.

e) The power spectral density of the partial response code concerned is given by (see Volume 1, Chapter 5):

$$\Gamma_{s_e}(f) = V^2 T_b \left[\frac{\sin(\pi f T_b)}{\pi f T_b}\right]^2 \times [\sin(2\pi f T_b)]^2$$

The zeros of $\Gamma_{s_e}(f)$ take place every $1/2T_b$, its bandwidth is then approximately $1/2T_b$.

Thus, there is a reduction of the bandwidth of the transmitted signal by a factor of 2, compared to question 4, because of the introduction of the correlation. In addition, one has no continuous component.

6) We have:

$$c_k = a_k - a_{k-2} = [(2b'_k - 1) - (2b'_{k-2} - 1)] = 2(b'_k - b'_{k-2})$$

$$= 2(b'_k \oplus b'_{k-2}) = 2b_k \Rightarrow \hat{b}_k = \frac{1}{2}\hat{c}_k \mod 2$$

hence, an instantaneous decoding (no recursion). Thus:

– if no decision error on $\hat{c}_k$, then, no decision error on $\hat{b}_k$;

– if decision error on $\hat{c}_k$, then decision error on $\hat{b}_k$ only (not on subsequent ones).

$\{b_k\}$	0	0	1	0	0	0	1	1	0	1	1	1	0	0	1	1	0	0	0	1	0	1
CODBip (V, 0, -V), 1 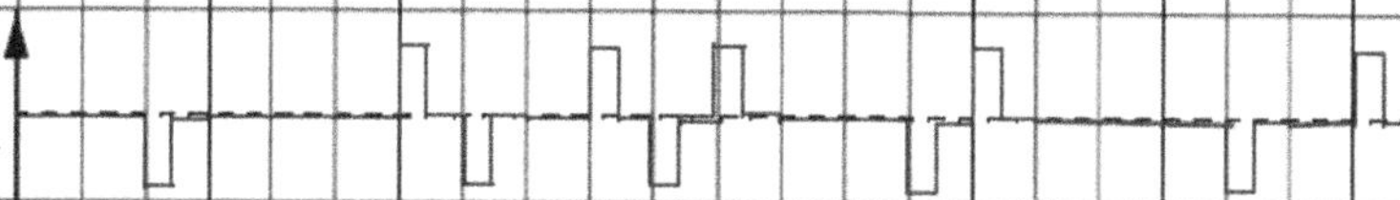																						
HDB-2 (V, 0, -V), 1 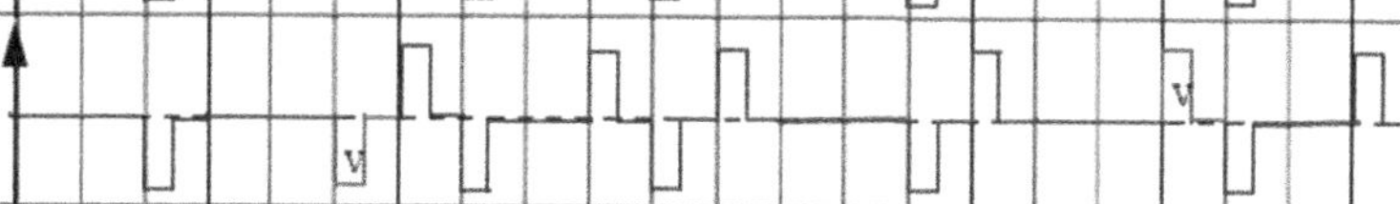																						
$\{b'_k\}$ 0 0	0	0	1	0	1	0	0	1	0	0	1	1	1	1	0	0	0	0	0	1	0	0
$\{a_k\}$ -1 -1	-1	-1	1	-1	1	-1	-1	1	-1	-1	1	1	1	1	-1	-1	-1	-1	-1	1	-1	-1
$\{c_k\}$	0	0	2	0	0	0	-2	2	0	-2	2	2	0	0	-2	-2	0	0	0	2	0	-2
PRC $s_e(t)$ (V, 0, -V) 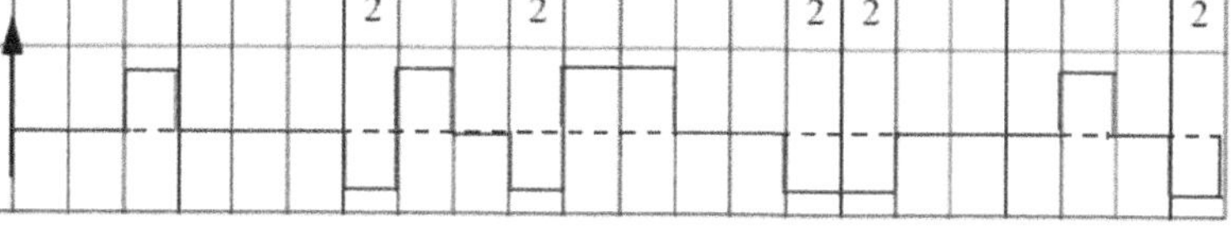																						

Table 2.4. *Chronograms of the signals and coded symbols. For a color version of this table, see www.iste.co.uk/assad/digital2.zip*

2.2. Problem 17 – Calculation of autocorrelation function and power spectral density by probabilistic approach of RZ and NRZ binary on-line codes

We consider the digital transmission signal defined by:

$$s(t) = \sum_{n=-\infty}^{\infty} a_n x(t - nT - t_0)$$

and:

$$x(t) = \begin{cases} V & \text{for} \quad 0 \le t < \theta \\ 0 & \text{for} \quad \theta \le t < T \end{cases}$$

with:

$$0 < \theta \le T \quad \text{and} \quad V > 0$$

The symbols a_n are independent random variables which can take only the values 0 and 1 with a probability equal to 1/2 and t_0 is random, of uniform law on the time interval $[0, T[$.

NOTE.– In this problem, we consider that the instant t belongs to the time interval $[t_0, t_0 + T[$. Without any loss of generalities, we will assign the index n to this interval where the time t is *a priori*.

Calculate the autocorrelation function $R_s(\tau)$ and the power spectral density $\Gamma_s(f)$ of $s(t)$ and carry out the two particular cases:

$$\theta = T/2 \text{ (binary RZ)} \qquad \text{and} \qquad \theta = T \text{ (binary NRZ)}$$

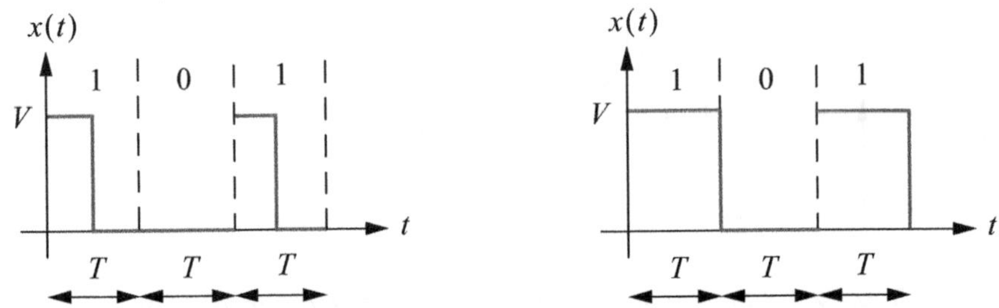

Figure 2.2. *Examples of binary RZ and NRZ codes. For a color version of this figure, see www.iste.co.uk/assad/digital2.zip*

Solution of problem 17

Let us make the following form of the signal $s(t)$ (with $\theta < T/2$ on Figure 2.3).

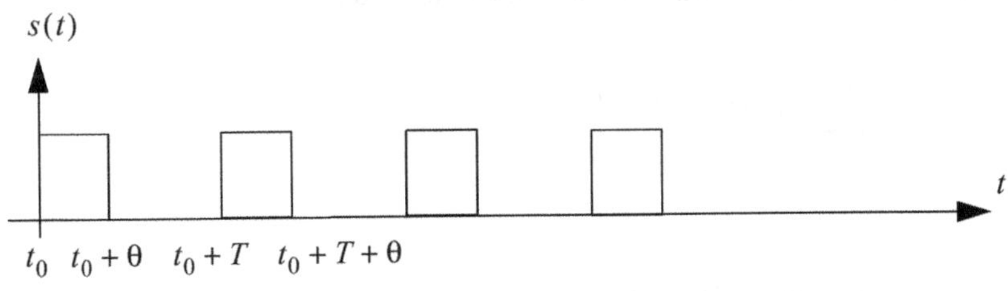

Figure 2.3. *Example of signal $s(t)$ waveform with $\theta < T/2$*

Such a signal can be represented by:

$$s(t) = \sum_{n=-\infty}^{\infty} a_n x(t - nT - t_0)$$

With the deterministic function (pulse):

$$x(t) = \begin{cases} V & \text{for } \ t \in [0, \theta[\\ 0 & \text{elsewhere} \end{cases}$$

and:

$$a_n = \begin{Bmatrix} 1 \\ 0 \end{Bmatrix} \quad \text{with} \quad Pr(a_n = 1) = Pr(a_n = 0) = 1/2$$

Since t_0 is uniform on time interval $[0, T[$, then $s(t)$ is a second-order stationary random signal.

Moreover, the autocorrelation function $R_s(\tau)$ is an even function: $R_s(-\tau) = R_s(\tau)$ and thus the calculation can be done with suitability with $\tau \geq 0$ or $\tau \leq 0$.

The autocorrelation function $R_s(\tau)$ is written:

$$R_s(\tau) = E[s(t) \times s(t - \tau)] = \sum_i \sum_j s_i s_j Pr\{s(t) = s_i \ \text{ and } \ s(t - \tau) = s_j\}$$

$$R_s(\tau) = \begin{cases} V \times V \times Pr\{V \ \text{at} \ t \ \text{ and } \ V \ \text{at} \ t - \tau\} + \\ 0 \times V \times Pr\{0 \ \text{at} \ t \ \text{ and } \ V \ \text{at} \ t - \tau\} + \\ V \times 0 \times Pr\{V \ \text{at} \ t \ \text{ and } \ 0 \ \text{at} \ t - \tau\} + \\ 0 \times 0 \times Pr\{0 \ \text{at} \ t \ \text{ and } \ 0 \ \text{at} \ t - \tau\} \end{cases}$$

hence:

$$R_s(\tau) = V^2 \times Pr\{V \ \text{at} \ t \ \text{ and } \ V \ \text{at} \ t - \tau\}$$

Moreover, using the theorem of compound probability, $R_s(\tau)$ is also written:

$$R_s(\tau) = V^2 \times Pr\{V \ \text{at} \ t\} \times Pr\{V \ \text{at} \ t - \tau / V \ \text{at} \ t\}$$

To calculate $R_s(\tau)$, it is sufficient to calculate $Pr\{V \ \text{at} \ t\}$ and $Pr\{V \ \text{at} \ t - \tau / V \ \text{at} \ t\}$.

Two cases can be encountered: $\begin{cases} \text{case}: 0 < \theta \leq T/2 \\ \text{case}: T/2 < \theta \leq T \end{cases}$

A. First case, where $0 < \theta \leq T/2$

There are three situations to consider.

– *First situation*: $0 \leq \tau \leq \theta$.

The only possibility is that t and $t-\tau$ belong to the same first part of time slice (hatched region).

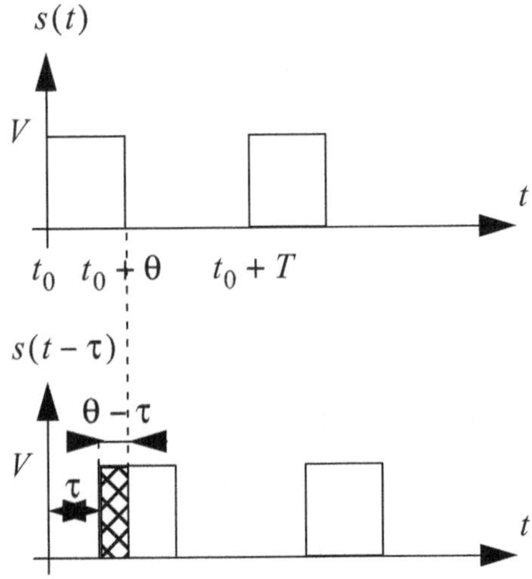

Figure 2.4. *First situation: $0 \leq \tau \leq \theta$ (with $\theta \leq T/2$)*

Let:

$$p = Pr\{V \text{ at } t\} = Pr\{t \in [t_0, t_0 + \theta[\} \times Pr\{a_n = 1\} = \frac{\theta}{T} \times \frac{1}{2}$$

and:

$$\begin{aligned} q &= Pr\{V \text{ at } t - \tau / V \text{ at } t\} \\ &= Pr\{t - \tau \in [t_0, t_0 + \theta[, a_n = 1 / t \in [t_0, t_0 + \theta[, a_n = 1\} \end{aligned}$$

or, since $a_n = 1$:

$$q = Pr\{t - \tau \in [t_0, t_0 + \theta[/ t \in [t_0, t_0 + \theta[\}$$

that is to say:

$$q = Pr\{t \in [t_0 + \tau, t_0 + \theta + \tau[/t \in [t_0, t_0 + \theta[\}$$

Or:

$$q = Pr\{t \in [t_0 + \tau, t_0 + \theta[\}$$

with random t_0, of *a posteriori* law uniform over a measurement interval θ, therefore with probability density $(ddp(t_0))$ equal to $1/\theta$.

Hence:

$$q = \int_{t_0+\tau}^{t_0+\theta} ddp(t_0)\, dt_0 = \frac{\theta - \tau}{\theta}$$

So finally:

$$R_s(\tau) = V^2 \times \frac{1}{2} \times \frac{\theta}{T} \times \left(\frac{\theta - \tau}{\theta}\right)$$

NOTE.– In general, if *a priori* t belongs to a time interval, of measurement T (t_0 is random, of uniform law on time interval $[0, T[$), we know *a posteriori* that t belongs to a first part of the time interval, of measure θ. It is therefore this posterior uniform law that is used in time conditional probabilities.

– *Second situation:* $\theta < \tau \leq T$.

The only possibility is that $s(t)$ and $s(t - \tau)$ come from two first parts of adjacent time slots, hence independence between the variables a considered.

In general, we will always have:

$$Pr\{a_{n-k} = \alpha_i / a_n = \alpha_j\}_{k \neq 0} = Pr\{a_{n-k} = \alpha_i\}$$

$$p = Pr\{V \text{ at } t\} = Pr\{t \in [t_0, t_0 + \theta[\} \times Pr\{a_n = 1\} = \frac{\theta}{T} \times \frac{1}{2}$$

$$q = Pr\{V \text{ at } t - \tau / V \text{ at } t\}$$

$$q = Pr\{t - \tau \in [t_0 - T, t_0 - T + \theta[, a_{n-1} = 1/t \in [t_0, t_0 + \theta[, a_n = 1\}$$

$$q = Pr\{t - \tau \in [t_0 - T, t_0 - T + \theta[/t \in [t_0, t_0 + \theta[\} \times Pr\{a_{n-1} = 1\}$$

and:

$$Pr\{t-\tau \in [t_0-T, t_0-T+\theta[/t \in [t_0, t_0+\theta[\}$$

turns into:

$$Pr\{t \in [t_0-T+\tau, t_0-T+\theta+\tau[/t \in [t_0, t_0+\theta[\}$$

Or:

$$Pr\{t \in [Sup[t_0-T+\tau, t_0], Inf[t_0-T+\theta+\tau, t_0+\theta\,]\,]/t \in [t_0, t_0+\theta[\}$$

Which depends on the value of τ with respect to θ and T.

Two situations can occur (see graph in Figure 2.5).

We know that if a real random variable follows a uniform law over a given interval $[c, d]$, then its probability density is equal to $1/Mes[c, d]$ and that:

$$Mes[c, d] = \begin{cases} d-c \text{ if } d \geq c \\ 0 \text{ otherwise} \end{cases}$$

and if $Mes[c, d] = 0$, then the probability density will be zero because we are faced here with a continuous random variable

if: $\theta < \tau \leq T$, then:

$$Sup[t_0-T+\tau, t_0] = t_0$$

$$Inf[t_0-T+\theta+\tau, t_0+\theta\,] = t_0-T+\theta+\tau$$

$$Inf[\quad] - Sup[\quad] = \tau+\theta-T = \tau-(T-\theta)$$

So if $\theta < \tau \leq T-\theta$, the measure of the interval is zero, thus $q = 0$.

Otherwise, if $T-\theta < \tau \leq T$, then the measure of the interval is given by: $\tau+\theta-T$, and:

$$q = \left[\frac{\tau+\theta-T}{\theta}\right] \times \frac{1}{2}$$

so if:

$$\theta < \tau \leq T-\theta: \rightarrow q = 0$$

and if:

$$T - \theta < \tau \leq T : \rightarrow q = \left[\frac{\tau + \theta - T}{\theta}\right] \times \frac{1}{2}$$

$- Third\ situation: \tau > T.$

$$T < \tau < T + \theta : \rightarrow q = \left[\frac{T + \theta - \tau}{\theta}\right] \times \frac{1}{2}$$

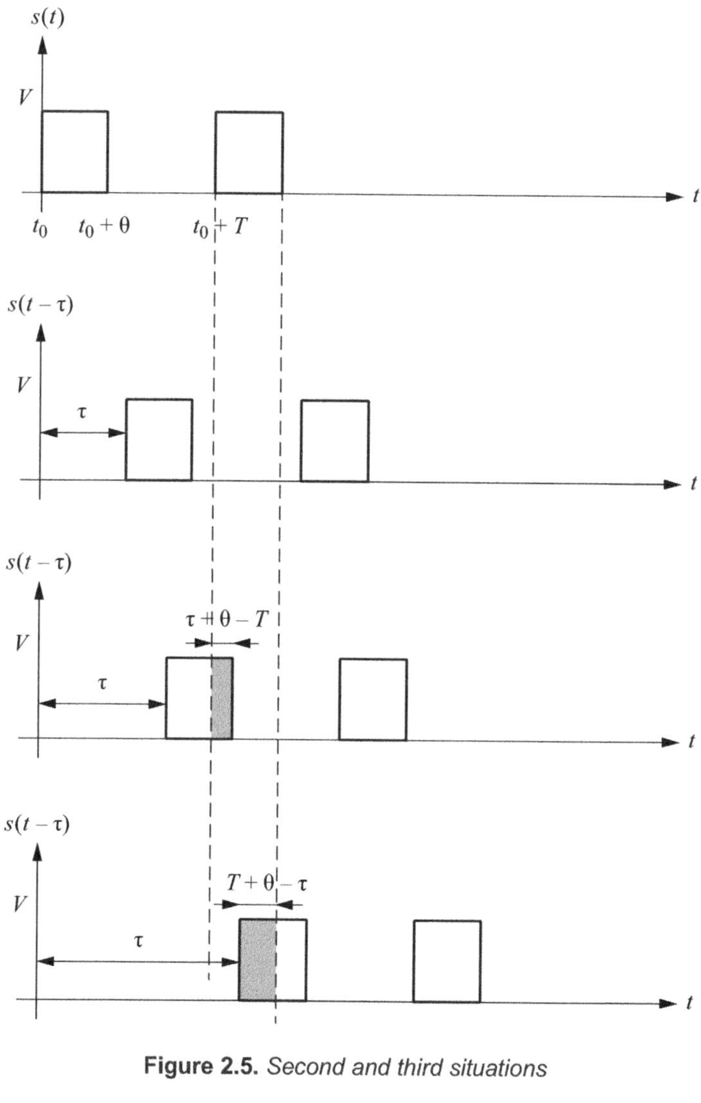

Figure 2.5. *Second and third situations*

In summary, for $0 < \theta \leq T/2$, if:

$$\begin{cases} 0 \leq \tau \leq \theta & \Rightarrow R_s(\tau) = V^2 \times \dfrac{\theta}{2T} \times \left[\dfrac{\theta - \tau}{\theta}\right] \\ \theta < \tau < T - \theta & \Rightarrow R_s(\tau) = 0 \\ T - \theta \leq \tau \leq T & \Rightarrow R_s(\tau) = V^2 \times \dfrac{\theta}{4T} \times \left[\dfrac{\tau + \theta - T}{\theta}\right] \\ T \leq \tau \leq T + \theta & \Rightarrow R_s(\tau) = V^2 \times \dfrac{\theta}{4T} \times \left[\dfrac{T + \theta - \tau}{\theta}\right] \end{cases}$$

For values $\tau > T + \theta$, we see that the intervals intervening in the conditional probability relating to instants $(t - \tau)$ and t are different and therefore $n - k < n$. This implies that conditional probability $Pr\{a_{n-k}/a_n\}_{k\neq 0} = Pr\{a_{n-k}\}$, and that the probability relating $(t - \tau)$ conditionally to t will evolve periodically, from T to T, strictly as we have just described it. More precisely, we have:

– for: $\theta + kT \leq \tau < (k+1)T - \theta$, then $q = 0$;

– for: $(k+1)T - \theta \leq \tau < (k+1)T$, then:

$$q = \frac{[\tau + \theta - (k+1)T]}{\theta} \times \frac{1}{2}$$

– for: $(k+1)T \leq \tau < (k+1)T + \theta$, then:

$$q = \frac{[(k+1)T + \theta - \tau]}{\theta} \times \frac{1}{2}$$

– for: $(k+1)T + \theta \leq \tau < (k+2)T - \theta$, then $q = 0$.

It is therefore a periodic function of period T.

Note that the autocorrelation function breaks down into a sum of two functions:

$$R_s(\tau) = R_1(\tau) + R_2(\tau)$$

with $R_1(\tau)$, a non-periodic function, and $R_2(\tau)$, a periodic function of period T.

Calculation of the power spectral density $\Gamma_s(f)$:

$$\Gamma_s(f) = F\{R_s(\tau)\} = \Gamma_1(f) + \Gamma_2(f)$$

Calculation of the power spectral density $\Gamma_1(f)$:

$$\Gamma_1(f) = \int_{-\infty}^{\infty} R_1(\tau)\exp[-j2\pi f\tau]d\tau$$

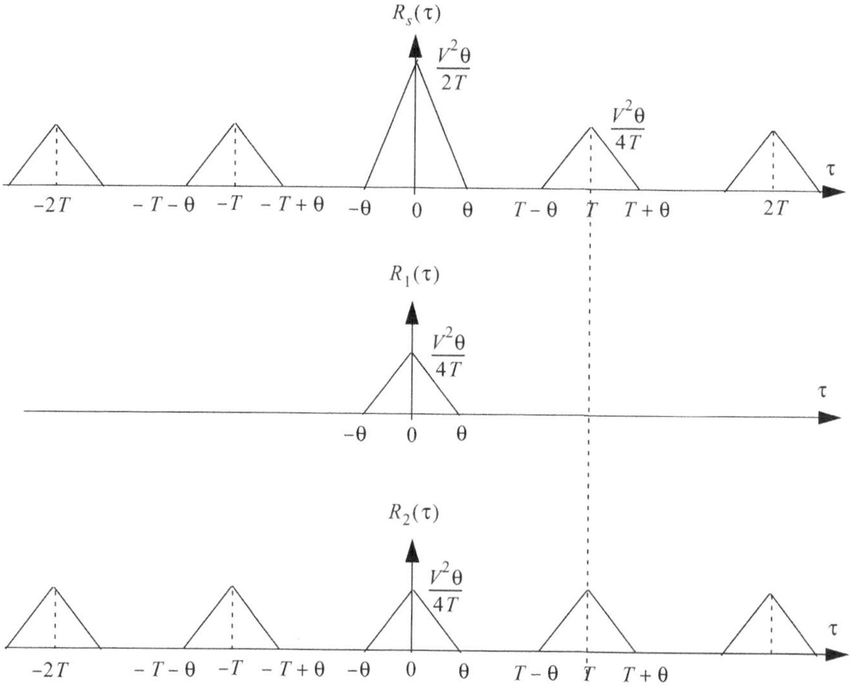

Figure 2.6. *Autocorrelation function $R_s(\tau)$ and its decomposition*

$R_1(\tau)$ is an even symmetric function, hence:

$$\Gamma_1(f) = 2\int_0^\theta \frac{V^2}{4T}[\theta - \tau]\cos(2\pi f\tau)\,d\tau$$

$$= \frac{V^2}{2T}\left\{\theta\int_0^\theta \cos(2\pi f\tau)\,d\tau - \int_0^\theta \tau\cos(2\pi f\tau)\,d\tau\right\}$$

but:

$$\int_0^\theta \cos(2\pi f\tau)\,d\tau = \frac{[\sin(2\pi f\tau)]_0^\theta}{2\pi f} = \frac{1}{2\pi f}\times\sin(2\pi f\theta)$$

and:

$$\int_0^\theta \tau\cos(2\pi f\tau)\,d\tau = \int_0^\theta u dv = uv - \int_0^\theta v du$$

Integration by part, with:

$$u = \tau \quad \text{and} \quad v = \cos(2\pi f\tau)\,d\tau$$

hence:

$$\int_0^\theta \tau\cos(2\pi f\tau)\,d\tau = \frac{[\tau\sin(2\pi f\tau)]_0^\theta}{2\pi f} - \int_0^\theta \frac{\sin(2\pi f\theta)}{2\pi f}d\tau$$

$$= \frac{1}{2\pi f}\theta \times \sin(2\pi f\theta) + \frac{1}{(2\pi f)^2}[\cos(2\pi f\tau)]_0^\theta$$

$$= \frac{\theta}{2\pi f} \times \sin(2\pi f\theta) + \frac{1}{(2\pi f)^2}[\cos(2\pi f\theta) - 1]$$

$$\Gamma_1(f) = \frac{V^2}{2T}\left\{\frac{\theta}{2\pi f}\sin(2\pi f\theta) - \frac{\theta}{2\pi f}\sin(2\pi f\theta) - \frac{1}{(2\pi f)^2}[\cos(2\pi f\theta) - 1]\right\}$$

$$\Gamma_1(f) = \frac{V^2}{2T}\left\{\frac{1}{(2\pi f)^2}[1 - \cos(2\pi f\theta)]\right\} = \frac{V^2}{2T}\left\{\frac{2[\sin(\pi f\theta)]^2}{(2\pi f)^2}\right\}$$

$$= \frac{V^2\theta^2}{4T}\left[\frac{\sin(\pi f\theta)}{\pi f\theta}\right]^2$$

which is a classic result. Indeed, recall that if:

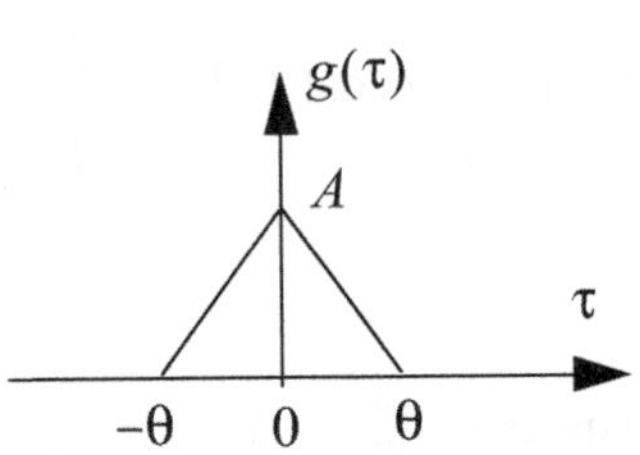

Figure 2.7. *Symmetrical triangular function*

then:

$$G(f) = A \times \theta \times \left[\frac{\sin(\pi f \theta)}{\pi f \theta}\right]^2$$

$\Gamma_1(f)$ has a continuous spectrum.

Calculation of $\Gamma_2(f)$.

The basic form of $R_2(\tau)$ is identical to that of $R_1(\tau)$ and in addition, it is periodic with period T, then:

$$R_2(\tau) = \sum_{n=-\infty}^{\infty} R_1(\tau - nT) = R_1(\tau) \otimes \sum_{n=-\infty}^{\infty} \delta(\tau - nT)$$

$$\Gamma_2(f) = F\{R_2(\tau)\} = F\{R_1(\tau)\} \times F\left\{\sum_{n=-\infty}^{\infty} \delta(\tau - nT)\right\}$$

$$= \Gamma_1(f) \times \frac{1}{T} \sum_{k=-\infty}^{\infty} \delta\left(f - \frac{k}{T}\right)$$

$$\Gamma_2(f) = \frac{V^2\theta^2}{4T}\left[\frac{\sin(\pi f \theta)}{\pi f \theta}\right]^2 \times \frac{1}{T} \sum_{k=-\infty}^{\infty} \delta\left(f - \frac{k}{T}\right)$$

$$= \frac{V^2\theta^2}{4T^2} \sum_{k=-\infty}^{\infty} \left[\frac{\sin(k\pi\theta/T)}{k\pi\theta/T}\right]^2 \delta\left(f - \frac{k}{T}\right)$$

It is a discrete spectrum, the discrete spectral components being spaced $1/T$ apart from each other.

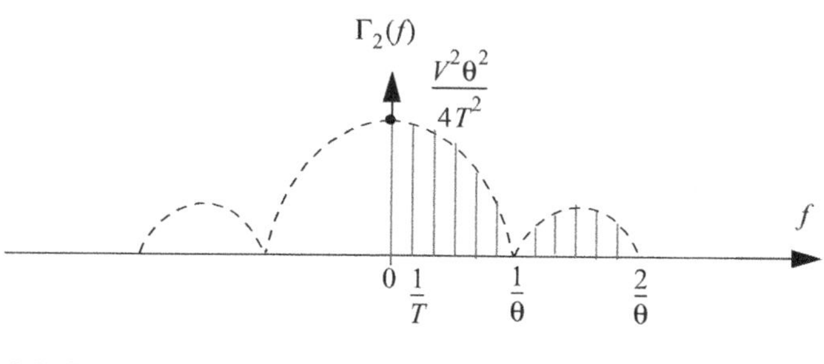

Figure 2.8. *Discrete spectral components of the power spectral density $\Gamma_2(f)$. For a color version of this figure, see www.iste.co.uk/assad/digital2.zip*

Special case where $\theta = T/2$: binary RZ code.

In this case, we then have:

$$\Gamma_1(f) = \frac{V^2 T}{16}\left[\frac{\sin(\pi f T/2)}{\pi f T/2}\right]^2$$

and:

$$\Gamma_2(f) = \frac{V^2}{16}\sum_{k=-\infty}^{\infty}\left[\frac{\sin(k\pi/2)}{k\pi/2}\right]^2 \delta\left(f - \frac{k}{T}\right)$$

The continuous component is such that:

$$\Gamma_2(0) = [\text{continuous component}]^2 \times \delta(f)$$

One has:

$$\Gamma_2(0) = \frac{V^2}{16}\left\{\lim_{k\to 0}\left[\frac{\sin(k\pi/2)}{k\pi/2}\right]^2\right\}\delta(f) = \left[\frac{V}{4}\right]^2 \delta(f)$$

Thus, the continuous component is then equal to $V/4$.

The function $\sin(k\pi/2)$ is non zero for odd integer k, so by setting $k = 2n + 1$, the expression of $\Gamma_2(f)$ is then written:

$$\Gamma_2(f) = \frac{V^2}{16}\delta(f) + \frac{V^2}{16}\sum_{n=-\infty}^{\infty}\left[\frac{\sin[(2n+1)\,\pi/2]}{(2n+1)\,\pi/2}\right]^2 \delta\left(f - \frac{(2n+1)}{T}\right)$$

$$\Gamma_2(f) = \frac{V^2}{16}\delta(f) + \frac{V^2}{16}\sum_{n=-\infty}^{\infty}\left[\frac{4}{(2n+1)^2\pi^2}\right]\delta\left(f - \frac{(2n+1)}{T}\right)$$

This is also written:

$$\Gamma_2(f) = \left[\frac{V}{4}\right]^2 \delta(f) + \left[\frac{V}{4}\right]^2 \sum_{n=-\infty}^{\infty}\left[\frac{2}{(2n+1)\pi}\right]^2 \delta\left(f - \frac{(2n+1)}{T}\right)$$

$\Gamma_2(f)$ has discrete components at odd frequencies and in particular at the frequency $f = 1/T$ which makes it possible to recover the clock.

B. Second case, the one where $T/2 < \theta \leq T$

There are also three situations to consider.

– *First situation*: $0 < \tau \leq \theta$.

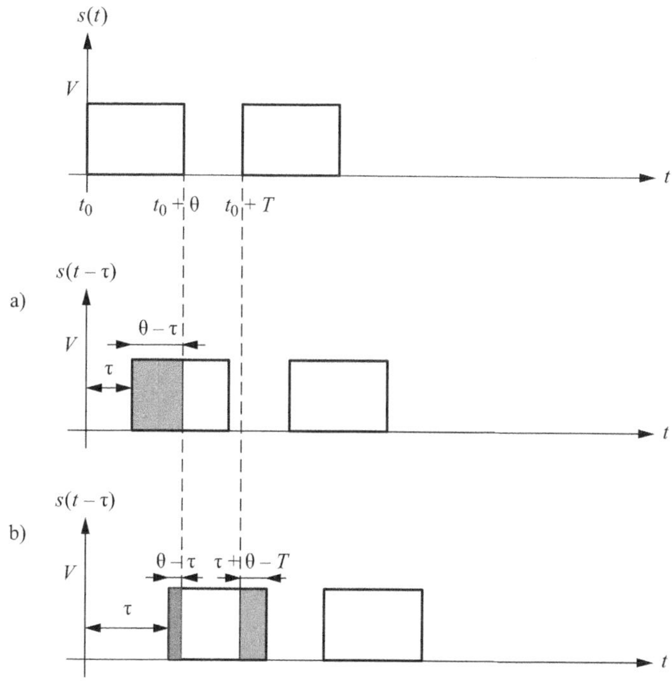

Figure 2.9. *First situation:* $0 < \tau \leq \theta$ *(with* $T/2 < \theta \leq T$*). For a color version of this figure, see www.iste.co.uk/assad/digital2.zip*

Calculation of p and q. There are two cases:

a) $\tau + \theta \leq T \rightarrow t$ and $t - \tau \in$ the same first part of time slice:

$$p_a = Pr\{V \text{ at } t\} = Pr\{t \in [t_0, t_0 + \theta[\} \times Pr\{a_n = 1\} = \frac{\theta}{T} \times \frac{1}{2}$$

$$q_a = Pr\{t - \tau \in [t_0, t_0 + \theta[/t \in [t_0, t_0 + \theta[\} = \frac{\theta - \tau}{\theta}$$

b) $\tau + \theta > T$, there are two possibilities:

$$\begin{cases} t \text{ and } t - \tau \notin \text{same first part of the time slice} & \rightarrow p_1, q_1 \\ \text{or } t \text{ and } t - \tau \in \text{same first part of the time slice} & \rightarrow p_2, q_2 \end{cases}$$

First possibility: t and $t - \tau$ do not belong to the same first part of the time slice, but to the first parts of adjacent time slices:

$$p_1 = Pr\{V \text{ at } t\} = Pr\{t \in [t_0, t_0 + \theta[\} \times Pr\{a_n = 1\} = \frac{\theta}{T} \times \frac{1}{2}$$

$$q_1 = Pr\{V \text{ at } t - \tau / V \text{ at } t\}$$

$$q_1 = Pr\{t - \tau \in [t_0 - T, t_0 - T + \theta[, a_{n-1} = 1 / t \in [t_0, t_0 + \theta[, a_n = 1\}$$

Or:

$$q_1 = Pr\{t - \tau \in [t_0 - T, t_0 - T + \theta[/ t \in [t_0, t_0 + \theta[\} \times Pr\{a_{n-1} = 1\}$$
$$= \left[\frac{\tau + \theta - T}{\theta}\right] \times \frac{1}{2}$$

Second possibility: t and $t - \tau$ belong to the same first part of the time slice, hence:

$$p_2 = p_a = p_1 = \frac{\theta}{T} \times \frac{1}{2} \quad \text{and} \quad q_2 = q_a = \frac{\theta - \tau}{\theta}$$

$$q_b = q_1 + q_2 = \frac{1}{2}\left[\frac{\tau + \theta - T}{\theta}\right] + \frac{\theta - \tau}{\theta} = \frac{1}{2}\left[\frac{3\theta - T - \tau}{\theta}\right]$$

NOTE.– Both hypotheses a) and b) exclude each other.

Thus, for:

$$0 < \tau \leq \theta \rightarrow \begin{cases} \text{a) } 0 < \tau \leq T - \theta & \rightarrow R_s(\tau) = \dfrac{V^2\theta}{2T}\left[\dfrac{\theta - \tau}{\theta}\right] = \dfrac{V^2}{2}\left[\dfrac{\theta - \tau}{T}\right] \\ \text{b) } T - \theta < \tau \leq \theta & \rightarrow R_s(\tau) = \dfrac{V^2\theta}{4T}\left[\dfrac{\tau - (T - \theta)}{\theta}\right] + \dfrac{V^2\theta}{2T}\left[\dfrac{\theta - \tau}{\theta}\right] \end{cases}$$

Note that in the case a), the expression of $R_s(\tau)$ is also valid throughout the time interval $\tau \in [0, \theta]$.

– *Second situation*: $\theta \leq \tau < T$.

The only possibility is that $s(t) = V$ and $s(t - \tau) = V$ come from the first two parts of adjacent time slices.

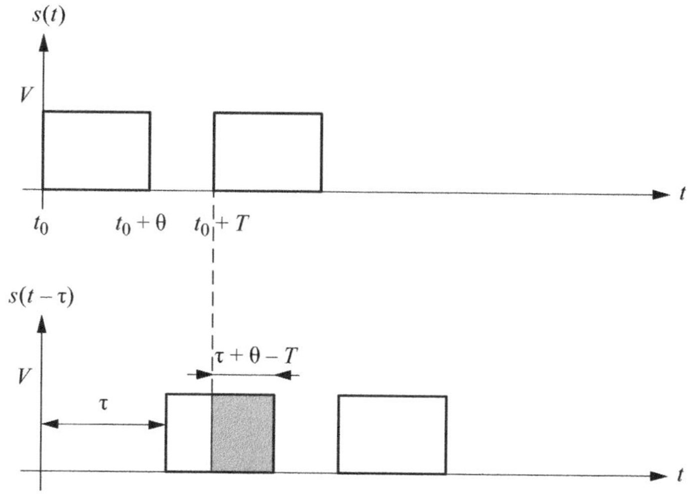

Figure 2.10. *Second situation: $\theta \le \tau < T$*

Let's set:

$$p = Pr\{V \text{ at } t\} = Pr\{t \in [t_0, t_0 + \theta[\} \times Pr\{a_n = 1\} = \frac{\theta}{T} \times \frac{1}{2}$$

and:

$$q = Pr\{V \text{ at } t - \tau / V \text{ at } t\}$$

$$q = Pr\{t - \tau \in [t_0 - T, t_0 - T + \theta[, a_{n-1} = 1 / t \in [t_0, t_0 + \theta[, a_n = 1\}$$

$$q = Pr\{t - \tau \in [[t_0 - T, t_0 - T + \theta[[/ t \in [t_0, t_0 + \theta[\} \times Pr\{a_{n-1} = 1\}$$
$$= \left[\frac{\tau + \theta - T}{\theta}\right] \times \frac{1}{2}$$

hence:

$$R_s(\tau) = \frac{V^2\theta}{4T}\left[\frac{\tau - (T - \theta)}{\theta}\right]$$

– *Third situation*: $T \le \tau < T + \theta$.

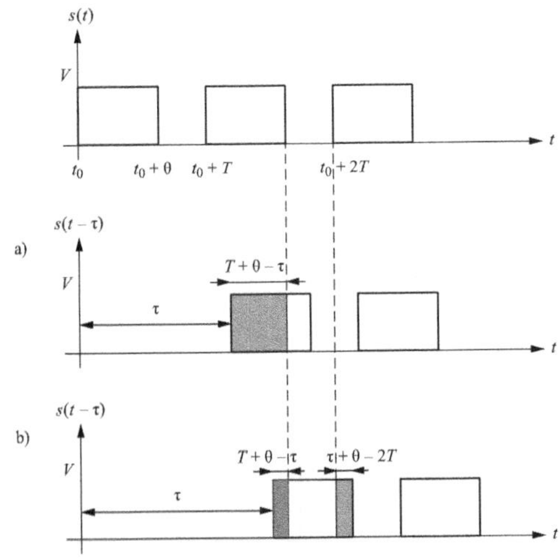

Figure 2.11. *Third situation: $T \leq \tau < T+\theta$. For a color version of this figure, see www.iste.co.uk/assad/digital2.zip*

Calculation of p and q. There are two cases:

a) $T < \tau \leq 2T - \theta \;\; \rightarrow t$ and $t - \tau \in$ to the first two parts of adjacent time slices:

$$p_a = Pr\{V \text{ at } t\} = Pr\{t \in [t_0, t_0 + \theta[\} \times Pr\{a_n = 1\} = \frac{\theta}{T} \times \frac{1}{2}$$

$$q_a = Pr\{V \text{ at } t - \tau / V \text{ at } t\}$$

Following the same approach as in the previous situation, we obtain:

$$q_a = Pr\{t - \tau \in [t_0 - T, t_0 - T + \theta[/t \in [t_0, t_0 + \theta[\} \times Pr\{a_{n-1} = 1\}$$
$$= \left[\frac{T + \theta - \tau}{\theta}\right] \times \frac{1}{2}$$

b) $2T - \theta < \tau < T + \theta$, there are two possibilities:

$$\begin{cases} t \text{ and } t - \tau \in \text{to the two first parts of time slices distant from } 2T & \rightarrow p_1, q_1 \\ \text{or } t \text{ and } t - \tau \in \text{ to the two first parts of adjacent time slices} & \rightarrow p_2, q_2 \end{cases}$$

$$p_1 = Pr\{V \text{ at } t\} = Pr\{t \in [t_0, t_0 + \theta[\} \times Pr\{a_n = 1\} = \frac{\theta}{T} \times \frac{1}{2}$$

$$q_1 = Pr\{V \text{ at } t - \tau / V \text{ at } t\}$$

$$q_1 = Pr\{t-\tau \in [t_0-2T, t_0-2T+\theta[/t \in [t_0, t_0+\theta[\} \times Pr\{a_{n-2}=1\}$$
$$= \left[\frac{\tau+\theta-2T}{\theta}\right] \times \frac{1}{2}$$

$$p_2 = p_a = \frac{\theta}{T} \times \frac{1}{2}$$

$$q_2 = q_a = \left[\frac{T+\theta-\tau}{\theta}\right] \times \frac{1}{2}$$

$$q_b = q_1 + q_2 = \frac{1}{2}\left[\frac{2\theta-T}{\theta}\right]$$

NOTE.– Both hypotheses a) and b) exclude each other.

Thus, for $T \leq \tau < T+\theta$:

$$\rightarrow \begin{cases} \text{a) } T < \tau \leq 2T-\theta & \rightarrow R_s(\tau) = \frac{V^2\theta}{4T}\left[\frac{T+\theta-\tau}{\theta}\right] = \frac{V^2}{4}\left[\frac{T+\theta-\tau}{T}\right] \\ \text{b) } 2T-\theta < \tau \leq T+\theta & \rightarrow R_s(\tau) = \frac{V^2\theta}{4T}\left[\frac{\tau+\theta-2T}{\theta}\right] + \frac{V^2\theta}{4T}\left[\frac{T+\theta-\tau}{\theta}\right] \\ & = \frac{V^2\theta}{4T}\left[\frac{2\theta-T}{\theta}\right] = \frac{V^2}{4}\left[\frac{2\theta-T}{T}\right] \end{cases}$$

Note that in case a), the expression of $R_s(\tau)$ is also valid throughout the time interval $[T \leq \tau \leq T+\theta]$.

In summary, for $T/2 < \theta \leq T$:

– for $0 < \tau \leq \theta$:

$$\rightarrow \begin{cases} \text{a) } \tau \leq T-\theta & \rightarrow R_s(\tau) = \frac{V^2}{2}\left[\frac{\theta-\tau}{T}\right] \\ \text{b) } T-\theta < \tau \leq \theta & \rightarrow R_s(\tau) = \frac{V^2\theta}{4T}\left[\frac{\tau-(T-\theta)}{\theta}\right] + \frac{V^2\theta}{2T}\left[\frac{\theta-\tau}{\theta}\right] = \frac{V^2}{4}\left[\frac{3\theta-T-\tau}{T}\right] \end{cases}$$

– for $\theta \leq \tau < T$:

$$\rightarrow R_s(\tau) = \frac{V^2\theta}{4T}\left[\frac{\tau-(T-\theta)}{\theta}\right] = \frac{V^2}{4}\left[\frac{\tau-(T-\theta)}{T}\right]$$

– for $T \leq \tau \leq T + \theta$:

$$\rightarrow \begin{cases} \text{a) } \tau \leq 2T - \theta & \rightarrow R_s(\tau) = \dfrac{V^2\theta}{4T}\left[\dfrac{T+\theta-\tau}{\theta}\right] = \dfrac{V^2}{4}\left[\dfrac{T+\theta-\tau}{T}\right] \\ \text{b) } 2T - \theta < \tau \leq T + \theta & \rightarrow R_s(\tau) = \dfrac{V^2\theta}{4T}\left[\dfrac{\tau+\theta-2T}{\theta}\right] + \dfrac{V^2\theta}{4T}\left[\dfrac{T+\theta-\tau}{\theta}\right] \\ & = \dfrac{V^2\theta}{4T}\left[\dfrac{2\theta-T}{\theta}\right] = \dfrac{V^2}{4}\left[\dfrac{2\theta-T}{T}\right] \end{cases}$$

As in the case $0 < \theta \leq T/2$, the probability relating to $(t - \tau)$ conditionally at t will evolve periodically from T to T, strictly as we have just described it above.

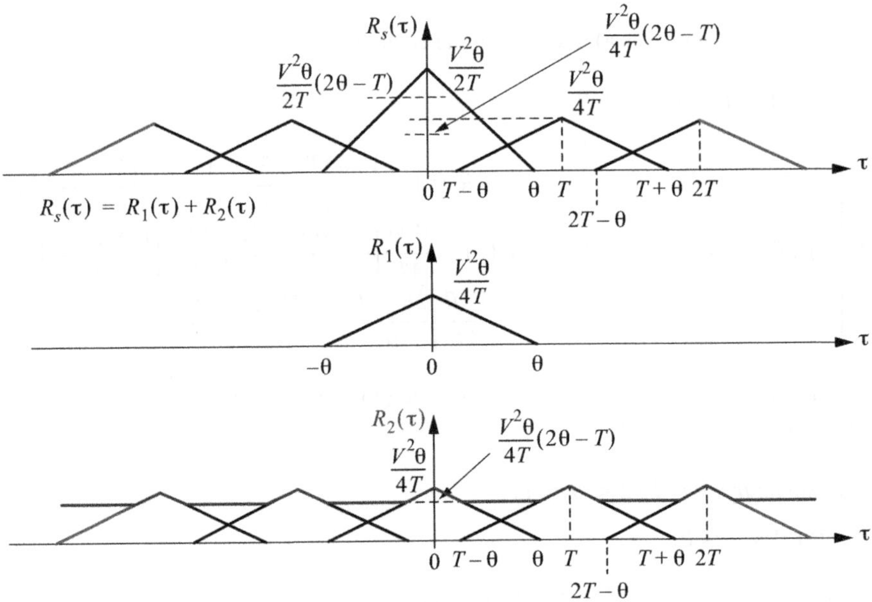

Figure 2.12. *Autocorrelation function $R_s(\tau)$ and its decomposition. For a color version of this figure, see www.iste.co.uk/assad/digital2.zip*

In the same way as for $0 < \theta \leq T/2$, the autocorrelation function breaks down into a sum of two functions:

$$R_s(\tau) = R_1(\tau) + R_2(\tau)$$

with $R_1(\tau)$, a non-periodic function, and $R_2(\tau)$, a periodic function, of period T. In addition, we note that the expression of $R_1(\tau)$ and $R_2(\tau)$ are the same as previously (case $0 < \theta \leq T/2$).

Calculation of the power spectral density, case where: $T/2 < \theta \leq T$.

The power spectral density of the signal $s(t)$ can be broken down into the sum of two functions:

$$\Gamma_s(f) = \Gamma_1(f) + \Gamma_2(f)$$

with:

$$\Gamma_1(f) = \frac{V^2\theta^2}{4T}\left[\frac{\sin(\pi f\theta)}{\pi f\theta}\right]^2$$

This is a continuous spectrum.

And:

$$\Gamma_2(f) = \Gamma_1(f) \times \frac{1}{T}\sum_{k=-\infty}^{\infty}\delta\left(f - \frac{k}{T}\right) = \frac{V^2\theta^2}{4T}\left[\frac{\sin(\pi f\theta)}{\pi f\theta}\right]^2 \times \frac{1}{T}\sum_{k=-\infty}^{\infty}\delta\left(f - \frac{k}{T}\right)$$

$$\Gamma_2(f) = \frac{V^2\theta^2}{4T^2}\sum_{k=-\infty}^{\infty}\left[\frac{\sin(\pi\theta\, k/T)}{\pi\theta\, k/T}\right]^2 \times \delta\left(f - \frac{k}{T}\right)$$

This is a discrete spectrum.

Special case: $\theta = T$ (NRZ code)

In this case, we obtain:

$$\Gamma_1(f) = \frac{V^2 T}{4}\left[\frac{\sin(\pi f T)}{\pi f T}\right]^2$$

$$\Gamma_2(f) = \frac{V^2}{4}\sum_{k=-\infty}^{\infty}\left[\frac{\sin(k\pi)}{k\pi}\right]^2 \times \delta\left(f - \frac{k}{T}\right)$$

So:

$$\Gamma_2(f) = \frac{V^2}{4}\delta(f) : \text{which gives the continuous component, equal to } V/2$$

Thus, finally we get:

$$\Gamma_s(f) = \frac{V^2T}{4}\left[\frac{\sin(\pi f T)}{\pi f T}\right]^2 + \frac{V^2}{4}\delta(f)$$

Notice that for $\theta = T$ (NRZ code), the signal does not have a discrete spectrum component at $1/T$ ($\Gamma_2(f)$ is zero for the other values of $k \neq 0$). Therefore, clock recovery is not easy with NRZ code.

NOTE.– One could easily generalize the problem to the situation where the symbols of information to be transmitted are not equiprobable: $Pr\{a = 0\} = \lambda$ and: $Pr\{a = 1\} = 1 - \lambda$.

2.3. Problem 18 – Calculation of the autocorrelation function and the power spectral density by probabilistic approach of the bipolar RZ code

The bipolar RZ code is a three-level code such as:

$$\text{if: } \begin{cases} b_n = 0 & \rightarrow a_n = 0 \\ b_n = 1 & \rightarrow a_n = \pm 1 \text{ alternately} \end{cases}$$

and:

$$x(t) = \begin{cases} V & \text{for } 0 \leq t \leq T/2 \\ 0 & \text{for } T/2 < t \leq T \end{cases}$$

The binary random variables are assumed to be equiprobable:

$$Pr\{b_n = 1\} = Pr\{b_n = 0\} = 1/2$$

We consider the digital transmission signal defined by:

$$s(t) = \sum_{n=-\infty}^{\infty} a_n x(t - nT - t_0)$$

The signal is a second-order stationary random signal assuming that the time of origin t_0 is random and uniformly distributed over the time interval $[0, T[$.

NOTE.– In this problem, we consider that the instant t belongs to the time interval $[t_0, t_0 + T[$. Without any loss of generalities, we will assign the index n to this interval where the time t is *a priori*.

Moreover, as for problem 17, one could immediately generalize to the situation where the information source does not generate equiprobable b symbols.

However, we preferred not to complicate the problem. For those who wish to do so, after having analyzed the situation in the equiprobable case, it will be easy to generalize the results to a situation with a non-equiprobable source.

Calculate the autocorrelation function $R_s(\tau)$ and the power spectral density $\Gamma_s(f)$ of the signal $s(t)$.

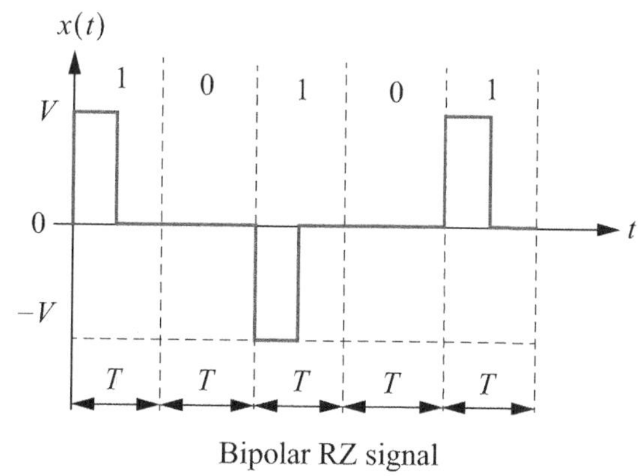

Figure 2.13. *Example of a bipolar RZ signal waveform. For a color version of this figure, see www.iste.co.uk/assad/digital2.zip*

Solution of problem 18

The signal $s(t)$ can be represented by:

$$s(t) = \sum_{n=-\infty}^{\infty} a_n x(t - nT - t_0)$$

where:

$$x(t) = \begin{cases} V & \text{for } \; t \in [t_0, t_0 + T/2] \\ 0 & \text{elsewhere} \end{cases}$$

is a deterministic function, and:

$$a_n = \{1, 0, -1\} \;\; \text{with} : \;\; Pr\{a_n = 0\} = Pr\{a_n = 1 \; \text{or} \; a_n = -1\} = 1/2$$

The autocorrelation $R_s(\tau)$ is written:

$$R_s(\tau) = E[s(t) \times s(t-\tau)] = \sum_i \sum_j s_i s_j Pr\{s(t) = s_i \; \text{and} \; s(t-\tau) = s_j\}$$

$$R_s(\tau) = \begin{cases} V \times V \times Pr\{V \text{ at } t \text{ and } V \text{ at } t-\tau\} + \\ V \times -V \times Pr\{V \text{ at } t \text{ and } -V \text{ at } t-\tau\} + \\ -V \times V \times Pr\{-V \text{ at } t \text{ and } V \text{ at } t-\tau\} + \\ -V \times -V \times Pr\{-V \text{ at } t \text{ and } -V \text{ at } t-\tau\} \end{cases}$$

Using the theorem of compound probabilities, $R_s(\tau)$ is written as:

$$R_s(\tau) = V^2 \times Pr\{V \text{ at } t\} \times Pr\{V \text{ at } t-\tau/V \text{ at } t\}$$
$$-V^2 \times Pr\{V \text{ at } t\} \times Pr\{-V \text{ at } t-\tau/V \text{ at } t\}$$
$$-V^2 \times Pr\{-V \text{ at } t\} \times Pr\{V \text{ at } t-\tau/-V \text{ at } t\}$$
$$+V^2 \times Pr\{-V \text{ at } t\} \times Pr\{-V \text{ at } t-\tau/-V \text{ at } t\}$$

Calculation of the simple probabilities $Pr\{V \text{ at } t\}$ and $Pr\{-V \text{ at } t\}$:

$$Pr\{V \text{ at } t\} = Pr\{t \in [t_0, t_0 + T/2], a_n = 1\}$$

Since there is independence between t_0 and the symbols of information, we have:

$$Pr\{V \text{ at } t\} = Pr\{t \in [t_0, t_0 + T/2]\} \times Pr\{a_n = 1\}$$

Likewise:

$$Pr\{-V \text{ at } t\} = Pr\{t \in [t_0, t_0 + T/2]\} \times Pr\{a_n = -1\}$$

and:

$$Pr\{t \in [t_0, t_0 + T/2]\} = \frac{T/2}{T} = \frac{1}{2}$$

Besides:

$$Pr\{a_n = 1 \text{ or } a_n = -1\}$$
$$= Pr\{a_n = 1\} + Pr\{a_n = -1\} - Pr\{a_n = 1 \text{ and } a_n = -1\}$$

And yet:

$$Pr\{a_n = 1 \text{ and } a_n = -1\} = 0$$

as we have two mutually exclusive hypotheses, hence:

$$Pr\{a_n = 1 \text{ or } a_n = -1\} = Pr\{a_n = 1\} + Pr\{a_n = -1\} = 1/2$$

thus:

$$Pr\{a_n = 1\} = Pr\{a_n = -1\} = 1/4$$

and:

$$Pr\{V \text{ at } t\} = Pr\{-V \text{ at } t\} = \frac{1}{2} \times \frac{1}{4} = \frac{1}{8}$$

Let's now calculate the other probabilities, representing four mutually exclusive hypotheses. Note that for reasons of symmetry, we have:

$$Pr\{V \text{ at } t-\tau/V \text{ at } t\} = Pr\{-V \text{ at } t-\tau/-V \text{ at } t\}$$

and:

$$Pr\{V \text{ at } t-\tau/-V \text{ at } t\} = Pr\{-V \text{ at } t-\tau/V \text{ at } t\}$$

– *First case*: $0 \leq \tau \leq T/2$.

1) Let:

$$q_1 = Pr\{V \text{ at } t-\tau/V \text{ at } t\}$$

The only possibility is that t and $t-\tau$ belong to the same first half of the time slice:

$$q_1 = Pr\{t-\tau \in [t_0, t_0 + T/2]/t \in [t_0, t_0 + T/2]\}$$

We have:

$$\left.\begin{array}{r} t_0 < t-\tau < t_0 + T/2 \ \rightarrow t_0 + \tau < t < t_0 + \tau + T/2 \\ \text{and } t_0 < t < t_0 + T/2 \end{array}\right\} \rightarrow t_0 + \tau < t < t_0 + T/2$$

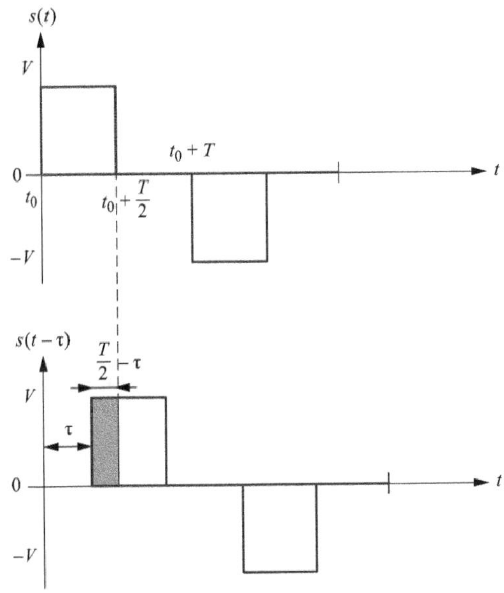

Figure 2.14. *First case: $0 < \tau < T/2$ and a positive impulsion at t. For a color version of this figure, see www.iste.co.uk/assad/digital2.zip*

therefore:

$$q_1 = Pr\{t \in [t_0 + \tau, t_0 + T/2]\} = \frac{T/2 - \tau}{T/2} = 1 - \frac{2\tau}{T}$$

2) Let:

$$q_2 = Pr\{-V \text{ at } t - \tau/V \text{ at } t\}$$

So, we have: $q_2 = 0$.

So, we have also:

$$q_3 = Pr\{-V \text{ at } t - \tau/-V \text{ at } t\} = q_1 = 1 - \frac{2\tau}{T}$$

and:

$$q_4 = Pr\{V \text{ at } t - \tau/-V \text{ at } t\} = q_2 = 0$$

– *Second case*: $T/2 < \tau \leq T$.

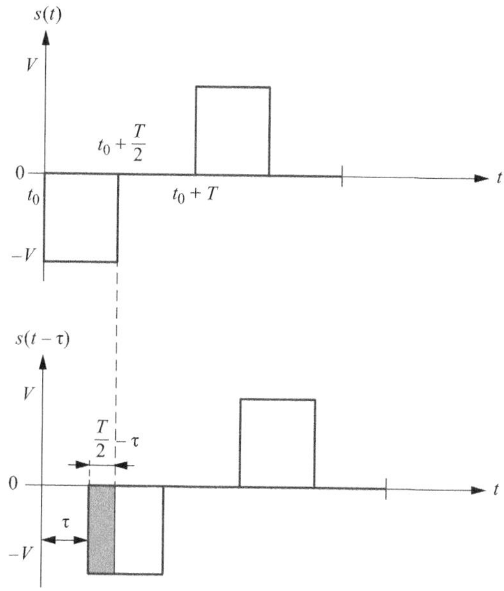

Figure 2.15. *First case:* $0 \leq \tau \leq T/2$ *and negative impulse at t. For a color version of this figure, see www.iste.co.uk/assad/digital2.zip*

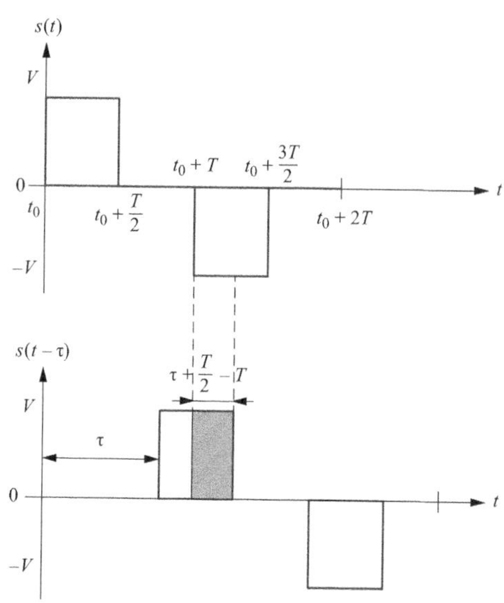

Figure 2.16. *Second case:* $T/2 < \tau \leq T$*. For a color version of this figure, see www.iste.co.uk/assad/digital2.zip*

The only possibility is that t and $t-\tau \in$ first halves of adjacent time slices.

Since two successive bits to 1 are issued in alternating polarity, then:

$$\begin{cases} q_1 = Pr\{V \text{ at } t-\tau/V \text{ at } t\} = 0 \\ q_3 = Pr\{-V \text{ at } t-\tau/-V \text{ at } t\} = 0 \end{cases}$$

$$q_4 = Pr\{V \text{ at } t-\tau/-V \text{ at } t\}$$

$$q_4 = Pr\{t-\tau \in [t_0-T, t_0-T+T/2], a_{n-1}=1/t \in [t_0, t_0+T/2], a_n=-1\}$$

$$q_4 = Pr\{t-\tau \in [t_0-T, t_0-T/2]/t \in [t_0, t_0+T/2]\} \times Pr\{a_{n-1}=1/a_n=-1\}$$

yet:

$$\begin{aligned} &Pr\{a_{n-1}=1/a_n=-1\} = Pr\{b_{n-1}=1/b_n=1\} \\ &= Pr\{b_{n-1}=1\} = Pr\{|a_{n-1}|=1\} = 1/2 \end{aligned}$$

(due to the independence between binary information symbols).

And on the other hand:

$$\left.\begin{array}{l} t_0-T < t-\tau < t_0-T/2 \rightarrow t_0-T+\tau < t < t_0-T/2+\tau \\ \text{and:} \qquad\qquad t_0 < t < t_0+T/2 \end{array}\right\}$$

$$\rightarrow t_0 < t < t_0-T/2+\tau$$

hence:

$$\begin{aligned} q_4 &= Pr\{t \in [t_0, t_0-T/2+\tau]\} \times Pr\{a_{n-1}=1/a_n=-1\} \\ &= \frac{\tau-T/2}{T/2} \times \frac{1}{2} = \frac{1}{2}\left[\frac{2\tau}{T}-1\right] \end{aligned}$$

and, by symmetry, we have:

$$q_2 = Pr\{-V \text{ at } t-\tau/V \text{ at } t\} = q_4 = \frac{1}{2}\left[\frac{2\tau}{T}-1\right]$$

– *Third case*: $T < \tau \leq T+T/2$.

The only possibility is that t and $t-\tau$ belong to the first two halves of adjacent time slices.

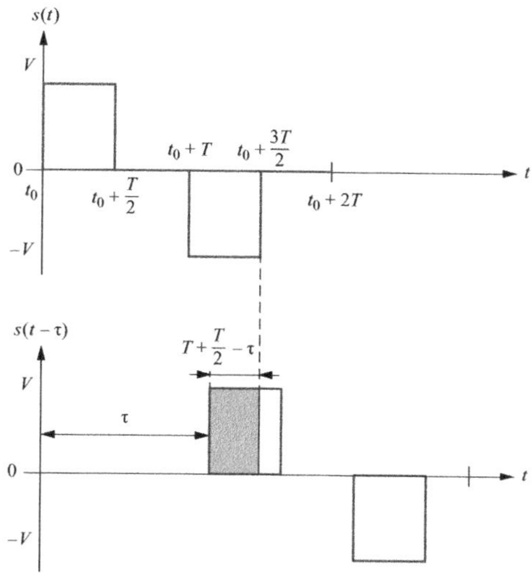

Figure 2.17. *Third case: $T < \tau \leq T + T/2$. For a color version of this figure, see www.iste.co.uk/assad/digital2.zip*

Since two successive bits at "1" are issued in alternating polarity, then as explained previously:

$$\begin{cases} q_1 = Pr\{V \text{ at } t - \tau/V \text{ at } t\} = 0 \\ q_3 = Pr\{-V \text{ at } t - \tau/-V \text{ at } t\} = 0 \end{cases}$$

$$q_4 = Pr\{V \text{ at } t - \tau/-V \text{ at } t\}$$

$$q_4 = Pr\{t - \tau \in [t_0 - T, t_0 - T + T/2], a_{n-1} = 1/t \in [t_0, t_0 + T/2], a_n = -1\}$$

hence:

$$q_4 = Pr\{t - \tau \in [t_0 - T, t_0 - T/2]/t \in [t_0, t_0 + T/2]\} \times Pr\{a_{n-1} = 1/a_n = -1\}$$

We have:

$$Pr\{a_{n-1} = 1/a_n = -1\} = Pr\{|a_{n-1}| = 1\} = 1/2$$

and, on the other hand:

$$\left.\begin{array}{c} t_0 - T < t - \tau < t_0 - T/2 \rightarrow t_0 - T + \tau < t < t_0 - T/2 + \tau \\ \text{and} \qquad\qquad\qquad\qquad t_0 < t < t_0 + T/2 \end{array}\right\}$$

$$\rightarrow t_0 - T + \tau < t < t_0 + T/2$$

hence:

$$q_4 = Pr\{t \in [t_0 - T + \tau, t_0 + T/2\,]\} \times Pr\{a_{n-1} = 1/a_n = -1\}$$
$$= \frac{3T/2 - \tau}{T/2} \times \frac{1}{2} = \frac{1}{2}\left[3 - \frac{2\tau}{T}\right]$$

and, by symmetry, we have:

$$q_2 = Pr\{-V \text{ at } t - \tau/V \text{ at } t\} = q_4 = \frac{1}{2}\left[3 - \frac{2\tau}{T}\right]$$

– *Fourth case*: $3T/2 < \tau \leq 2T$.

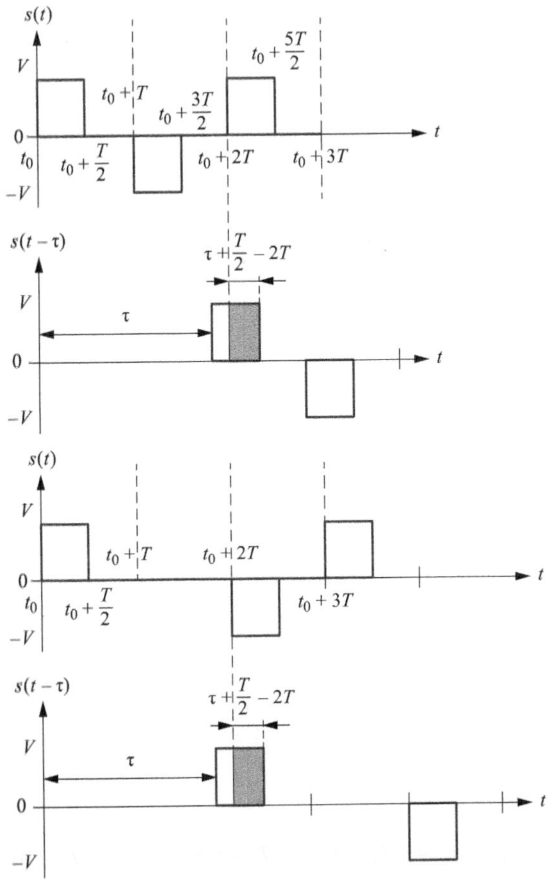

Figure 2.18. *Fourth case:* $3T/2 < \tau \leq 2T$. *For a color version of this figure, see www.iste.co.uk/assad/digital2.zip*

The only possibility is that $(t-\tau)$ and t belong to two first time slices separated by a time slot of intermediate duration T:

$$q_1 = Pr\{V \text{ at } t-\tau/V \text{ at } t\}$$

Moreover, since we have the conditional event: $\{V \text{ at } (t-\tau)/V \text{ at } t\}$, we must also have in the intermediate time slot a negative impulse, and therefore a symbol $a_{n-1} = -1$.

Therefore:

$$q_1 = Pr\{t-\tau \in [t_0-2T, t_0-2T+T/2], a_{n-2} = 1, a_{n-1} = -1/t \in [t_0, t_0 + T/2],\ a_n = 1\}$$

Because of the statistical independence between the pairs of instants considered and the values of the symbols considered, we have:

$$q_1 = Pr\{t-\tau \in [t_0-2T, t_0-3T/2]/t \in [t_0, t_0 + T/2]\} \times Pr\{a_{n-2} = 1, a_{n-1} = -1\ /a_n = 1\}$$

The first conditional probability on the instants gives:

$$Pr\{t \in [t_0-2T+\tau, t_0-3T/2+\tau\ ]/t \in [t_0, t_0 + T/2]\} = Pr\{t \in [t_0, t_0-3T/2+\tau\ ]\} = \frac{\tau - 3T/2}{T/2} = \left[\frac{2\tau}{T} - 3\right]$$

The second conditional probability on the symbols gives (because of the independence between the symbols):

$$Pr\{a_{n-2} = 1, a_{n-1} = -1\ /a_n = 1\} = Pr\{b_{n-2} = 1/b_n = 1\} \times Pr\{b_{n-1} = 1/b_n = 1\}$$

and due to the independence between the symbols b:

$$Pr\{b_{n-2} = 1/b_n = 1\} \times Pr\{b_{n-1} = 1/b_n = 1\} = Pr\{b_{n-2} = 1\} \times Pr\{b_{n-1} = 1\} = \frac{1}{2} \times \frac{1}{2} = \frac{1}{4}$$

hence:

$$q_1 = \frac{1}{4}\left[\frac{2\tau}{T} - 3\right]$$

and, by symmetry, we have: $q_3 = q_1$ and therefore:

$$q_3 = Pr\{-V \text{ at } t-\tau/-V \text{ at } t\} = \frac{1}{4}\left[\frac{2\tau}{T} - 3\right]$$

$$q_4 = Pr\{V \text{ at } t-\tau/-V \text{ at } t\}$$

As before, the only possibility is that $t-\tau$ and t each belong to two first time slices separated by a time slice of intermediate duration T but, unlike in the previous case, in the intermediate time slice, it must not have transmitted a pulse, therefore a symbol: $a_{n-1} = 0$. The conditional probability for the instants remains the same. It is the same for conditional probabilities on symbols a. As a result, we have:

$$q_4 = \frac{1}{4}\left[\frac{2\tau}{T} - 3\right]$$

and therefore:

$$q_4 = q_1 = q_3$$

And by symmetry, we also have $q_2 = q_4$ and therefore:

$$q_2 = Pr\{-V \text{ at } t-\tau/V \text{ at } t\} = \frac{1}{4}\left[\frac{2\tau}{T} - 3\right]$$

Consequently, we finally get:

$$R_s(\tau) = \frac{1}{8}V^2 \times [q_1 + q_3 - q_2 - q_4] = 0$$

Actually, we show that for:

$$\tau > kT, \text{ with} : k \geq 2 \rightarrow R_s(\tau) = 0$$

Indeed, let's take for example the case: $2T \leq \tau < 5T/2$.

From the study of the previous case, we see that:

$$q_1 = q_3 = \frac{1}{4} \times Pr\{t \in [t_0 - 2T + \tau, t_0 - 2T + T/2 + \tau\,]/t \in [t_0, t_0 + T/2]\}$$

$$q_1 = q_3 = \frac{1}{4} \times Pr\{t \in [t_0 - 2T + \tau, t_0 + T/2\,]\} = \frac{1}{4} \times \left[\frac{5T/2 - \tau}{T/2}\right] = \frac{1}{4} \times \left[5 - \frac{2\tau}{T}\right]$$

Likewise:

$$q_4 = q_2 = q_1 = q_3 \rightarrow R_s(\tau) = 0$$

So in the general case where: $kT < \tau \leq (k + 1/2)T$ and $k \geq 2$, we have: $R_s(\tau) = 0$.

Indeed, (for clarity, see the case: $3\,T/2 < \tau \leq 2T$), we have:

$$q_1 = Pr\{V \text{ at } t - \tau / V \text{ at } t\}$$

That is:

$$q_1 = Pr\{t - \tau \in [t_0 - kT, t_0 - (k - 1/2)T], a_{n-k} = 1 / t \in [t_0, t_0 + T/2], a_n = 1\}$$

Because of the independence between t_0 and the information symbols a, we therefore have:

$$q_1 = Pr\{t - \tau \in [t_0 - kT, t_0 - (k - 1/2)T] / t \in [t_0, t_0 + T/2]\}$$
$$\times Pr\{a_{n-k} = 1 / a_n = 1\}$$

or:

$$q_1 = q_{11} \times q_{13}$$

with:

$$q_{11} = Pr\{t - \tau \in [t_0 - kT, t_0 - (k - 1/2)T] / t \in [t_0, t_0 + T/2]\}$$

and:

$$q_{13} = Pr\{a_{n-k} = 1 / a_n = 1\}$$

This last conditional probability implies implicitly the fact that between $n - k$ and n, we have an odd number of bits $b_{n-k'}$ at 1 $(0 < k' < k)$.

Thus:

$$q_{13} = Pr\{\, a_{n-k} = 1, \text{odd number of bits } b_{n-k'} \text{ at } 1 \,/ a_n = 1\}$$

Because of the independence between the symbols, then one obtains successively:

$$q_{13} = Pr\{\, a_{n-k} = 1/a_n = 1\} \\ \times Pr\{\text{ odd number of bits } b_{n-k'} \text{ at } 1\ /a_n = 1\}$$

$$q_{13} = Pr\{a_{n-k} = 1\,\} \times Pr\{\text{ odd number of bits } b_{n-k'} \text{ at } 1\} = \frac{1}{2} \times \frac{1}{2} = \frac{1}{4}$$

because it is obvious that on a given bit length k, the probability of having an even number of bits at 1 is identical to having an odd number. Indeed, for each given configuration of $k-1$ bits, there are two configurations of the additional bit (whatever its position in the k bits): one with 0, the other with 1.

So we have: $q_1 = \frac{1}{4} \times q_{11}$.

By symmetry, we also have:

$$q_3 = Pr\{-V \text{ at } t-\tau/-V \text{ at } t\} = q_1$$

In addition, we also have:

$$q_4 = Pr\{V \text{ at } t-\tau/-V \text{ at } t\}$$

i.e.: $q_4 = q_{11} \times q_{42}$

with:

$$q_{42} = Pr\{a_{n-k} = 1/\ a_n = -1\}.$$

For the same reasons as previously with the calculation of q_{13}, it is easy to show that:

$$q_{42} = Pr\{\, a_{n-k} = 1, \text{even number of bits } b_{n-k'} \text{ at } 1\ /a_n = -1\}$$

and with the independence of the information symbols:

$$q_{42} = Pr\{a_{n-k} = 1\,\} \times Pr\{\text{ even number of bits } b_{n-k'} \text{ at } 1\} = \frac{1}{2} \times \frac{1}{2} = \frac{1}{4}$$

By symmetry, we also have:

$$q_2 = Pr\{-V \text{ at } t-\tau/V \text{ at } t\} = q_4$$

because:

$$q_{44} = Pr\{a_{n-k} = -1/\ a_n = 1\} = q_{42} = \frac{1}{4}$$

Thus, we obtain:

$$R_s(\tau) = \frac{1}{8}V^2 \times [q_1 + q_3 - q_2 - q_4] = 0$$

In summary:

$$R_s(\tau) = \frac{1}{4}V^2 \times [q_1 - q_4]$$

for:

$$0 \leq \tau \leq T/2 \quad \rightarrow R_s(\tau) = \frac{V^2}{4}\left[1 - \frac{2\tau}{T}\right]$$

$$T/2 \leq \tau \leq T \quad \rightarrow R_s(\tau) = \frac{-V^2}{8}\left[\frac{2\tau}{T} - 1\right]$$

$$T \leq \tau \leq 3T/2 \quad \rightarrow R_s(\tau) = \frac{-V^2}{8}\left[3 - \frac{2\tau}{T}\right]$$

$$3T/2 \leq \tau \leq 2T \quad \rightarrow R_s(\tau) = 0$$

The autocorrelation function $R_s(\tau)$ of the bipolar RZ code represented in the Figure 2.19a shows that it can be decomposed into the sum of two functions $R_1(\tau)$ and $R_2(\tau)$ represented in the Figure 2.19b.

The power spectral density $\Gamma_s(f)$ of the bipolar code is therefore given by:

$$\Gamma_s(f) = \Gamma_1(f) + \Gamma_2(f)$$

with:

– $\Gamma_1(f)$, power spectral density of the signal whose autocorrelation function is $R_1(\tau)$ of triangular shape. It is given by:

$$\Gamma_1(f) = \frac{V^2}{2} \times T \times \left[\frac{\sin(\pi f T)}{\pi f T}\right]^2$$

– $\Gamma_2(f)$, power spectral density of the signal whose autocorrelation function is $R_2(\tau)$ of trapezoidal shape. And if $g(\tau)$ has a trapezoidal form (see Figure 2.20):

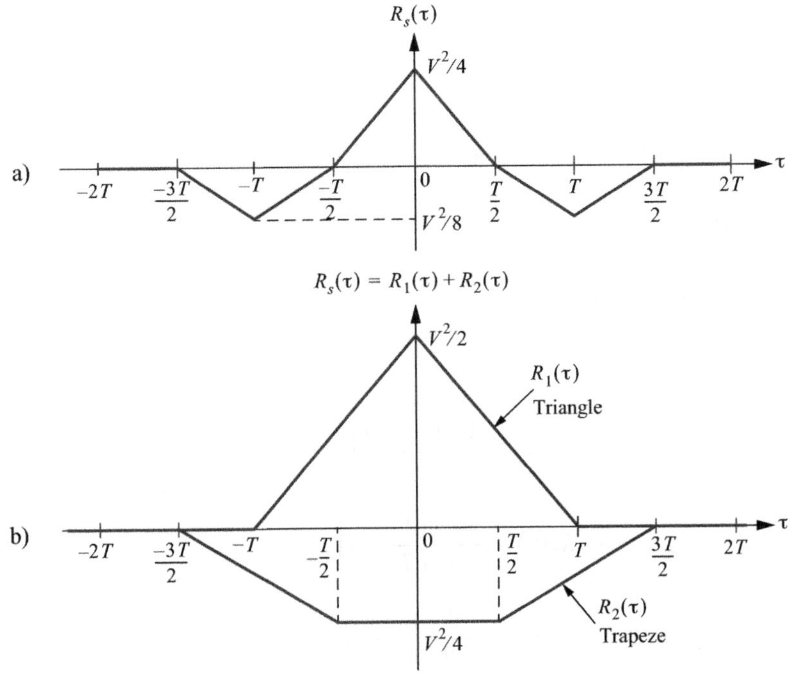

Figure 2.19. *(a) Autocorrelation function of the bipolar RZ code and (b) its decomposition into the sum of two functions. For a color version of this figure, see www.iste.co.uk/assad/digital2.zip*

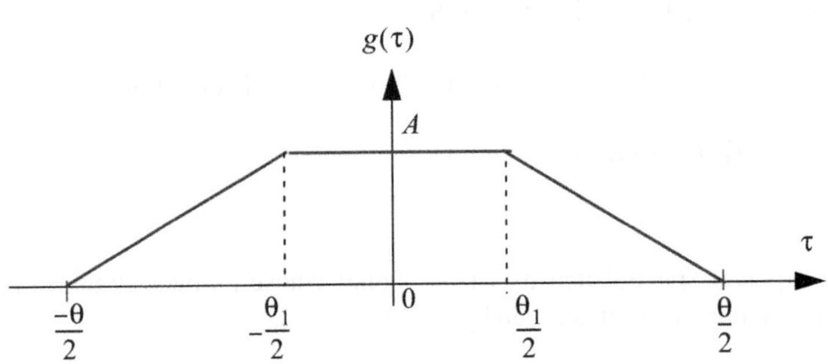

Figure 2.20. *Trapezoidal function. For a color version of this figure, see www.iste.co.uk/assad/digital2.zip*

It is obtained by convolution between two even rectangular functions that do not have the same support in general. Hence:

$$G(f) = \frac{8A}{(2\pi f)^2(\theta - \theta_1)} \times \sin\left[\frac{\pi f(\theta + \theta_1)}{2}\right] \times \sin\left[\frac{\pi f(\theta - \theta_1)}{2}\right]$$

Let's apply this to $R_2(\tau)$, with:

$$A = -V^2/4, \qquad \theta = 3T, \qquad \theta_1 = T$$

Thus, we have successively:

$$\Gamma_2(f) = \frac{8[-V^2/4]}{(2\pi f)^2 2T} \times \sin[2\pi fT] \times \sin[\pi fT]$$

$$\Gamma_2(f) = \frac{-V^2}{T(2\pi f)^2} \times 2\sin[\pi fT]\cos[\pi fT] \times \sin[\pi fT]$$

$$\Gamma_2(f) = \frac{-V^2}{4} T \times [\sin(\pi fT)]^2 \times \frac{2}{(\pi fT)^2}\cos[\pi fT]$$

$$\Gamma_2(f) = \frac{V^2}{4} T \times [\sin(\pi fT)]^2 \times \left[\frac{-2}{(\pi fT)^2}\{1 - 2[\sin(\pi f\,T/2)]^2\}\right]$$

$$\Gamma_2(f) = \frac{V^2}{4} T \times [\sin(\pi fT)]^2 \times \left[\frac{-2}{(\pi fT)^2} + \frac{[\sin(\pi f\,T/2)]^2}{(\pi f\,T/2)^2}\right]$$

Hence:

$$\Gamma_s(f) = \Gamma_1(f) + \Gamma_2(f)$$

$$\Gamma_s(f) = \frac{V^2}{4} T \times [\sin(\pi fT)]^2 \left\{\left[\frac{\sin(\pi f\,T/2)}{(\pi f\,T/2)}\right]^2 - \frac{2}{(\pi fT)^2} + \frac{2}{(\pi fT)^2}\right\}$$

Finally, it gives:

$$\Gamma_s(f) = \frac{V^2}{4} T \times \left[\frac{\sin(\pi f\,T/2)}{(\pi f\,T/2)}\right]^2 \times [\sin(\pi fT)]^2$$

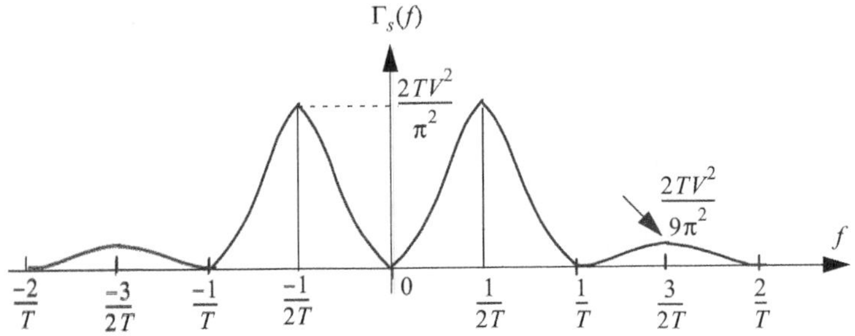

Figure 2.21. *Power spectral density of the bipolar RZ code. For a color version of this figure, see www.iste.co.uk/assad/digital2.zip*

Properties of $\Gamma_s(f)$:

– no continuous component;

– no energy at frequency $f = 1/T$, however a double alternation rectification of the bipolar RZ code gives a RZ code which has a discrete component at frequency $f = 1/T$ in its power spectral density and therefore, a rather easy clock recovery;

– more than 90% of the energy is located in the physical frequency band $[0, 1/T]$.

2.4. Problem 19 – Transmission using a partial response linear coding

We consider the transmission system using the partial response linear coding of Figure 2.22.

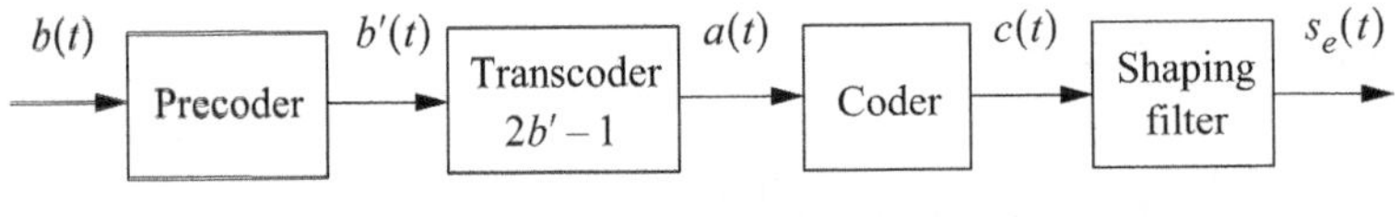

Figure 2.22. *Transmission system with partial response linear encoder*

$$b(t) = \sum_n b_n \delta(t - nT_b) \qquad b_n \in \{0, 1\}$$

$$b'(t) = \sum_n b'_n \delta(t - nT_b) \qquad b'_n \in \{0, 1\}$$

$$a(t) = \sum_n a_n \delta(t - nT_b) \qquad a_n = 2b'_n - 1 \qquad a_n \in \{-1, 1\}$$

$$c(t) = \sum_n c_n \delta(t - nT_b) \qquad c_n \text{ positive, negative or null integer}$$

1) Explain in no more than two sentences the interest of the use of a partial response linear encoder for the baseband transmission of independent binary symbols b_n.

2) Why is it advantageous to use a precoder associated with the partial response linear coder at the transmitter side? Is this precoder necessary for reception? Why?

A partial response linear encoder of the form $1 - D^2$ where D is the delay operator of T_b (time slice dedicated to the transmission of symbol c_n) is used.

3) Give the construction rule of c_n from a_n. Deduce the associated precoder: you will give the construction rule of b'_n as a function of b_n.

We consider the following 21-bit sequence $\{b_n\}$ (time running from left to right):

000 010 110 100 011 101 000

4) Determine:

– the associated time sequence $\{b'_n\}$ at the output of the pre-coder (the latter is considered initialized to zero);

– the time sequence $\{a_n\}$ at the output of the transcoder;

– the time sequence $\{c_n\}$ at the output of the encoder.

5) Give the decoding relationship providing the $\hat{b}_n$ as a function of $\hat{c}_n$.

Let $x(t)$ be the following deterministic pulse (return to zero code, RZ):

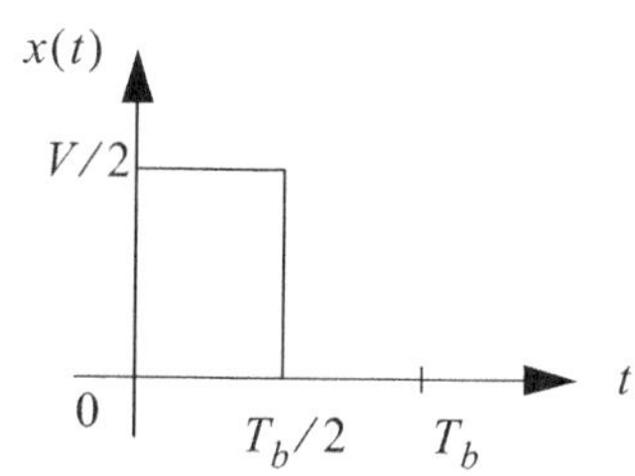

Figure 2.23. *Basic pulse shape (type RZ)*

And the signal transmitted (without pre-filtering) is $s_e(t)$.

6) How is it related to the sequence $\{c_n\}$? Represent on a time diagram the signal $s_e(t)$ transmitted, for the sequence of binary information $\{b_n\}$ in question 4. In this example, does the signal transmitted have a continuous component? On statistical average, does this signal have a continuous component?

7) If the b_n are equiprobable and independent, determine the probabilities of achievement of each element of the alphabet of symbol c_n.

8) Determine the power spectral density function $\Gamma_{s_e}(f)$.

9) We now consider the following 21-bit time sequence $\{b_n\}$ (time running from left to right):

100 001 100 000 110 000 010

Use the HDB-3 coding scheme (high density bipolar pulse code of order 3) to represent the signal $s_{HDB-3}(t)$ carrying the information and represent it on a time diagram.

Solution of problem 19

1) It makes a spectrum shaping. This is performed by the introduction of a certain correlation. The effect is a reduction of the frequency bandwidth of the signal transmitted.

2) A precoder is associated with the encoder to ensure that the decoding of the symbols b_n is instantaneous. The precoder is not necessary in reception because the decoding is instantaneous.

3) Since we have:

$$H(z) = \frac{C(z)}{A(z)} = 1 - z^{-2} \rightarrow C(z) = A(z) - A(z) \times z^{-2} \rightarrow c_n = a_n - a_{n-2}$$

then:

$$P(z) = \frac{B'(z)}{B(z)} = \frac{1}{H(z)} = \frac{1}{1 - z^{-2}} \rightarrow B'(z) = B(z) + B'(z) \times z^{-2}$$

$$\rightarrow b'_n = b_n \oplus b'_{n-2}$$

4) Time sequence.

$\{b_n\}$	0	0	0	0	1	0	1	1	0	1	0	0	0	1	1	1	0	1	0	0	0
$\{b'_n\}$ 0 0	0	0	0	0	1	0	0	1	0	0	0	0	0	1	1	0	1	1	1	1	1
$\{a_n\}$ -1 -1	-1	-1	-1	-1	1	-1	-1	1	-1	-1	-1	-1	-1	1	1	-1	1	1	1	1	1
$\{c_n\}$	0	0	0	0	2	0	-2	2	0	-2	0	0	0	2	2	-2	0	2	0	0	0
$s_e(t)$ V 0 -V	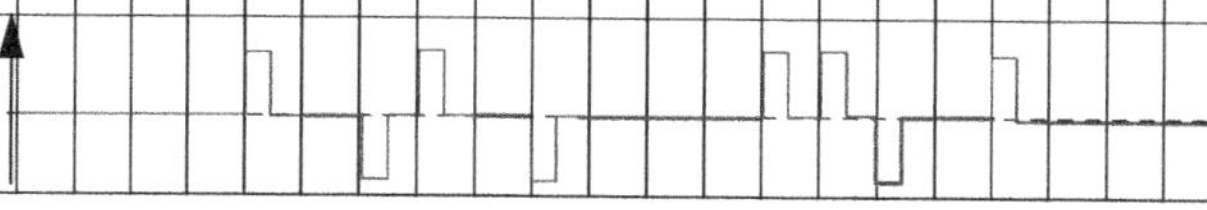																				

Table 2.5. *Chronogram of the time sequence and of the transmitted signal. For a color version of this table, see www.iste.co.uk/assad/digital2.zip*

5) From the previous relation, we have:

$$c_n = a_n - a_{n-2} = [2b'_n - 1 - (2b'_{n-2} - 1)] = 2[b'_n - b'_{n-2}] = 2b_n$$

$$\rightarrow \hat{b}_n = \frac{1}{2}\hat{c}_n \mod 2 \quad \rightarrow \hat{b}_n = \frac{1}{2}|\hat{c}_n| \mod 2$$

6) We have:

$$s_e(t) = \sum_n c_n x(t - nT_b)$$

The chronogram of the signal $s_e(t)$ is shown in Table 2.5 above (answer to question 4).

In this sequence, we have a continuous component, since there are five $+V$ pulses and three $-V$ pulses. This continuous component has the value:

$$2V \times \frac{1}{21 \times 2} = \frac{V}{21}$$

Value 2 in the denominator comes from the fact that the signal $x(t)$ is of type RZ.

On a statistical average: $E\{s_e(t)\} = 0$. (i.e. no continuous component)

7) Since $\{b_n\}$ are equiprobable, then:

$$Pr\{c_n = 0\} = 1/2, \quad Pr\{c_n = 2\} = Pr\{c_n = -2\} = 1/4$$

8) The power spectral density $\Gamma_{s_e}(f)$ is given by:

$$\Gamma_{s_e}(f) = \frac{1}{T_b} \times |H(f)|^2 \times |X(f)|^2$$

$$H(z) = 1 - z^{-2} \rightarrow H(f) = H(z = \exp(j2\pi f T_b) = 1 - \exp(-j4\pi f T_b)$$

$$H(f) = 1 - [\cos(4\pi f T_b) - j\sin(4\pi f T_b)]$$

hence:

$$|H(f)|^2 = [1 - \cos(4\pi f T_b)]^2 + [\sin(4\pi f T_b)]^2$$

$$|H(f)|^2 = 1 + [\cos(4\pi f T_b)]^2 - 2\cos(4\pi f T_b) + [\sin(4\pi f T_b)]^2$$
$$= 2[1 - \cos(4\pi f T_b)]$$

since:

$$1 - \cos(2x) = 2[\sin(x)]^2$$

$$\rightarrow |H(f)|^2 = 2 \times 2[\sin(2\pi f T_b)]^2 = 4[\sin(2\pi f T_b)]^2$$

Furthermore:

$$|X(f)| = \frac{V}{2} \times \frac{T_b}{2} \times \frac{\sin(2\pi f T_b/4)}{2\pi f T_b/4} = \frac{VT_b}{4} \times \frac{\sin(\pi f T_b/2)}{\pi f T_b/2}$$

hence:

$$\Gamma_{s_e}(f) = \frac{1}{T_b} \times 4[\sin(2\pi f T_b)]^2 \times \left[\frac{VT_b}{4} \times \frac{\sin(\pi f T_b/2)}{\pi f T_b/2}\right]^2$$

finally, we have:

$$\Gamma_{s_e}(f) = \frac{V^2 T_b}{4} \times [\sin(2\pi f T_b)]^2 \times \left[\frac{\sin(\pi f T_b/2)}{\pi f T_b/2}\right]^2$$

This code is well suited to long distance cable transmissions because:

– it has no continuous component, and no power spectral density at very low frequencies;

– from a practical point of view, most of its power is distributed in the frequency band $[0, 1/2T_b]$.

9) High density pulse code: HDB-3 ($s_{HDB\text{-}3}(t)$ signal).

$\{b_n\}$	1	0	0	0	0 V +	1	1	0 B -	0	0	0 V -	0	1	1	0 B +	0	0	0 V +	0	1	0

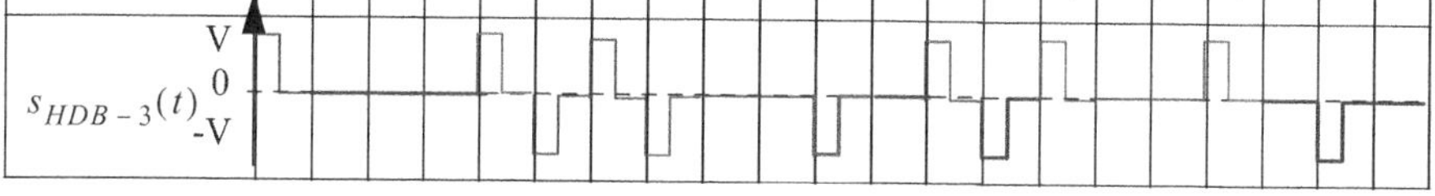

With V: polarity alternation violation bit (bit of violation); B: stuffing bit.

Table 2.6. *Generation of the HDB-3 signal and chronogram. For a color version of this table, see www.iste.co.uk/assad/digital2.zip*

2.5. Problem 20 – Signal information coding and digital transmissions with partial response linear encoder

We consider the problem of long-distance transmission (d > 1 Km) over an electrical cable of a source S, of equiprobable binary symbol information, and delivering a binary symbol every T_b seconds. To illustrate this problem, it will be considered that a limited (20-element) length realization of the binary symbol sequence produced by S is:

$$\{b_n\} = \cdots 101\ 000\ 110\ 000\ 001\ 001\ 01\cdots$$

The amplitude of the modulated pulses is equal to V except for the partial response coding where the amplitude will be $V/2$.

NOTE.– For a better comparison, time representations of transmitted signals $s_i(t)$, $i = \{1,\cdots,4\}$ will be drawn on the same sheet, as well as the sequences $\{b'_n\}, \{a_n\}$ and $\{c_n\}$.

In order to construct the signal $s_1(t)$ carrying the information, a binary return to zero code (RZ code) is used.

1) Represent the signal $s_1(t)$ carrying the sequence $\{b_n\}$. Is this RZ code of interest for long distance digital transmissions? Justify precisely the reasons for this (you can rely on the properties of the power spectral density $\Gamma_1(f)$ of $s_1(t)$ to argue).

A bipolar code of the RZ type is now used to construct the signal $s_2(t)$ carrying the information to be transmitted.

2) Represent the signal $s_2(t)$ encoding the sequence $\{b_n\}$. Is this code more interesting than the first one? What are the qualities and defects of long-range transmissions (argue based on the properties of power spectral density $\Gamma_2(f)$).

We want to use a code with a high density of pulses of type HDB-2.

3) Represent the signal $s_3(t)$ carrying the sequence $\{b_n\}$ of information transmitted. What are the characteristics of such a code compared to the RZ bipolar code? What do you conclude about its suitability for long-range transmissions over an electric cable of binary information?

We want to further reduce the bandwidth of the transmitted signal. For this, we use a partial response linear coding as shown in the diagram of Figure 2.24, with:

$$b(t) = \sum_n b_n \delta(t - nT_b) \qquad b_n \in \{0, 1\}$$

$$b'(t) = \sum_n b'_n \delta(t - nT_b) \qquad b'_n \in \{0, 1\}$$

$$a(t) = \sum_n a_n \delta(t - nT_b) \qquad a_n = 2b'_n - 1 \qquad a_n \in \{-1, 1\}$$

$$c(t) = \sum_n c_n \delta(t - nT_b) \qquad c_n \text{ positive, negative or null integer}$$

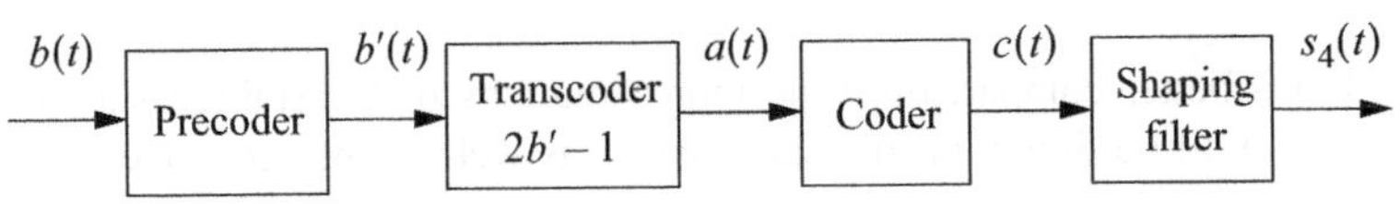

Figure 2.24. *Block diagram of the partial response linear coder*

The structure of the partial response coder is given in Figure 2.25.

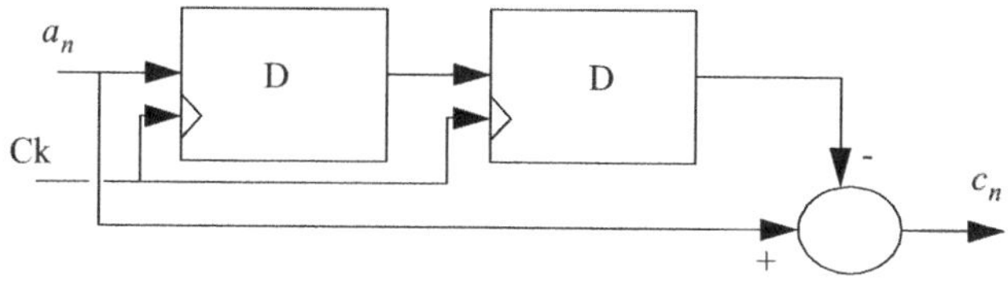

Figure 2.25. *Structure of the partial response linear coder (D: D flip-flop synchronized on the binary symbol clock)*

4) Describe the relationship between the c_n output of the encoder and its input a_n. Why does the encoder have to be preceded by a precoder?

5) Describe the relationship connecting the output of the precoder b'_n to its input b_n (the precoder used is obviously that associated with the partial response linear coder).

6) For the sequence $\{b_n\}$, give successively the sequences obtained:

– at the output of the precoder;

– at the output of the transcoder;

– at the output of the linear partial response coder (it will be considered that the $\{b'_n\}$ are zero for the two instants preceding the beginning of the sequence $\{b_n\}$).

7) Represent the signal $s_4(t)$ coding the sequence $\{b_n\}$ obtained at the output of the partial response coder for a RZ shaping signal $x(t)$ of amplitude $V/2$.

8) Determine the power spectral density $\Gamma_4(f)$ of this partial response linear code. Is it adequate for long distance transmission?

9) For this partial response linear code, how does the decoding produce $\hat{b}_n$ from $\hat{c}_n$? Justify your answer.

10) What happens to the reconstructed binary information $\hat{b}_n$ if an error (due to transmission) occurs for one of the $\hat{c}_n$ symbols reconstructed on reception?

11) Considering that the symbols b_n are independent (besides being equiprobable), determine the probabilities of realization of each of the possible values of the symbols c_n.

Solution of problem 20

Chronograms of the different signals:

$\{b_k\}$	1	0	1	0	0	0	1	1	0	0	0	0	0	0	1	0	0	1	0	1
$s_1(t)$ V 0	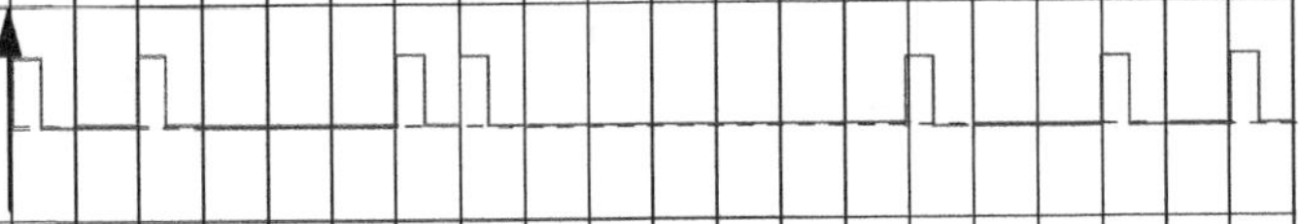																			
$s_2(t)$ V 0 -V	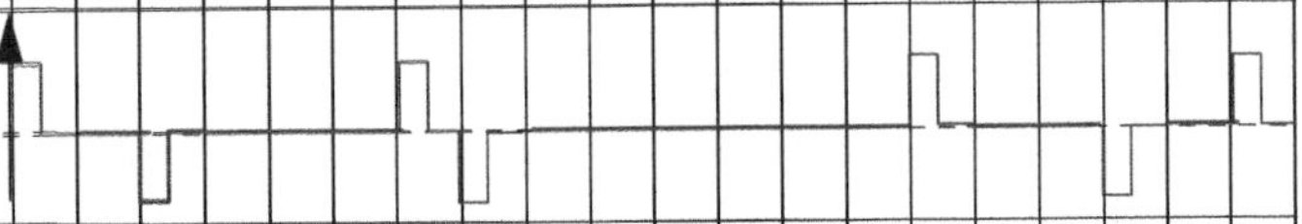																			
$s_3(t)$ V 0 -V	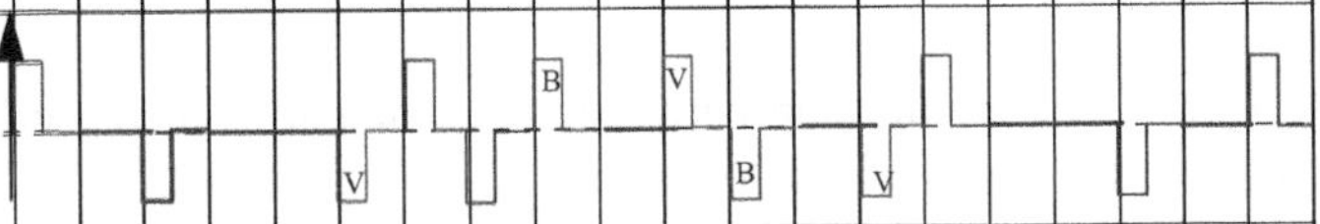																			
$\{b'_k\}$ 0 0	1	0	0	0	0	0	1	1	1	1	1	1	1	1	0	1	0	0	0	1
$\{a_k\}$ -1 -1	1	-1	-1	-1	-1	-1	1	1	1	1	1	1	1	1	-1	1	-1	-1	-1	1
$\{c_k\}$	2	0	-2	0	0	0	2	2	0	0	0	0	0	0	-2	0	0	-2	0	2
$s_4(t)$ V 0 -V	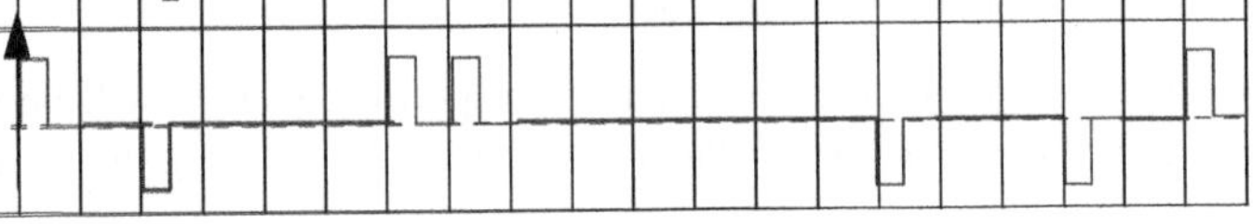																			

Table 2.7. *Temporal representations of the different signals. For a color version of this table, see www.iste.co.uk/assad/digital2.zip*

1) Look at chronograms of the different signals.

The power spectral density $\Gamma_1(f)$ of the RZ code is:

$$\Gamma_1(f) = \frac{V^2 T_b}{16}\left[\frac{sin(\pi f\, T_b/2)}{\pi f\, T_b/2}\right]^2 + \frac{V^2}{16}\delta(f)$$

$$+\frac{V^2}{4\pi^2}\sum_{n=-\infty}^{\infty}\left[\frac{1}{(2n+1)^2}\right]\delta\left(f-\frac{(2n+1)}{T_b}\right)$$

This code is not interesting for transmission over long distance cable because:

– it has a continuous component equal to $V/4$ and a high power spectral density at low frequencies;

– the spectral occupancy of the code is practically $2/T_b$, twice that of the NRZ code.

However, the presence of a discrete component at the frequency $1/T_b$ in the power spectral density facilitates the recovery of the clock rate in reception.

2) See the graph of the temporal representations of the different signals in Table 2.7. The power spectral density $\Gamma_2(f)$ of the bipolar code RZ is:

$$\Gamma_2(f) = \frac{V^2}{4} T_b \times \left[\frac{\sin(\pi f\, T_b/2)}{(\pi f\, T_b/2)}\right]^2 \times [\sin(\pi f T_b)]^2$$

Benefits of this code:

– it has no continuous component, and no power spectral density at very low frequencies;

– its spectral occupancy is only $1/T_b$.

Disadvantage of this code: it does not produce pulses to encode a sequence of consecutive 0, therefore the receiver may lose synchronization.

3) See the graph of the temporal representations of the different signals in Table 2.7. The HDB code has the same advantages as the bipolar code RZ, but in addition we always have pulses even if a long sequence of 0 is presented. So, it has a good match for the transmission of binary information over long distance cable.

4) We have:

$$H(z) = \frac{C(z)}{A(z)} = 1 - z^{-2} \rightarrow C(z) = A(z) - A(z) \times z^{-2} \rightarrow c_n = a_n - a_{n-2}$$

The precoding makes it possible to perform in reception (after transmission) an instantaneous decoding.

5) We have:

$$P(z) = \frac{B'(z)}{B(z)} = \frac{1}{H(z)} = \frac{1}{1 - z^{-2}} \rightarrow B'(z) = B(z) + B'(z) \times z^{-2}$$

$$\rightarrow b'_n = b_n \oplus b'_{n-2}$$

6) and 7) See the graph of the temporal representations of the different signals in Table 2.7.

8) The power spectral density $\Gamma_4(f)$ is:

$$\Gamma_4(f) = \frac{1}{T_b} \times |H(f)|^2 \times |X(f)|^2$$

$$H(z) = 1 - z^{-2} \rightarrow H(f) = H(z = \exp(j2\pi f T_b) = 1 - \exp(-j4\pi f T_b)$$

$$H(f) = 1 - [\cos(4\pi f T_b) - j\sin(4\pi f T_b)]$$

$$|H(f)|^2 = [1 - \cos(4\pi f T_b)]^2 + [\sin(4\pi f T_b)]^2$$

$$|H(f)|^2 = 1 + [\cos(4\pi f T_b)]^2 - 2\cos(4\pi f T_b) + [\sin(4\pi f T_b)]^2$$
$$= 2[1 - \cos(4\pi f T_b)]$$

Since:

$$1 - \cos(2x) = 2[\sin(x)]^2$$

$$\rightarrow |H(f)|^2 = 2 \times 2[\sin(2\pi f T_b)]^2 = 4[\sin(2\pi f T_b)]^2$$

Furthermore:

$$|X(f)| = \frac{V}{2} \times \frac{T_b}{2} \times \frac{\sin(2\pi f T_b/4)}{2\pi f T_b/4} = \frac{VT_b}{4} \times \frac{\sin(\pi f T_b/2)}{\pi f T_b/2}$$

hence:

$$\Gamma_4(f) = \frac{1}{T_b} \times 4[\sin(2\pi f T_b)]^2 \times \left[\frac{VT_b}{4} \times \frac{sin(\pi f T_b/2)}{\pi f T_b/2}\right]^2$$

$$\Gamma_4(f) = \frac{V^2 T_b}{4} \times [\sin(2\pi f T_b)]^2 \times \left[\frac{sin(\pi f T_b/2)}{\pi f T_b/2}\right]^2$$

This code is well suited for long distance cable transmission because:

– it has no continuous component, and no power spectral density at very low frequencies;

– its spectral occupancy is only $1/2T_b$.

9) We have:

$$c_n = a_n - a_{n-2} = [2b'_n - 1 - (2b'_{n-2} - 1)] = 2[b'_n - b'_{n-2}] = 2b_n$$

$$\rightarrow \hat{b}_n = \frac{1}{2}\hat{c}_n \mod 2 \quad \rightarrow \hat{b}_n = \frac{1}{2}|\hat{c}_n| \mod 2$$

10) If error on $\hat{c}_n$, we have:

$$\hat{c}_n = c_n + 2e_n \qquad e_n = \begin{cases} 0 : \text{even error number} \\ 1 : \text{odd error number} \end{cases}$$

$$\hat{b}_n = \frac{1}{2}|\hat{c}_n| + e_n \mod 2$$

Thus, decision error on b_n if e_n is odd ($e_n = 1$) and no decision error on b_n if e_n is even ($e_n = 0$).

11) We have:

$$Pr\{b_n = 0\} = Pr\{b_n = 1\} = 1/2$$

Independence:

$$Pr\{b_n b_m\} = Pr\{b_n\} \times Pr\{b_m\} \quad m \neq n$$

For the reasoning that follows, see the graph of the temporal representations of the different signals:

$$Pr\{c_n = 0\} = Pr\{[b'_n - b'_{n-2}] = 0\} = Pr\{b'_n = b'_{n-2}\} = Pr\{b_n = 0\} = 1/2$$

$$Pr\{c_n = 2\} = Pr\{b'_n = 1, b'_{n-2} = 0\} = Pr\{b_n = 1, b'_{n-2} = 0\}$$
$$= Pr\{b_n = 1\} \times Pr\{b'_{n-2} = 0\} = 1/2 \times 1/2 = 1/4$$

$$Pr\{c_n = -2\} = 1 - Pr\{c_n = 0\} - Pr\{c_n = 2\} = 1/4$$

2.6. Problem 21 – Baseband digital transmission system (1)

A baseband digital transmission system of binary information is considered. It transmits coded digital images (with information compression) on a reduced capacity transmission channel (cable). The characteristics of the system in Figure 2.26 will be studied.

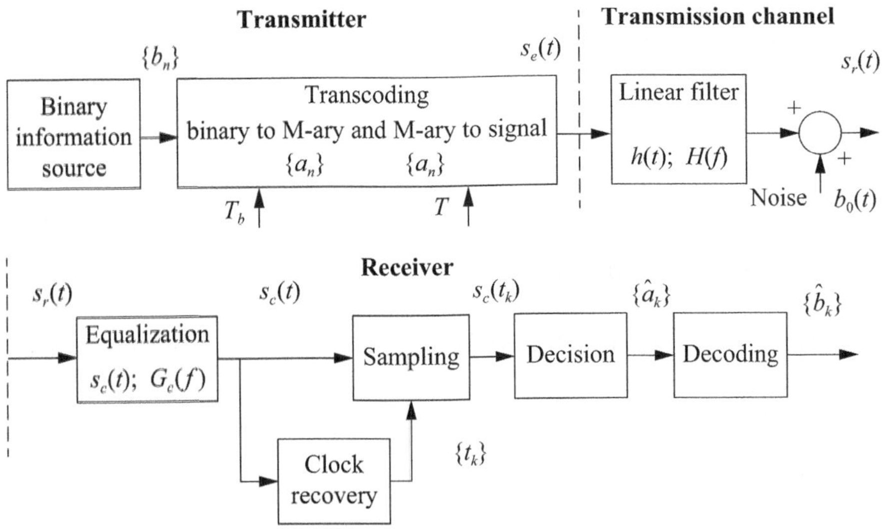

Figure 2.26. *Block diagram of the baseband transmission system*

The random sequence $\{b_n\}$ is of a given probability law and b_n are assumed to be independent. The transcoding of binary information sequence $\{b_n\}$ into symbol sequence $\{a_n\}$ corresponds to the following assignment:

$$\text{if } b = 1 \text{ then } a = 1\,; \quad \text{if } b = 0 \text{ then } a = -1$$

with the following probability law:

$$Pr\{a_n = 1\} = p_1 = Pr\{b_n = 1\} = 4/5$$

$$Pr\{a_n = -1\} = p_{-1} = Pr\{b_n = 0\} = 1/5$$

The symbols a_n of information to be transmitted are supplied to the transmitter at a rate of $1/T = 13.5$ MHz which corresponds to the sampling frequency of the luminance signal in television (standard CCIR 4: 2: 2).

The transmitted signal $s_e(t)$ is given by:

$$s_e(t) = \sum_n a_n x(t - nT)$$

where $x(t)$ is a rectangular signal of amplitude V and duration:

$$\theta = T \text{ (NRZ: non return to zero), or } \theta = T/2 \text{ (RZ: return to zero)}$$

The transmission channel is modeled by a linear filter whose impulse response is denoted by $h(t)$ (the propagation delay is not taken into account) and an additive degradation noise $b_0(t)$ at the output of the channel. The noise $b_0(t)$ is assumed to be a second-order stationary Gaussian random process, independent of the useful signal. It has a zero mean value, a power $\sigma_{b_0}^2$, and its power spectral density $\Gamma_{b_0}(f)$ is modeled by a rectangular function of support Δf_b (on positive frequency axis) as shown on Figure 2.27.

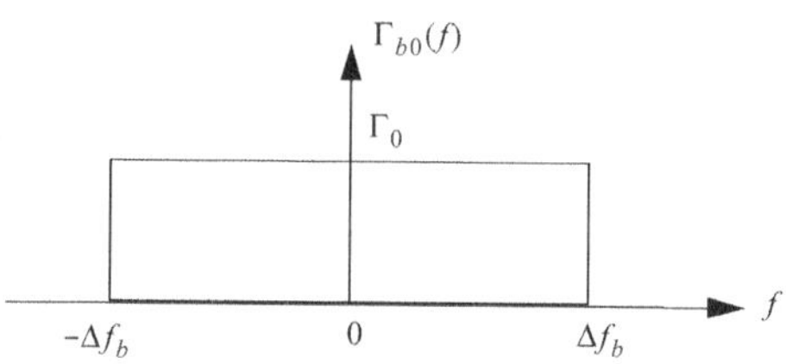

Figure 2.27. *Power spectral density $\Gamma_{b_0}(f)$ of noise $b_0(t)$*

The equalization of the channel is performed by a linear filtering of the signal received $s_r(t)$: impulse response filter $g_c(t)$, frequency gain $G_c(f)$ such that its support is fully included in the frequency band of the noise $b_0(t)$.

The clock regeneration system is assumed to be flawless, and thus provides the decision system with a sequence of decision instants $\{t_k\}$ with $t_k = t_0 + kT$ (thereafter, t_0 is assumed to be zero).

The decision system uses a given decision threshold μ_0 and the decision rule is as follows:

$$d[s_c(t_k)] = \hat{a}_k = 1 \text{ if } s_c(t_k) \geq \mu_0$$

$$d[s_c(t_k)] = \hat{a}_k = -1 \text{ if } s_c(t_k) < \mu_0$$

Decoding $\{\hat{a}_k\} \rightarrow \{b_k\}$ is obvious.

We denote successively:

$$y(t) = x(t) \otimes h(t) \quad \text{and} \quad p(t) = y(t) \otimes g_c(t)$$

where $\otimes$ is the convolution product.

The frequency gain of the equalizer $G_c(f)$ is assumed to be equal to 1 at zero frequency: $G_c(0) = 1$.

1) Determine the noise characteristics $b_1(t)$ at the output of the equalization: the power spectral density $\Gamma_{b_1}(f)$ and the power $\sigma_{b_1}^2$ of the noise $b_1(t)$, as a function of that $\sigma_{b_0}^2$ of $b_0(t)$, its bandwidth Δf_b and the energy bandwidth Δf_c of the equalization filter.

From now on, the equalization filter is assumed such that the amplitude spectrum $P(f)$ of $p(t)$ is constant, equal to VT on the frequency domain $\left[\frac{-(1+\alpha)}{2T}, \frac{(1+\alpha)}{2T}\right]$, and equal to zero otherwise.

2) Give the expression of the signal $s_c(t_k)$ (made of the useful signal + intersymbol interference + noise) at the instants of the form: $t_k = kT$.

Subsequently, for sake of simplification, it will be considered that only the two symbols adjacent to a given symbol a_k can interfere with it (namely symbols a_{k-1} and a_{k+1}), and that $\alpha = 1/6$.

3) Calculate the probability p_{m_l} and the intersymbol interference $I_{m_l}(t_k)$ for each possible message m_l interfering with a_k. (For the sake of simplification, take $\pi \cong 3$ in the rest of this problem).

Assuming that at the output of the equalizer the signal-to-noise ratio obtained is:

$$\left[\frac{S}{b}\right]_{c,dB} = 20 \times \log_{10}\left[\frac{p(0)}{\sigma_{b_1}}\right] = 10.88 \text{ dB}$$

4) Give the expressions of the conditional probabilities of error:

$$Pr\{\hat{a}_k = a_j / a_k = a_i, m_l\}; \quad i \neq j; \quad a_j = \begin{Bmatrix}-1\\1\end{Bmatrix}; \quad a_i = \begin{Bmatrix}-1\\1\end{Bmatrix}$$

5) For each possible message m_l, calculate these probabilities.

6) Finally, deduce the average probability of error $P_{e,1}$.

NOTE.– For a Gaussian random variable X centered ($m = 0$) and reduced ($\sigma = 1$), we will assume that we have approximately:

$$Pr\{|X| > 2.3\} = 2.14 \times 10^{-2}\,; Pr\{|X| > 2.7\} = 7 \times 10^{-3}$$

$$Pr\{|X| > 3.3\} = 9.6 \times 10^{-4}\,; Pr\{|X| > 3.7\} = 2.2 \times 10^{-4}$$

$$Pr\{|X| > 4.3\} = 3 \times 10^{-5}\,; Pr\{|X| > 4.7\} \cong 0$$

7) Calculate the average probability of error $P_{e,0}$ that we would have had (with the same ratio $\left[\frac{S}{b}\right]_C$ at the output of the equalizer) if the intersymbol interference had been canceled.

Solution of problem 21

1) The power spectral density is:

$$\Gamma_{b_1}(f) = \Gamma_{b_0}(f) \times |G_c(f)|^2 = \Gamma_0 \times |G_c(f)|^2$$

The noise power $b_0(t)$ is $\sigma_{b_0}^2$ given by:

$$\sigma_{b_0}^2 = \int_{-\infty}^{\infty} \Gamma_{b_0}(f) df = \Gamma_0 \int_{-\Delta f_b}^{\Delta f_b} df = 2\Gamma_0 \Delta f_b \rightarrow \Gamma_0 = \frac{\sigma_{b_0}^2}{2\Delta f_b}$$

The power of the noise $b_1(t)$ is $\sigma_{b_1}^2$ given by:

$$\sigma_{b_1}^2 = \int_{-\infty}^{\infty} \Gamma_{b_1}(f) df = \Gamma_0 \int_{-\Delta f_c}^{\Delta f_c} |G_c(f)|^2 df = \Gamma_0 2\,\Delta f_c |G_c(0)|^2 = 2\Gamma_0 \Delta f_c$$

$$\rightarrow \sigma_{b_1}^2 = \sigma_{b_0}^2 \times \frac{\Delta f_c}{\Delta f_b}$$

2) The signal transmitted $s_e(t)$ is:

$$s_e(t) = \sum_{n=-\infty}^{\infty} a_n x(t - nT)$$

The signal received $s_r(t)$ is:

$$s_r(t) = \sum_{n=-\infty}^{\infty} a_n y(t - nT) + b_0(t)$$

with:

$$y(t) = x(t) \otimes h(t)$$

The signal $y(t)$ is the response of the channel to the basic pulse (rectangular shape) $x(t)$, of duration θ, in the noiseless case.

The equalized signal (corrected) $s_c(t)$ is:

$$s_c(t) = \sum_{n=-\infty}^{\infty} a_n p(t - nT) + b_1(t)$$

with:

$$p(t) = y(t) \otimes g_c(t) = x(t) \otimes h(t) \otimes g_c(t)$$

The noise $b_1(t)$ is the result of noise $b_0(t)$ filtering by the equalizer whose impulse response is $g_c(t)$.

At sampling times $t_k = kT$, the signal $s_c(t)$ is written:

$$s_c(kT) = a_k p(0) + \sum_{n=-\infty, n\neq k}^{\infty} a_n\, p[(k-n)T] + b_1(kT)$$

The term $a_k p(0)$ represents the useful response of the system (channel + equalization) to the transmission of the symbol a_k associated with the time interval kT.

The term $I_{m_l}(kT) = \sum_{n=-\infty, n\neq k}^{\infty} a_n\, p[(k-n)T]$ is the intersymbol interference. It is a disturbing signal depending on all of the symbols transmitted $\{a_n\}$, except for the symbol a_k which is related to the time interval considered.

The term $b_1(kT)$ is the noise at the time of decision.

3) The messages of only the form $m_l = [a_{k-1}, a_{k+1}]$ interferes with symbol a_k, thus:

$$I_{m_l}(kT) = \sum_{n=k-1, n\neq k}^{k+1} a_n\, p[(k-n)T] = a_{k-1} p(T) + a_{k+1} p(-T)$$

$I_{m_l}(kT)$ depends on $p[(k-n)T]$, here on $p(\pm T)$. We must first calculate $p(t)$ from $P(f)$.

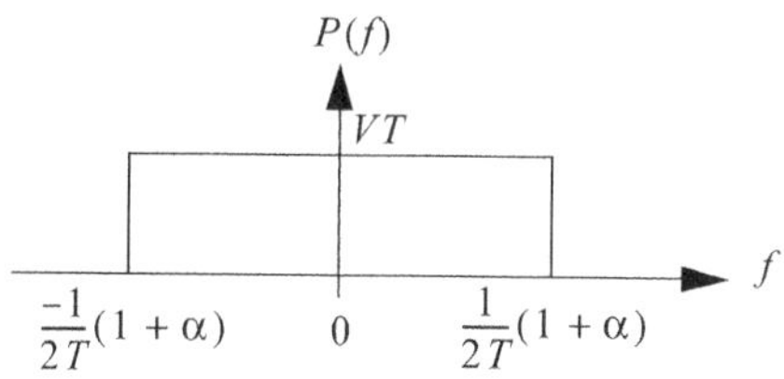

Figure 2.28. *Amplitude spectrum $P(f)$ of $p(t)$*

By definition, we have: $p(t) = F^{-1}\{P(f)\}$, hence:

$$p(t) = VT \times \frac{1}{T}(1+\alpha) \times \frac{\sin\left[2\pi t \times \frac{1}{2T}(1+\alpha)\right]}{2\pi t \times \frac{1}{2T}(1+\alpha)}$$

$$= V(1+\alpha) \times \frac{\sin[\pi(1+\alpha)\,t/T]}{\pi(1+\alpha)\,t/T}$$

This gives:

$$p(0) = V(1+\alpha) = V\,7/6$$

and:

$$p(\pm T) = \frac{7V}{6} \times \frac{\sin[\pi + \pi/6]}{7\pi/6} = \frac{-V}{\pi}\sin[\pi/6] = \frac{-V}{2\pi} \cong \frac{-V}{6} \quad (\text{with } \pi \cong 3)$$

$\boldsymbol{m_l = [a_{k-1}, a_{k+1}]}$	$\boldsymbol{I_{m_l}(kT) \cong \frac{-V}{6}[a_{k-1} + a_{k+1}]}$	$\boldsymbol{p_{m_l}}$
-1 -1	$V/3$	$\frac{1}{5} \times \frac{1}{5} = \frac{1}{25}$
-1 1	0	$\frac{1}{5} \times \frac{4}{5} = \frac{4}{25}$
1 -1	0	$\frac{4}{5} \times \frac{1}{5} = \frac{4}{25}$
1 1	$-V/3$	$\frac{4}{5} \times \frac{4}{5} = \frac{16}{25}$

Table 2.8. *Intersymbol interference: amplitudes and probabilities*

4) The two expressions of conditional probabilities of error are:

$$P_{e_{-1}} = Pr\{\hat{a}_k = 1/a_k = -1, m_l\} = \frac{1}{\sigma_{b_1}\sqrt{2\pi}} \int_{\mu_0+p(0)-I_{m_l}(kT)}^{\infty} \exp\left[\frac{-1}{2}\frac{b_1^2}{\sigma_{b_1}^2}\right] db_1$$

$$P_{e_1} = Pr\{\hat{a}_k = -1/a_k = 1, m_l\} = \frac{1}{\sigma_{b_1}\sqrt{2\pi}} \int_{-\infty}^{\mu_0-[p(0)+I_{m_l}(kT)]} \exp\left[\frac{-1}{2}\frac{b_1^2}{\sigma_{b_1}^2}\right] db_1$$

5) To calculate these conditional probabilities for each message m_l, it is necessary to express the integration domains as a function of σ_{b_1}. As:

$$\left[\frac{S}{b}\right]_{c,dB} = 20 \times \log_{10}\left[\frac{p(0)}{\sigma_{b_1}}\right] = 20 \times \log_{10}\left[\frac{7V/6}{\sigma_{b_1}}\right] = 10.88 \text{ dB}$$

$$\rightarrow V \cong 3\,\sigma_{b_1} \text{ and } p(0) = 7V/6 \cong 3.5\,\sigma_{b_1}$$

Furthermore:

$$\mu_0 = \frac{\sigma_{b_1}^2}{2p(0)} \log_e\left[\frac{p_{-1}}{p_1}\right] \cong -0.2\,\sigma_{b_1}$$

hence the following Table 2.9.

$m_l = [a_{k-1}, a_{k+1}]$	$\mu_0 + p(0) - I_{m_l}(kT)$	$P_{e_{-1}}$	$\mu_0 - p(0) - I_{m_l}(kT)$	P_{e_1}
-1 -1	$2.3\,\sigma_{b_1}$	$\cong 1.07 \times 10^{-2}$	$-4.7\,\sigma_{b_1}$	$\cong 0$
-1 1	$3.3\,\sigma_{b_1}$	$\cong 4.8 \times 10^{-4}$	$-3.7\,\sigma_{b_1}$	$\cong 1.1 \times 10^{-4}$
1 -1	$3.3\,\sigma_{b_1}$	$\cong 4.8 \times 10^{-4}$	$-3.7\,\sigma_{b_1}$	$\cong 1.1 \times 10^{-4}$
1 1	$4.3\,\sigma_{b_1}$	$\cong 1.5 \times 10^{-5}$	$-2.7\,\sigma_{b_1}$	$\cong 3.5 \times 10^{-3}$

Table 2.9. *Conditional probabilities of error with intersymbol interference*

6) The average probability of error $P_{e,1}$ is given by:

$$P_{e,1} = p_{-1}\left[\sum_{l=1}^{2^2} p_{m_l} \times P_{e_{-1}}\right] + p_1\left[\sum_{l=1}^{2^2} p_{m_l} \times P_{e_1}\right]$$

$$P_{e,1} = \frac{1}{5}\left[\frac{1}{25} \times 1.07 \times 10^{-2} + 2 \times \frac{4}{25} \times 4.8 \times 10^{-4} + \frac{16}{25} \times 1.5 \times 10^{-5}\right]$$
$$+\frac{4}{5}\left[\frac{1}{25} \times 0 + 2 \times \frac{4}{25} \times 1.1 \times 10^{-4} + \frac{16}{25} \times 3.5 \times 10^{-3}\right] \cong 1.938 \times 10^{-3}$$

7) As there is no intersymbol interference, this means that the first Nyquist frequency criterion is verified ($\alpha = 0$). So, from the previous results we get:

$$p(t) = V \times \frac{\sin[\pi\, t/T]}{\pi\, t/T}; \quad p(0) = V; \quad p(\pm lT) = 0; \text{ with } l \neq 0$$

At sampling times $t_k = kT$, the equalized signal (corrected) $s_c(t)$ is now:

$$s_c(kT) = a_k p(0) + b_1(kT)$$

The two simplified expressions of conditional probabilities of error are now:

$$P_{e_{-1}} = Pr\{\hat{a}_k = 1/a_k = -1\} = \frac{1}{\sigma_{b_1}\sqrt{2\pi}} \int_{\mu_0+p(0)}^{\infty} \exp\left[\frac{-1}{2}\frac{b_1^2}{\sigma_{b_1}^2}\right] db_1$$

$$P_{e_1} = Pr\{\hat{a}_k = -1/a_k = 1\} = \frac{1}{\sigma_{b_1}\sqrt{2\pi}} \int_{-\infty}^{\mu_0-p(0)} \exp\left[\frac{-1}{2}\frac{b_1^2}{\sigma_{b_1}^2}\right] db_1$$

As we have the same signal-to-noise ratio as before, it means that:

$$p(0) \cong 3.5\, \sigma_{b_1}$$

And we can keep (as the approximation remains rather good) the value of the optimal threshold μ_0 as a function of the noise power $\sigma_{b_1}^2$:

$$\mu_0 = \frac{\sigma_{b_1}^2}{2p(0)} \log_e\left[\frac{p_{-1}}{p_1}\right] \cong -0.2\, \sigma_{b_1}$$

Thus, in this case, the previous Table 2.9 is replaced by Table 2.10:

$\mu_0 + p(0)$	$P_{e_{-1}}$	$\mu_0 - p(0)$	P_{e_1}
$3.3\ \sigma_{b_1}$	$\cong 4.8 \times 10^{-4}$	$-3.7\ \sigma_{b_1}$	$\cong 1.1 \times 10^{-4}$

Table 2.10. *Conditional probabilities of error without intersymbol interference*

Finally, we get:

$$P_{e,1} = p_{-1} \times P_{e_{-1}} + p_1 \times P_{e_1}$$

$$P_{e,0} = \frac{1}{5}[4.8 \times 10^{-4}] + \frac{4}{5}[1.1 \times 10^{-4}] = 9.6 \times 10^{-5} + 8.8 \times 10^{-5}$$

$$\cong 1.84 \times 10^{-4}$$

Thus, in this case (with the same signal-to-noise ratio), the probability of transmission error without intersymbol interference is approximately 10 times lower than it was in the presence of intersymbol interference.

2.7. Problem 22 – Baseband digital transmission (2)

The following baseband digital transmission system (Figure 2.29) is considered for the transmission of binary information.

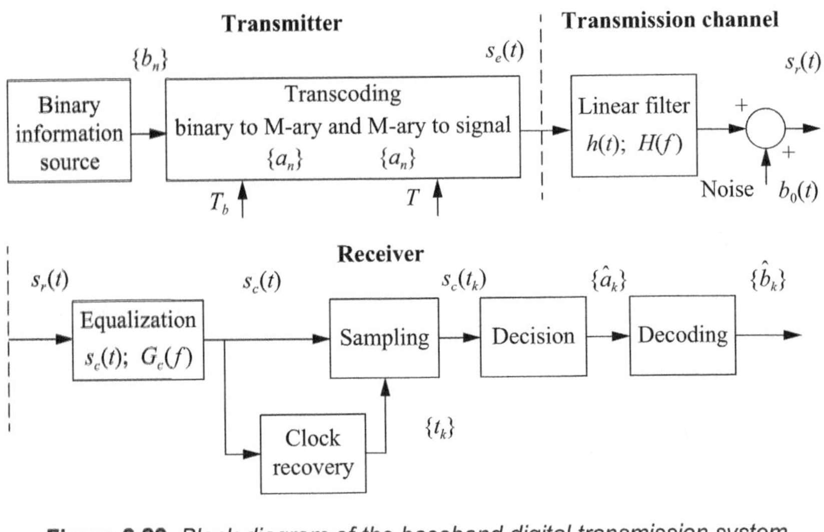

Figure 2.29. *Block diagram of the baseband digital transmission system*

The source of information produces a random sequence $\{b_n\}$ of equiprobable and independent binary variables. The coding of binary information $\{b_n\}$ into information symbol $\{a_n\}$ corresponds to the following assignment:

$$\text{if } b = 1 \text{ then } a = 1\,; \text{if } b = 0 \text{ then } a = -1$$

The symbols a_n of information to be transmitted are supplied to the transmitter at a rate of $1/T$. The coder information to signal generates a transmitted signal $s_e(t)$ given by:

$$s_e(t) = \sum_n a_n x(t - nT)$$

where $x(t)$ is a rectangular signal of amplitude V and duration $T/2$.

The transmission channel is modeled by a linear filter whose impulse response is denoted by $h(t)$ (the propagation delay here is not taken into account) and an additive degradation noise $b_0(t)$ at the transmission channel output.

The noise $b_0(t)$ is modeled by the low pass filtering of a white noise, of constant power spectral density Γ_0. This low pass filter is considered as a first-order low pass R-C filter whose frequency gain is denoted by $L(f)$. The noise $b_0(t)$ is assumed to be a second-order stationary Gaussian random process with zero mean, and independent of the useful signal. We called $\sigma_{b_0}^2$ its average power and $\Gamma_{b_0}(f)$ its power spectral density.

A receiver makes it possible to retrieve the binary information from the signal received at the output of the channel according to the block diagram from Figure 2.29. The channel equalization is produced by a linear filter of impulse response $g_c(t)$ and complex gain $G_c(f)$. The clock recovery, supposed to be faultless, produces a sequence of decision instants $\{t_k\}$ of the form $t_k = t_0 + kT$ (thereafter, t_0 is assumed to be zero).

The decision system uses a given decision threshold μ_0 and the decision rule is as follows:

$$d[s_c(t_k)] = \hat{a}_k = 1 \text{ if } s_c(t_k) \geq \mu_0$$

$$d[s_c(t_k)] = \hat{a}_k = -1 \text{ if } s_c(t_k) < \mu_0$$

Decoding $\{\hat{a}_k\} \rightarrow \{b_k\}$ is obvious.

We denote successively ($\otimes$: convolution product):

$$y(t) = x(t) \otimes h(t) \text{ and } p(t) = y(t) \otimes g_c(t)$$

1) Determine the energy bandwidth Δf_b of the noise $b_0(t)$. This will allow us to consider in the following problem that its spectrum $\Gamma_{b_0}(f)$ is constant on the frequency band $[-\Delta f_b, \Delta f_b]$, and zero otherwise (see Figure 2.30).

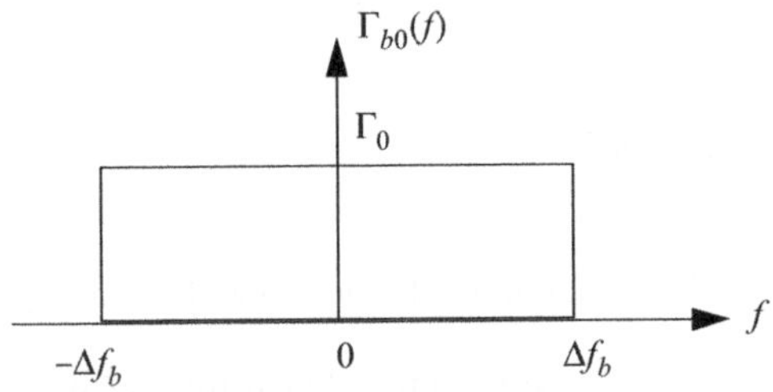

Figure 2.30. *Equivalent power spectral density of noise $b_0(t)$*

It is assumed that the equalization is of gain $G_c(f)$ on a support fully included in the frequency band $[-\Delta f_b, \Delta f_b]$ and $G_c(0) = 1$. We denoted Δf_c as its equivalent energy bandwidth.

2) Determine the noise characteristics $b_1(t)$ at the output of the equalization: the power spectral density $\Gamma_{b_1}(f)$ and the power $\sigma_{b_1}^2$ of the noise $b_1(t)$, as a function of that $\sigma_{b_0}^2$ of $b_0(t)$, its equivalent energy bandwidth Δf_b and the energy bandwidth Δf_c of the equalization filter.

The equalization filter is set so that the amplitude spectrum $P(f)$ of $p(t)$ is constant, equal to VT/α in the frequency band $\left[\frac{-\alpha}{2T}, \frac{\alpha}{2T}\right]$, and equal to zero otherwise (see Figure 2.31).

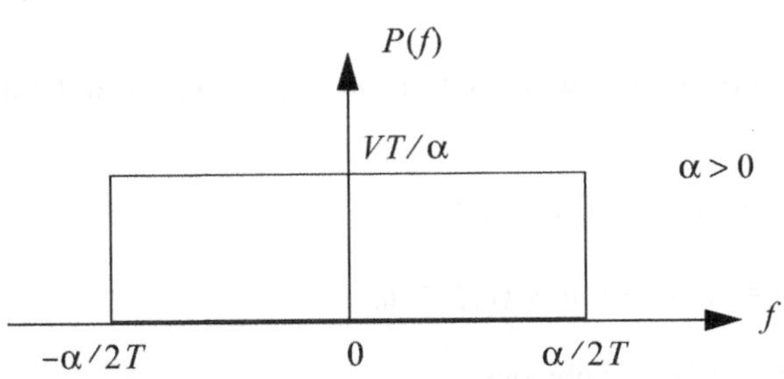

Figure 2.31. *Amplitude spectrum $P(f)$ of $p(t)$*

3) Give the expression of the signal $s_c(t_k)$ (made of the useful signal + intersymbol interference + noise) at the instants of the form: $t_k = kT$.

Subsequently, for sake of simplification, it will be considered that only the two symbols adjacent to a given symbol a_k can interfere with it (namely symbols a_{k-1} and a_{k+1}).

4) Which minimum value α_0 of the parameter α ensures no intersymbol interference?

We then adjust the equalizer with the value α_0. Under these conditions, the signal-to-noise ratio obtained at the output of the equalizer is equal to 6 dB with:

$$\left[\frac{S}{b}\right]_{c,dB} = 20 \times \log_{10}\left[\frac{p(0)}{\sigma_{b_1}}\right]$$

5) Calculate the conditional probabilities of error:

$$Pr\{\hat{a}_k = a_j/a_k = a_i\}\,;\; i \neq j\,;\; a_j = \begin{Bmatrix}-1\\1\end{Bmatrix};\; a_i = \begin{Bmatrix}-1\\1\end{Bmatrix}$$

knowing that we have:

$$\int_{-2\sigma}^{2\sigma} p(x)dx = 0.95 \quad \text{if} \quad p(x) = \frac{1}{\sigma\sqrt{2\pi}}\exp\left[\frac{-x^2}{2\sigma^2}\right]$$

Solution of problem 22

1) The calculation of energy bandwidth Δf_b of noise $b_0(t)$ is made from the expression of the power $\sigma_{b_0}^2$:

$$\sigma_{b_0}^2 = \int_{-\infty}^{\infty} \Gamma_{b_0}(f)df$$

On one hand:

$$\sigma_{b_0}^2 = \int_{-\infty}^{\infty} \Gamma_{b_0}(f)df = \Gamma_0 \int_{-\Delta f_b}^{\Delta f_b} df = 2\Gamma_0\,\Delta f_b$$

On the other hand, we have to calculate $\sigma_{b_0}^2$ from the expression of $\Gamma_{b_0}(f)$, which is the result of filtering Γ_0 by the first-order RC low pass filter:

$$\Gamma_{b_0}(f) = \Gamma_0 \times |L(f)|^2$$

Calculation of the transfer function of a 1st order low pass R-C filter:

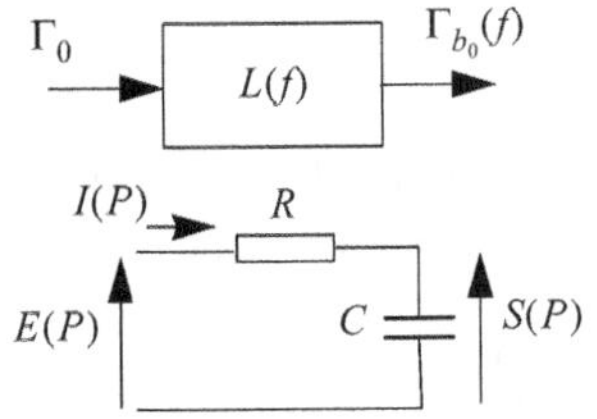

Figure 2.32. *First-order R-C low pass filter*

$$Z_1(P) = R, \quad Z_2(P) = \frac{1}{CP} \quad \text{and } P\text{: Laplace variable}$$

$$\left.\begin{array}{l} E(P) = R \times I(P) + S(P) \\ S(P) = I(P)/CP \end{array}\right\} \rightarrow [RCP + 1]S(P) = E(P)$$

$$\rightarrow L(P) = \frac{S(P)}{E(P)} = \frac{1}{1 + RCP} = \frac{1}{1 + \tau P}$$

For:

$$\left.\begin{array}{l} E(P) = 1 \\ P = j2\pi f \end{array}\right\} \rightarrow L(f) = \frac{1}{1 + j2\pi f\tau} \rightarrow |L(f)|^2 = \frac{1}{1 + 4\pi^2 f^2 \tau^2}$$

and:

$$\Gamma_{b_0}(f) = \Gamma_0 \times |L(f)|^2 = \frac{\Gamma_0}{1 + 4\pi^2 f^2 \tau^2}$$

then:

$$\sigma_{b_0}^2 = \int_{-\infty}^{\infty} \Gamma_{b_0}(f) df = \Gamma_0 \int_{-\infty}^{\infty} \frac{1}{1 + 4\pi^2 f^2 \tau^2} df$$

Recall:

$$\int \frac{dx}{a^2+x^2} = \frac{1}{a}\tan^{-1}\left[\frac{x}{a}\right]$$

$$\rightarrow \sigma_{b_0}^2 = \frac{\Gamma_0}{4\pi^2\tau^2}\int_{-\infty}^{\infty} \frac{1}{\frac{1}{4\pi^2\tau^2}+f^2}df = \frac{\Gamma_0}{4\pi^2\tau^2}\times 2\pi\tau\times[\tan^{-1}(2\pi\tau f)]_{-\infty}^{\infty}$$

$$\rightarrow \sigma_{b_0}^2 = \frac{\Gamma_0}{2\pi\tau}\left[\frac{\pi}{2}-\left(-\frac{\pi}{2}\right)\right] = \frac{\Gamma_0}{2\tau}$$

Finally, we get:

$$\sigma_{b_0}^2 = \frac{\Gamma_0}{2\tau} = 2\Gamma_0\Delta f_b \rightarrow \Delta f_b = \frac{1}{4\tau} = \frac{1}{4RC}$$

2) Power spectral density $\Gamma_{b_1}(f)$ and power $\sigma_{b_1}^2$ of noise $b_1(t)$.

We have:

$$\Gamma_{b_1}(f) = \Gamma_{b_0}(f)\times|G_c(f)|^2$$

and:

$$\sigma_{b_1}^2 = \int_{-\infty}^{\infty}\Gamma_{b_1}(f)df = \int_{-\infty}^{\infty}\Gamma_{b_0}(f)\times|G_c(f)|^2 df = \Gamma_0\int_{-\Delta f_b}^{\Delta f_b}|G_c(f)|^2 df$$

Since the support of $G_c(f)$ is included in $[-\Delta f_b, \Delta f_b]$, then:

$$\sigma_{b_1}^2 = \Gamma_0\int_{-\Delta f_c}^{\Delta f_c}|G_c(f)|^2 df = \Gamma_0\int_{-\Delta f_c}^{\Delta f_c}|G_c(0)|^2 df = \Gamma_0\times 2\Delta f_c$$

hence:

$$\sigma_{b_1}^2 = \sigma_{b_0}^2\times\frac{\Delta f_c}{\Delta f_b}$$

3) The equalized signal (corrected) $s_c(t)$ is:

$$s_c(t) = \sum_{n=-\infty}^{\infty} a_n p(t-nT) + b_1(t)$$

with:

$$p(t) = y(t) \otimes g_c(t) = x(t) \otimes h(t) \otimes g_c(t)$$

The noise $b_1(t)$ is the result of filtering $b_0(t)$ by the equalizer filter whose impulse response is $g_c(t)$.

At the sampling times $t_k = kT$, the signal $s_c(t)$ is written:

$$s_c(kT) = a_k p(0) + \sum_{n=-\infty, n \neq k}^{\infty} a_n \, p[(k-n)T] + b_1(kT)$$

The term $a_k p(0)$ represents the response of the system (channel + equalization) to the transmission of the symbol a_k associated with the time interval kT.

The term $I_{m_l}(kT) = \sum_{n=-\infty, n \neq k}^{\infty} a_n \, p[(k-n)T]$ is the intersymbol interference.

It is a disturbing signal depending on all the transmitted symbols $\{a_n\}$, except for the symbol a_k which is related to the time interval considered.

The term $b_1(kT)$ is the noise at the output of the equalizer at the decision instant.

4) Only messages in the form $m_l = [a_{k-1}, a_{k+1}]$ interfere with symbol a_k, so:

$$I_{m_l}(kT) = \sum_{n=k-1, n \neq k}^{k+1} a_n \, p[(k-n)T] = a_{k-1} p(T) + a_{k+1} p(-T)$$

$I_{m_l}(kT)$ depends on $p[(k-n)T]$, here on $p(\pm T)$. We have to calculate $p(t)$ from its Fourier transform $P(f)$ which is defined (see Figure 2.33).

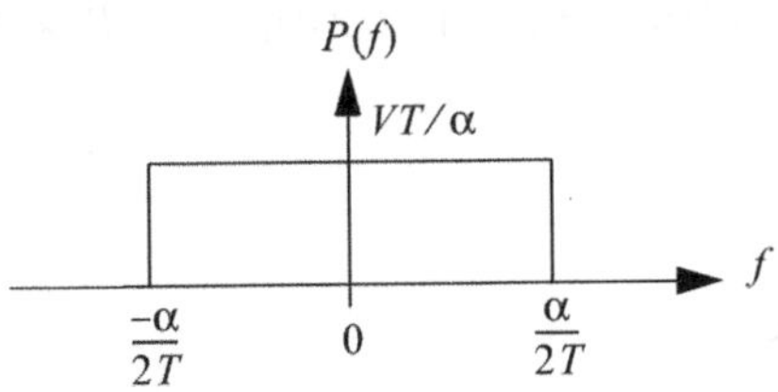

Figure 2.33. *Amplitude spectrum $P(f)$ of $p(t)$*

We have: $p(t) = F^{-1}\{P(f)\}$. This gives:

$$p(t) = \frac{VT}{\alpha} \times \frac{\alpha}{T} \times \frac{\sin\left[2\pi t \times \frac{\alpha}{2T}\right]}{2\pi t \times \frac{\alpha}{2T}} = V \times \frac{\sin[\alpha\pi\, t/T]}{\alpha\pi\, t/T}$$

Finally, we get:

$$p(0) = V, \quad p(\pm T) = V \times \frac{\sin[\alpha\pi]}{\alpha\pi}$$

So:

$$I_{m_l}(kT) = 0 \;\; if \;\; p(\pm T) = 0$$

This must be true for α non null integer.

$$\rightarrow \alpha_0 = min(\alpha) = 1 \quad \text{and} \quad p(t) = V \times \frac{sin[\pi\, t/T]}{\pi\, t/T}$$

5) Conditional probabilities of error:

Relation between $V\ and\ \sigma_{b_1}$. We have

$$\left[\frac{S}{b}\right]_{c,dB} = 20 \times \log_{10}\left[\frac{p(0)}{\sigma_{b_1}}\right] = 20 \times \log_{10}\left[\frac{V}{\sigma_{b_1}}\right] = 6\text{ dB} \rightarrow V = 2\sigma_{b_1}$$

Since $\mu_0 = 0$ (equiprobable symbols) and $I_{m_l}(kT) = 0$, then:

$$P_{e_{-1}} = Pr\{\hat{a}_k = 1/a_k = -1, m_l\} = \frac{1}{\sigma_{b_1}\sqrt{2\pi}} \int_{2\sigma_{b_1}}^{\infty} \exp\left[\frac{-1}{2}\frac{b_1^2}{\sigma_{b_1}^2}\right] db_1$$

$$P_{e_1} = Pr\{\hat{a}_k = -1/a_k = 1, m_l\} = \frac{1}{\sigma_{b_1}\sqrt{2\pi}} \int_{-\infty}^{-2\sigma_{b_1}} \exp\left[\frac{-1}{2}\frac{b_1^2}{\sigma_{b_1}^2}\right] db_1$$

Knowing that:

$$\int_{-\infty}^{\infty} p(x)dx = 1 \quad \text{and} \quad \int_{-2\sigma_{b_1}}^{2\sigma_{b_1}} p(x)dx = 0.95 = 1 - 0.05$$

and:

$$\int_{-\infty}^{\infty} p(x)dx = \int_{-\infty}^{-2\sigma_{b_1}} p(x)dx + \int_{-2\sigma_{b_1}}^{2\sigma_{b_1}} p(x)dx + \int_{2\sigma_{b_1}}^{\infty} p(x)dx = 1$$

$$\rightarrow \int_{-\infty}^{-2\sigma_{b_1}} p(x)dx = \int_{2\sigma_{b_1}}^{\infty} p(x)dx = \frac{0.05}{2} = 0.025$$

Finally we then get:

$$P_{e_{-1}} = P_{e_1} = 0.025$$

2.8. Problem 23 – M-ary baseband digital transmission

This problem deals with the baseband transmission of coded digital images over a transmission channel (cable) with reduced capacity. The different characteristics of the transmitter and receiver system in Figure 2.34 below will be analyzed together with its performances.

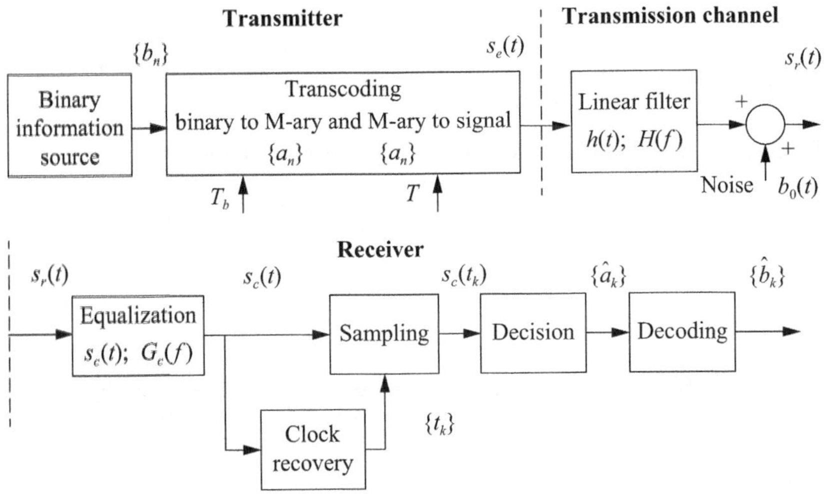

Figure 2.34. *Block diagram of the baseband transmission system on a cable*

The symbols b_n are delivered by the binary source every T_b second (with the use of a buffer memory). Moreover, they are supposed to be independent and equiprobable.

The coding of binary information $\{b_n\}$ into information symbols $\{a_n\}$ is done by grouping 2 bits b_n to form a quaternary symbol $a_n = \{-3, -1, 1, 3\}$ of period $T = 2T_b$.

The symbols a_n of information are provided to the transmitter at a rate of: $1/T = 10$ MHz.

The transmitted signal $s_e(t)$ is given by $s_e(t) = \sum_n a_n x(t - nT)$ where $x(t)$ is a rectangular signal of amplitude V over the time interval $[0, T[$, zero elsewhere.

The transmission channel is modeled by a linear filtering whose impulse response is denoted $h(t)$ with an additive degradation noise $b_0(t)$ at the output of the channel. The latter is supposed to be a second-order stationary Gaussian random noise, with zero mean value, having a broad frequency bandwidth, and a mean power $\sigma_{b_0}^2$.

The equalization filter of the transmission channel works in a frequency band totally included in that of the noise. The clock regeneration is assumed to be perfect and provides a sequence of decision instants of the form: $t_k = t_0 + kT$.

The decision system uses three thresholds, denoted μ_{-1}, μ_0, μ_1, to separate the equalized signal $s_c(t_k)$ into four classes. They are given by:

$$\mu_m = 2m \times p(0) \text{ with } m \in [-1, 0, 1]$$

We have:

$$d[s_c(t_k)] = \hat{a}_k = -3 \text{ if } s_c(t_k) < \mu_{-1}$$

$$d[s_c(t_k)] = \hat{a}_k = -1 \text{ if } \mu_{-1} \leq s_c(t_k) < \mu_0$$

$$d[s_c(t_k)] = \hat{a}_k = 1 \text{ if } \mu_0 \leq s_c(t_k) < \mu_1$$

$$d[s_c(t_k)] = \hat{a}_k = 3 \text{ if } s_c(t_k) \geq \mu_1$$

We denote successively ($\otimes$ is the convolution product):

$$y(t) = x(t) \otimes h(t) \quad \text{and} \quad p(t) = y(t) \otimes g_c(t)$$

In a first phase, the equalization filter is such that the frequency gain $P(f)$ of $p(t)$ is constant, equal to $2VT$ in the frequency band $[-1/4T, 1/4T]$, and zero elsewhere.

1) Give the expression of the signal $s_c(t_k)$ (composed of the useful signal + intersymbol interference + noise) at the decision instants of the form: $t_k = kT$.

2) From the expression of the intersymbol interference $I_{m_l}(kT)$, show that only symbols a of odd-rank index $[k \pm (2i+1)]$ interfere with symbol a_k (i positive or negative integer).

Afterwards, for sake of simplification, it is considered that only the two symbols adjacent to the symbol a_k interfere with it (namely a_{k-1} and a_{k+1}).

3) By listing the different possible combinations of the message $m_l = [a_{k-1}, a_{k+1}]$ interfering with a_k, show that $I_{m_l}(kT)$ can only take seven possible values that will be determined. (Afterwards, you will take $\pi \cong 3$ as a simplification for the calculation).

Also calculate the different probabilities, each of them associated to one of the seven different values of the intersymbol interference. Here the a_k will be considered equiprobable.

4) Show that, even without noise at the input of the receiver, the probability of error is very high.

So, we decide to perform a better equalization of the cable distortion. This second equalization is such that the frequency spectrum $P(f)$ of $p(t)$ is constant on the frequency band $[-1/2T, 1/2T]$, zero elsewhere, and $P(0) = 2VT$.

5) Show that the intersymbol interference is now cancelled.

Under these new conditions, we will assume that at the output of the equalizer the signal-to-noise ratio is:

$$\left[\frac{S}{b}\right]_{c,dB} = 20 \times \log_{10}\left[\frac{p(0)}{\sigma_{b_1}}\right] = 12 \text{ dB}$$

6) Calculate the following 16 conditional probabilities:

$$Pr\{\hat{a}_k = a_j / a_k = a_i\}; \quad i,j = 1,\cdots,4$$

and show that the 4x4 conditional probability matrix:

$$Pr\{\hat{a}_k = a_j / a_k = a_i\}; \quad i,j = 1,\cdots,4$$

is quasi of the form given in Table 2.11:

$a_k \backslash \widehat{a}_k$	- 3	- 1	+ 1	+ 3
-3	$1-p$	p	0	0
-1	p	$1-2p$	p	0
+1	0	p	$1-2p$	p
+3	0	0	p	$1-p$

Table 2.11. *Form of the conditional probability matrix*

NOTE.– We will consider here that if $p(x)$ is the probability density function of a Gaussian random variable with zero mean value:

$$p(x) = \frac{1}{\sigma\sqrt{2\pi}} \exp\left[\frac{-x^2}{2\sigma^2}\right], \text{ then } \int_{-4\sigma}^{4\sigma} p(x)dx = 1 - 2 \times 10^{-5} \text{ with } p = 10^{-5}$$

Solution of problem 23

1) The transmitted signal $s_e(t)$ is:

$$s_e(t) = \sum_{n=-\infty}^{\infty} a_n x(t - nT)$$

The received signal $s_r(t)$ is:

$$s_r(t) = \sum_{n=-\infty}^{\infty} a_n y(t - nT) + b_0(t)$$

with: $y(t) = x(t) \otimes h(t)$.

The signal $y(t)$ is the output of the channel when its input is the basic impulse (rectangular shape) $x(t)$ of period T and without noise.

The equalized (corrected) signal $s_c(t)$ is:

$$s_c(t) = \sum_{n=-\infty}^{\infty} a_n p(t - nT) + b_1(t)$$

with: $p(t) = y(t) \otimes g_c(t)$, that is: $p(t) = x(t) \otimes h(t) \otimes g_c(t)$.

The noise $b_1(t)$ is the result of filtering the noise $b_0(t)$ by the equalizer whose impulse response is $g_c(t)$.

At the sampling times $t_k = kT$, the signal $s_c(t)$ is written:

$$s_c(kT) = a_k p(0) + \sum_{n=-\infty, n\neq k}^{\infty} a_n\, p[(k-n)T] + b_1(kT)$$

The term $a_k p(0)$ represents the useful response of the system (channel + equalization) to the transmission of the symbol a_k associated with the time interval kT.

The term $I_{m_l}(kT) = \sum_{n=-\infty, n\neq k}^{\infty} a_n\, p[(k-n)T]$ is the intersymbol interference.

It is a disturbing signal which depends on all the symbol $\{a_n\}$ transmitted, except for the symbol a_k which is related to the time interval considered.

The term $b_1(kT)$ is the noise at the decision instant.

2) The intersymbol interference $I_{m_l}(kT)$ is given by:

$$I_{m_l}(kT) = \sum_{n=-\infty, n\neq k}^{\infty} a_n\, p[(k-n)T]$$

It depends on $p[(k-n)T]$, so we have to calculate $p(t)$ from $P(f)$.

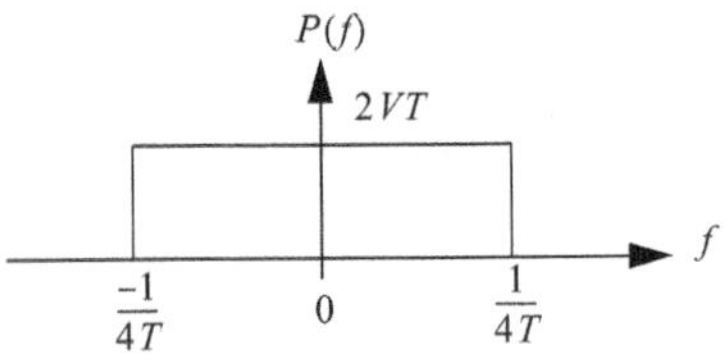

Figure 2.35. *Amplitude spectrum $P(f)$ of $p(t)$*

By definition, we have: $p(t) = F^{-1}\{P(f)\}$, hence:

$$p(t) = 2VT \times \frac{1}{2T} \times \frac{\sin\left[2\pi t \times \frac{1}{4T}\right]}{2\pi t \times \frac{1}{4T}} = V \times \frac{\sin[\pi\, t/2T]}{\pi\, t/2T}$$

$$p(0) = V,\ p(\pm T) = \frac{2V}{\pi},\ p(\pm 2T) = 0,\ p(\pm 3T) = \frac{-2V}{3\pi},\quad p(\pm 4T) = 0$$

From these values, we can conclude that:

$$p[\pm(2iT)] = 0 \rightarrow I_{m_l}(kT) = 0 \quad \text{for} \;\; n = (k \pm 2i)$$

$$p[\pm(2i+1)T] \neq 0 \rightarrow I_{m_l}(kT) \neq 0 \quad \text{for} \;\; n = [k \pm (2i+1)]$$

3) Possible values of $I_{m_l}(kT)$.

In the case where the messages only of the form $m_l = [a_{k-1}, a_{k+1}\,]$ interfere with the symbol a_k, we have (with $\pi \cong 3$):

$$I_{m_l}(kT) = \sum_{n=k-1, n\neq k}^{k+1} a_n\, p[(k-n)T] = a_{k-1}p(T) + a_{k+1}p(-T)$$

$$= [a_{k-1} + a_{k+1}]\frac{2V}{3}$$

$m_l = [a_{k-1}, a_{k+1}\,]$	$I_{m_l}(kT) = [a_{k-1} + a_{k+1}]\frac{2V}{3}$
- 3 - 3	- 4 V
- 3 - 1	- 2.6 V
- 3 1	- 1.3 V
- 3 3	0
- 1 - 3	- 2.6 V
- 1 - 1	- 1.3 V
- 1 1	0
- 1 3	1.3 V
1 - 3	- 1.3 V
1 - 1	0
1 1	1.3 V
1 3	2.6 V
3 - 3	0
3 - 1	1.3 V
3 1	2.6 V
3 3	4 V

Table 2.12. *Interfering messages and intersymbol interference amplitude*

Thus, there are seven distinct values of $I_{m_l}(kT)$.

$I_{m_l}(kT)$	$Pr\{I_{m_l}(kT)\}$
4 V	1/16
2.6 V	1/8
1.3 V	3/16
0	1/4
- 1.3 V	3/16
- 2.6 V	1/8
- 4 V	1/16

Table 2.13. *Probabilities of amplitude of intersymbol interference*

4) Neglecting the noise, after equalization the signal $s_c(kT)$ is written:

$$s_c(t_k) = a_k p(0) + I_{m_l}(kT) = a_k V + I_{m_l}(kT)$$

The decision thresholds are given by:

$$\mu_m = 2m \times p(0) \;\; \text{with} \;\; m \in [-1, 0, 1]$$

$$\rightarrow \mu_{-1} = -2p(0) = -2V\,; \quad \mu_0 = 0\,; \quad \mu_1 = 2p(0) = 2V$$

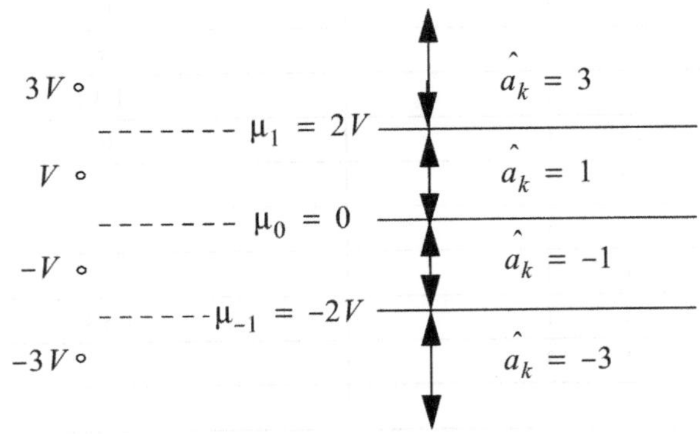

Figure 2.36. *Sample values $a_k V$, optimal thresholds and decision classes of a_k: $\hat{a}_k$*

It can easily be seen that to change the decision class, it is sufficient that

$I_{m_l}(kT) = \pm[V + \varepsilon]$. More precisely:

– an erroneous decision on the transmitted symbol $a_k = 3$ is made for all the values of $I_{m_l}(kT) < -V$, that is in 6/16 of cases;

– similarly, an erroneous decision on the transmitted symbol $a_k = -3$ is made for all the values of $I_{m_l}(kT) > V$, which is also in 6/16 cases;

– An erroneous decision on the symbols transmitted $a_k = \pm 1$ is made for all the values of $|I_{m_l}(kT)| > V$, that is in 12/16 of the cases.

In view of these results, even without noise, the probability of error is extremely high.

5) Null value of the intersymbol interference:

Under these new conditions, we have:

$$p(t) = F^{-1}\{P(f)\} = 2V \times \frac{\sin[\pi\, t/T]}{\pi\, t/T}$$

$$p(0) = 2V,\ p(kT) = 0\ \forall\ k \text{ non null integer } \rightarrow I_{m_l}(kT) = 0$$

6) Calculation of the 16 following conditional probabilities:

$$Pr\{\hat{a}_k = a_j / a_k = a_i\};\quad i, j = 1, \cdots, 4$$

Let us first express V as a function of $\sigma_{b_1}^2$ and calculate the new values of the decision thresholds μ_{-1}, μ_0, μ_1:

$$\left[\frac{S}{b}\right]_{c,dB} = 20 \times \log_{10}\left[\frac{p(0)}{\sigma_{b_1}}\right] = 20 \times \log_{10}\left[\frac{2V}{\sigma_{b_1}}\right] = 12 \text{ dB} \rightarrow V = 2\sigma_{b_1}$$

At the output of the equalizer, the signal is then written:

$$s_c(kT) = a_k p(0) + b_1(kT) = a_k 2V + b_1(kT) = a_k 4\sigma_{b_1} + b_1(kT)$$

and the threshold values of the decision blocks are:

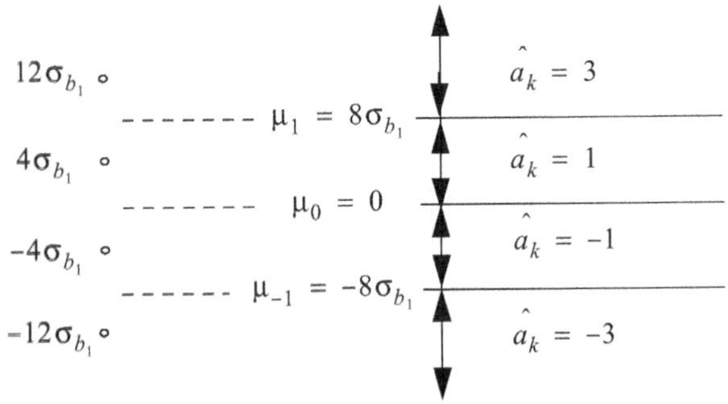

Figure 2.37. *Sample values $a_k 2V$, optimal thresholds and decision classes of a_k: $\hat{a}_k$*

The calculation of the conditional error probabilities is based on the knowledge of the noise intervals given by the course formulas (see the relations in Chapter 6 of Volume 1). For each value of the symbol a_k transmitted, we have four possible decisions (three erroneous, and a correct one) on the estimated value $\hat{a}_k$ of the symbol a_k.

Recall that for the intermediate values of the symbol a_k transmitted, the conditional error probability is given by:

$$P_{e_{2i+1}} = Pr\{\hat{a}_k = 2m + 1/a_k = 2i + 1, m_l\}$$

for $m \neq i$ and $m \neq -M/2\,;\ (M/2) - 1$

and for the extreme values of the transmitted symbol a_k, the two conditional error probabilities are:

$$P_{e_{2i+1}} = Pr\{\hat{a}_k = (M-1)/a_k = 2i + 1 \neq (M-1), m_l\}$$

$$P_{e_{2i+1}} = Pr\{\hat{a}_k = -(M-1)/a_k = 2i + 1 \neq -(M-1), m_l\}$$

The conditional probability matrix is then obtained like this:

– for the symbol transmitted $a_k = -1 \to i = -1 \to P_{e_{-1}}$, and the four decisions in reception are:

$$\hat{a}_k = 1 \to m = 0 \to \text{from (6.90)}, b_1(kT) \in [p(0), 3p(0)[$$

$$\to b_1(kT) \in \left[4\sigma_{b_1}, 12\sigma_{b_1}\right[$$

$\hat{a}_k = -1 \rightarrow m = -1 \rightarrow$ from (6.90), $b_1(kT) \in [-p(0), p(0)[$

$\rightarrow b_1(kT) \in \left[-4\sigma_{b_1}, 4\sigma_{b_1}\right[$

$\hat{a}_k = 3 \rightarrow$ from (6.96), $b_1(kT) \in [3p(0), \infty[\rightarrow b_1(kT) \in \left[12\sigma_{b_1}, \infty\right[$

$\hat{a}_k = -3 \rightarrow$ from (6.103), $b_1(kT) \in [-\infty, -p(0)[\rightarrow b_1(kT) \in \left[-\infty, -4\sigma_{b_1}\right[$

– for the symbol transmitted $a_k = 1 \rightarrow i = 0 \rightarrow P_{e_1}$, and the four decisions in reception are:

$\hat{a}_k = -1 \rightarrow m = -1 \rightarrow$ from (6.90), $b_1(kT) \in [-3p(0), -p(0)[$

$\rightarrow b_1(kT) \in \left[-12\sigma_{b_1}, -4\sigma_{b_1}\right[$

$\hat{a}_k = 1 \rightarrow m = 0 \rightarrow$ from (6.90), $b_1(kT) \in [-p(0), p(0)[$

$\rightarrow b_1(kT) \in \left[-4\sigma_{b_1}, 4\sigma_{b_1}\right[$

$\hat{a}_k = 3 \rightarrow$ from (6.96), $b_1(kT) \in [p(0), \infty[\rightarrow b_1(kT) \in \left[4\sigma_{b_1}, \infty\right[$

$\hat{a}_k = -3 \rightarrow$ from (6.103), $b_1(kT) \in [-\infty, -3p(0)[$

$\rightarrow b_1(kT) \in \left[-\infty, -12\sigma_{b_1}\right[$

– for the symbol transmitted $a_k = 3 \rightarrow i = 1 \rightarrow P_{e_3}$, and the four decisions in reception are:

$\hat{a}_k = -1 \rightarrow m = -1 \rightarrow$ from (6.90), $b_1(kT) \in [-5p(0), -3p(0)[$

$\rightarrow b_1(kT) \in \left[-20\sigma_{b_1}, -12\sigma_{b_1}\right[$

$\hat{a}_k = 1 \rightarrow m = 0 \rightarrow$ from (6.90), $b_1(kT) \in [-3p(0), -p(0)[$

$\rightarrow b_1(kT) \in \left[-12\sigma_{b_1}, -4\sigma_{b_1}\right[$

$\hat{a}_k = -3 \rightarrow$ from (6.103), $b_1(kT) \in [-\infty, -5p(0)[$

$\rightarrow b_1(kT) \in \left[-\infty, -20\sigma_{b_1}\right[$

$\hat{a}_k = 3 \rightarrow$ from (6.96), $b_1(kT) \in [-p(0), \infty[\rightarrow b_1(kT) \in \left[-4\sigma_{b_1}, \infty\right[$

– for the symbol transmitted $a_k = -3 \rightarrow i = -2 \rightarrow P_{e_{-3}}$, and the four decisions in reception are:

$$\hat{a}_k = -1 \rightarrow m = -1 \rightarrow \text{from (6.90)}, b_1(kT) \in [p(0), 3p(0)[$$

$$\rightarrow b_1(kT) \in \left[4\sigma_{b_1}, 12\sigma_{b_1}\right[$$

$$\hat{a}_k = 1 \rightarrow m = 0 \rightarrow \text{from (6.90)}, b_1(kT) \in [3p(0), 5p(0)[$$

$$\rightarrow b_1(kT) \in \left[12\sigma_{b_1}, 20\sigma_{b_1}\right[$$

$$\hat{a}_k = -3 \rightarrow \text{from (6.103)}, b_1(kT) \in [-\infty, p(0)[$$

$$\rightarrow b_1(kT) \in \left[-\infty, 4\sigma_{b_1}\right[$$

$$\hat{a}_k = 3 \rightarrow \text{from (6.96)}, b_1(kT) \in [5p(0), \infty[\rightarrow b_1(kT) \in \left[20\sigma_{b_1}, \infty\right[$$

Furthermore, we have:

$$\int_{-4\sigma}^{4\sigma} p(x)dx = 1 - 2 \times 10^{-5} = 1 - 2p$$

$$\int_{-\infty}^{\infty} p(x)dx = \int_{-\infty}^{-4\sigma} + \int_{-4\sigma}^{4\sigma} + \int_{4\sigma}^{\infty} = 1 \rightarrow \int_{-\infty}^{-4\sigma} = \int_{4\sigma}^{\infty} = 10^{-5} = p$$

$$\int_{-\infty}^{-20\sigma} = \int_{-\infty}^{-12\sigma} = \int_{20\sigma}^{\infty} = \int_{12\sigma}^{\infty} = 0 \rightarrow \int_{4\sigma}^{12\sigma} = \int_{4\sigma}^{\infty} = 10^{-5} = \int_{-12\sigma}^{-4\sigma}$$

$$\int_{-\infty}^{4\sigma} = \int_{-\infty}^{-4\sigma} + \int_{-4\sigma}^{4\sigma} = 10^{-5} + 1 - 2 \times 10^{-5} = 1 - 10^{-5} = \int_{-4\sigma}^{\infty}$$

$$\int_{-\infty}^{12\sigma} = \int_{-\infty}^{-4\sigma} + \int_{-4\sigma}^{4\sigma} + \int_{4\sigma}^{12\sigma} = 10^{-5} + 1 - 2 \times 10^{-5} + 10^{-5} = 1 \rightarrow \int_{-\infty}^{20\sigma} = 1$$

Hence see the matrix of conditional decision probabilities in Table 2.14 below.

$a_k \backslash \hat{a}_k$	-3	-1	+1	+3
-3	$1-p$	p	0	0
-1	p	$1-2p$	p	0
+1	0	p	$1-2p$	p
+3	0	0	p	$1-p$

Table 2.14. *Conditional decision probability matrix*

2.9. Problem 24 – Baseband digital transmission of bipolar coded information

We consider the transmission of information (speech) in digital form on a two-wire cable transmission channel. The on-line code used is the bipolar code. The block diagram of the transmission system is shown in Figure 2.38.

The source produces a series of independent but not equiprobable binary sequence $\{b_n\}$, with:

$$Pr\{b_k = 0\} = 2/5\,; \quad Pr\{b_k = 1\} = 3/5$$

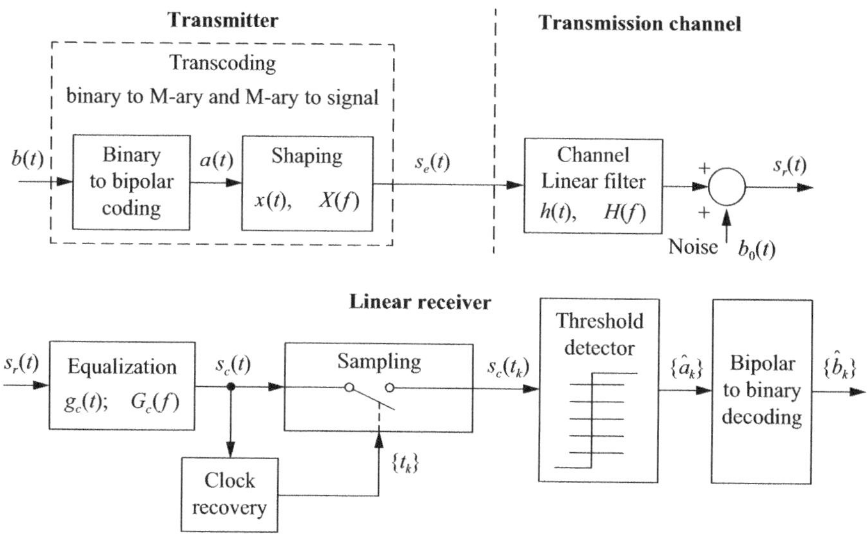

Figure 2.38. *Practical chain of a digital baseband communication system with bipolar code*

The transmitted signal is given by: $s_e(t) = \sum_n a_n x(t - nT)$.

The signal $x(t)$ is a pulse of amplitude V on the time interval $[0, T/2[$. The additive noise $b_0(t)$ is assumed to be stationary, Gaussian and centered, with a very broad power spectral density $\Gamma_{b_0}(f)$ compared to that of the signal (energy bandwidth equal to Δf_b) and an average power $\sigma_{b_0}^2$.

1) Cite two major reasons that digital information transmissions are superior to analog transmissions.

2) What is the bandwidth of the standard telephone channel? And what is the maximum possible bitrate of a digital signal transmitted through this channel?

3) Give the transformation rule which allows us to transcode the binary symbol b_n into the ternary symbol a_n (bipolar code).

4) Considering that the first non-zero symbol transmitted is always positive, what is the sequence $\{a_n\}$ resulting from the bipolar coding of the following binary sequence $\{b_n\}$ of length 16 in Table 2.15.

$\{b_n\}$	1	0	0	1	1	0	1	0	1	1	1	0	0	0	1	0
$\{a_n\}$																

Table 2.15. *Generation of the sequence $\{a_n\}$ of the bipolar code*

5) What is the advantage of using a bipolar code in baseband transmission (cite at least two reasons)?

6) What is the limitation? How do we bypass this limitation to prevent the coding of four consecutive null symbols b from leading to the absence of pulses in the on-line code, while ensuring that the system is operating correctly on reception?

7) Draw the diagram of realization of the RZ bipolar encoder and decoder.

8) What is the purpose of equalization? What does the Nyquist frequency criterion express?

9) Assuming that the equalization filter $G_c(f)$ has a unit gain at zero frequency, determine the power $\sigma_{b_1}^2$ of the noise at the decision level based on that of the observation noise and the equivalent energy bandwidth Δf_c of the equalization filter.

10) Give the number and value(s) of the decision threshold(s) μ of the bipolar code.

In the rest of the problem, it will be considered that the equalization is not perfect and that actually, the amplitude spectrum of the signal $p(t)$ at the output of the equalizer, denoted $P(f)$, when a single impulse $x(t)$ is sent by the transmitter and without considering the noise, is constant, equal to VT on the frequency domain $[-(1+\alpha)/2T, (1+\alpha)\ /\ 2T\ ]$, and zero elsewhere, with $\alpha = 0.1$.

So we have: $p(t) = x(t) \otimes h(t) \otimes g_c(t)$.

11) Give the expression of the signal $p(t)$ and its particular values at times $t = 0$ and $t = \pm T$ (We shall consider for simplification that $\sin[\pi(1-\alpha)]/\pi \cong 0.1$).

Similarly and for simplicity, intersymbol interference will only be considered as that resulting from the two symbols a_{k-1} and a_{k+1} adjacent to symbol a_k.

12) Give the expression of the equalized signal $s_c(kT)$ for the kth instant of decision by showing the different contributions to the amplitude of this signal.

13) Determine for each possible value of a_k the possible messages $m_l = [a_{k-1}, a_{k+1}\ ]$ and their conditional probabilities $p_{(m_l/i)} = Pr\{m_l/a_k = i\}$, according to Tables 2.16 and 2.17.

14) Give the expression of the probability p_{m_l} and its value for each message m_l. Give also the amplitude of the intersymbol interference I_{m_l} of each message.

15) Deduce the different possible values of intersymbol interference and the associated probabilities.

$m_l = [a_{k-1}, a_{k+1}\ ]$							
$a_k = -1$							
$a_k = 0$							
$a_k = 1$							

Table 2.16. *Possible messages m_l for each a_k*

$m_l = [a_{k-1}, a_{k+1}]$	$Pr\{m_l/a_k = -1\}$	$Pr\{m_l/a_k = 0\}$	$Pr\{m_l/a_k = 1\}$
$m_1(,)$			
$m_2(,)$			
$m_3(,)$			
$m_4(,)$			
$m_5(,)$			
$m_6(,)$			
$m_7(,)$			
$m_8(,)$			
$m_9(,)$			

Table 2.17. *Conditional probabilities $Pr\{m_l/a_k\}$ of messages m_l*

$m_l = [a_{k-1}, a_{k+1}]$	p_{m_l}	I_{m_l}
$m_1(,)$		
$m_2(,)$		
$m_3(,)$		
$m_4(,)$		
$m_5(,)$		
$m_6(,)$		
$m_7(,)$		
$m_8(,)$		
$m_9(,)$		

Table 2.18. *Probability p_{m_l} and value of the intersymbol interference for each message m_l*

I_{m_l}	I_{-2}	I_{-1}	I_0	I_1	I_2
Value of I_{m_l}					
$Pr\{I_{m_l}\}$					

Table 2.19. *Possible values of inter-symbol interference and associated probabilities*

16) Give the expression of the probability of error $P_{e,a}$ on the symbols of the bipolar code.

To simplify, it is considered that only the errors $\hat{a}_k$ of the following type: "the decided values are adjacent to the prior value a_k", are of non-zero probability:

$$a_k = -1 \rightarrow \hat{a}_k = 0\,; \quad a_k = 1 \rightarrow \hat{a}_k = 0\,; \quad a_k = 0 \rightarrow \hat{a}_k = \pm 1$$

17) Give the expression of each of these four conditional probabilities of possible errors (by specifying the intervals of the noise amplitude):

$$P_{e(-1/0,m_l)} = Pr\{\hat{a}_k = -1/\ a_k = 0\,, m_l\}$$

$$P_{e(1/0,m_l)} = Pr\{\hat{a}_k = 1/\ a_k = 0\,, m_l\}$$

$$P_{e(0/-1,m_l)} = Pr\{\hat{a}_k = 0/\ a_k = -1\,, m_l\}$$

$$P_{e(0/1,m_l)} = Pr\{\hat{a}_k = 0/\ a_k = 1\,, m_l\}$$

It is assumed that at the output of the equalization, the signal-to-noise ratio is:

$$\left[\frac{S}{b}\right]_{c,dB} = 20 \times \log_{10}\left[\frac{p(0)}{\sigma_{b_1}}\right] = 13.064 \text{ dB}$$

18) For each message m_l, calculate the intervals of the noise dynamics and the values of the conditional probability of errors according to Table 2.20 (based on the table of a centered and reduced Gaussian law).

NOTE.–

– *ul*: upper limit of the interval of the noise amplitude.

– *ll*: lower limit of the interval of the noise amplitude.

m_l	m_1	m_2	m_3	m_4	m_5	m_6	m_7	m_8	m_9
$ul =$ $ll =$									
$P_{e(-1/0,m_l)}$									
$ul =$ $ll =$									
$P_{e(1/0,m_l)}$									
$ul =$ $ll =$									
$P_{e(0/-1,m_l)}$									
$ul =$ $ll =$									
$P_{e(0/1,m_l)}$									

Table 2.20. *Noise dynamics interval and conditional error probabilities*

19) Deduce the value of the error probability $P_{e,a}$.

20) What would be the value of the probability of error $P_{e1,a}$, if we kept, for each of the possible values of a_k, only the configuration m_l leading to the most unfavorable value of the intersymbol interference I_{m_l}?

21) Give the expression and the value of the probability of error $P_{e,b}$ on the binary symbols decoded, under the same conditions of question 20.

22) What would be the value of the probability of error $P_{e2,a}$, if there was no more intersymbol interference?

Solution of problem 24

1) Major reasons:

– integration of services, therefore lower costs;

– performance in terms of error / distortion not cumulable, because the regeneration of signals can be exactly performed.

2) The frequency bandwidth of the standard telephone channel is:

$$B = f_h - f_b = 3{,}400 - 300 = 3{,}100 \text{ Hz}$$

The maximum possible symbol rate (according to Nyquist criterion) is:

$$D_s = 2B = 6{,}200 \text{ symbol/s}$$

The maximum possible bitrate is:

$$D_b = D_s \times \log_2 M$$

If coding on two levels, then:

$$D_b = D_s = 6{,}200 \text{ bit/s}$$

NOTE.– Usually, one uses a M-ary coding system with $M \gg 2$ where the number M is a function of the signal-to-noise ratio at the input of the decision block which ensures a probability of a wrong decision lower than a given level of admissible errors.

3) Rule of transformation of a binary symbol into a ternary symbol (bipolar code).

The bipolar code is a three-level code such as:

$$b_n = 0 \rightarrow a_n = 0\,;\; b_n = 1 \rightarrow a_n = \pm 1 \text{ alternately.}$$

4) Generation of the sequence $\{a_n\}$ (bipolar code).

$\{b_n\}$	1	0	0	1	1	0	1	0	1	1	1	0	0	0	1	0
$\{a_n\}$	1	0	0	-1	1	0	-1	0	1	-1	1	0	0	0	-1	0

Table 2.21. *Generation of the sequence* $\{a_n\}$

5) The interests in using a bipolar code in baseband transmission are:

– no continuous component;

– the spectrum of the transmitted signal vanishes at all the multiples of $1/T_b$ and limitation of the spectral occupation.

6) Bipolar code limitation:

If we have a long series of bits at zero, then there are no pulses issued over a period that can be significant. This causes the loss of clock synchronization at the receiver side. To overcome this drawback, the bipolar code with a high pulse density is used.

If on a period of $4T_b$, there are no pulses, we have to use the HDB-3 code.

7) Block diagram of the RZ bipolar encoder and decoder.

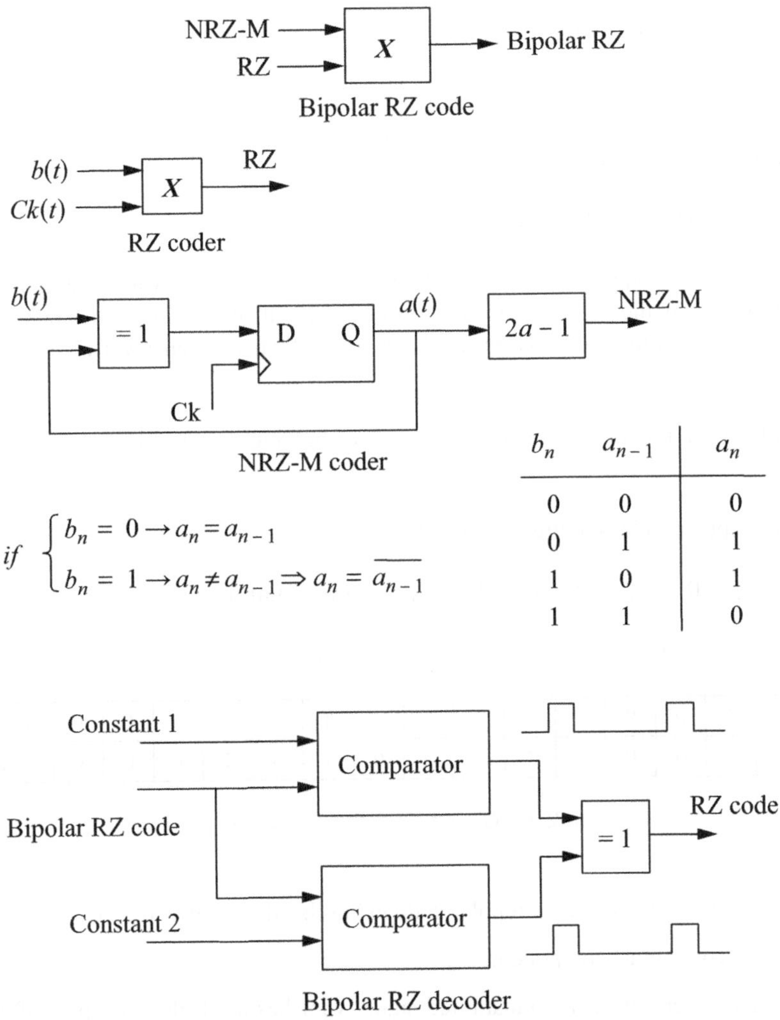

$$if \begin{cases} b_n = 0 \rightarrow a_n = a_{n-1} \\ b_n = 1 \rightarrow a_n \neq a_{n-1} \Rightarrow a_n = \overline{a_{n-1}} \end{cases}$$

b_n	a_{n-1}	a_n
0	0	0
0	1	1
1	0	1
1	1	0

Figure 2.39. *Block diagram of RZ bipolar coder and decoder*

8) The ultimate goal of equalization is to cancel the influence of the transmission channel in order to have an ISI as small as possible (and even null).

The Nyquist frequency criterion states that the equalization must ensure that we have:

$$\sum_{k=-\infty}^{\infty} P\left(f-\frac{k}{T}\right) = Tp(0) = \text{Constant}$$

9) The noise power $b_1(t)$ is:

$$\sigma_{b_1}^2 = \int_{-\infty}^{\infty} \Gamma_{b_1}(f)\, df$$

The power spectral density $b_1(t)$ is given by:

$$\Gamma_{b_1}(f) = \Gamma_{b_0}(f) \times |G_c(f)|^2 = \Gamma_0 \times |G_c(f)|^2$$

The noise power $b_0(t)$ is:

$$\sigma_{b_0}^2 = \int_{-\infty}^{\infty} \Gamma_{b_0}(f)\, df = \Gamma_0 \int_{-\Delta f_b}^{\Delta f_b} df = 2\,\Gamma_0 \times \Delta f_b$$

hence:

$$\Gamma_0 = \frac{\sigma_{b_0}^2}{2\Delta f_b}$$

and:

$$\sigma_{b_1}^2 = \int_{-\infty}^{\infty} \Gamma_{b_1}(f)\, df = \frac{\sigma_{b_0}^2}{2\Delta f_b} \int_{-\Delta f_c}^{\Delta f_c} |G_c(f)|^2 df = \frac{\sigma_{b_0}^2}{2\Delta f_b}$$

$$\times\, 2\Delta f_c |G_c(0)|^2$$

$$\sigma_{b_1}^2 = \sigma_{b_0}^2 \times \frac{\Delta f_c}{\Delta f_b}$$

10) Number and value(s) of the decision threshold(s) of the bipolar code: two decision thresholds, denoted: $\mu^+ = -\mu^- = p(0)/2$.

11) Expression of the signal $p(t)$ obtained by the inverse Fourier transform of the amplitude spectrum $P(f)$.

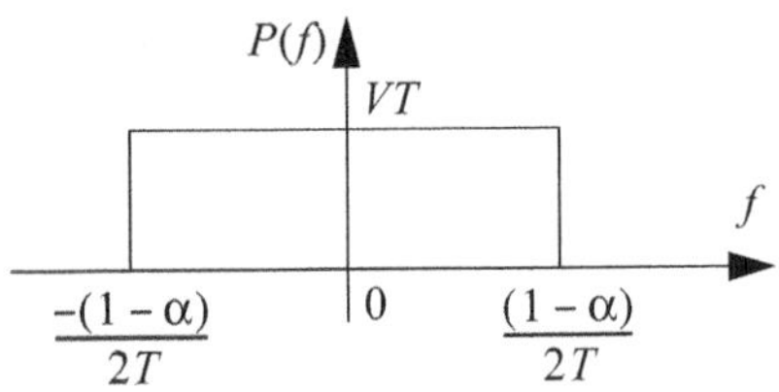

Figure 2.40. *Amplitude spectrum $P(f)$ of $p(t)$*

$$p(t) = (1-\alpha)V \times \frac{\sin[\pi\,(1-\alpha)t/T]}{\pi(1-\alpha)t/T}$$

$$p(0) = (1-\alpha)V = 0.9V = \frac{9V}{10}$$

$$p(\pm T) = V\frac{\sin[\pi(1-\alpha)]}{\pi} \cong 0.1\,V = V/10$$

12) Expression of the received and corrected signal:

$$s_c(kT) = a_k p(0) + I_{m_l}(kT) + b_1(kT)$$

with:

$$I_{m_l}(kT) = a_{k-1}p(T) + a_{k+1}p(-T) = \frac{V}{10}[a_{k-1} + a_{k+1}]$$

hence:

$$s_c(kT) = \frac{9}{10}Va_k + \frac{V}{10}[a_{k-1} + a_{k+1}] + b_1(kT)$$

13) For each possible value of a_k, determination of possible interfering messages and their conditional probabilities: $p_{(m_l/i)} = Pr\{m_l/a_k = i\}$.

$\boldsymbol{m_l = [a_{k-1}, a_{k+1}]}$							
$\boldsymbol{a_k = -1}$	0 0	0 1	1 0	1 1			
$\boldsymbol{a_k = 0}$	0 0	0 -1	0 1	-1 0	-1 1	1 0	1 -1
$\boldsymbol{a_k = 1}$	0 0	0 -1	-1 0	-1 -1			

Table 2.22. *Possible messages interfering with a_k*

$\{m_l/a_k = -1\}$ $m_l = [a_{k-1}, a_{k+1}]$	$p_{(m_l/-1)} = Pr\{m_l/a_k = -1\}$
0 0	$Pr\{b_{k-1} = 0, b_{k+1} = 0\} = 4/25$
0 1	$Pr\{b_{k-1} = 0, b_{k+1} = 1\} = 6/25$
1 0	$Pr\{b_{k-1} = 1, b_{k+1} = 0\} = 6/25$
1 1	$Pr\{b_{k-1} = 1, b_{k+1} = 1\} = 9/25$

Table 2.23. *Conditional probabilities:* $p_{(m_l/-1)} = Pr\{m_l/a_k = -1\}$

$\{m_l/a_k = 0\}$ $m_l = [a_{k-1}, a_{k+1}]$	$p_{(m_l/0)} = Pr\{m_l/a_k = 0\}$
0 0	$Pr\{b_{k-1} = 0, b_{k+1} = 0\} = 4/25$
0 1	$Pr\{b_{k-1} = 0, b_{k+1} = 1 \text{ and } a_{k+1} > 0\} = 6/50$
0 -1	$Pr\{b_{k-1} = 0, b_{k+1} = 1 \text{ and } a_{k+1} < 0\} = 6/50$
1 0	$Pr\{b_{k-1} = 1, b_{k+1} = 0 \text{ and } a_{k-1} > 0\} = 6/50$
1 -1	$Pr\{b_{k-1} = 1, b_{k+1} = 1 \text{ and } a_{k-1} > 0\} = 9/50$
-1 0	$Pr\{b_{k-1} = 1, b_{k+1} = 0 \text{ and } a_{k-1} < 0\} = 6/50$
-1 1	$Pr\{b_{k-1} = 1, b_{k+1} = 1 \text{ and } a_{k-1} < 0\} = 9/50$

Table 2.24. *Conditional probabilities:* $p_{(m_l/0)} = Pr\{m_l/a_k = 0\}$

$\{m_l/a_k = 1\}$ $m_l = [a_{k-1}, a_{k+1}]$	$p_{(m_l/1)} = Pr\{m_l/a_k = 1\}$
0 0	$Pr\{b_{k-1} = 0, b_{k+1} = 0\} = 4/25$
0 -1	$Pr\{b_{k-1} = 0, b_{k+1} = 1\} = 6/25$
-1 0	$Pr\{b_{k-1} = 1, b_{k+1} = 0\} = 6/25$
-1 -1	$Pr\{b_{k-1} = 1, b_{k+1} = 1\} = 9/25$

Table 2.25. *Conditional probabilities:* $p_{(m_l/1)} = Pr\{m_l/a_k = 1\}$

The summary of all these values is given in Table 2.26.

$\boldsymbol{m_l = [a_{k-1}, a_{k+1}]}$	$\boldsymbol{Pr\{m_l/a_k = -1\}}$	$\boldsymbol{Pr\{m_l/a_k = 0\}}$	$\boldsymbol{Pr\{m_l/a_k = 1\}}$
$m_1 = (0, \quad 0)$	$4/25$	$4/25$	$4/25$
$m_2 = (0, \quad 1)$	$6/25$	$6/50$	
$m_3 = (0, \quad -1)$		$6/50$	$6/25$
$m_4 = (1, \quad 0)$	$6/25$	$6/50$	
$m_5 = (1, \quad -1)$		$9/50$	
$m_6 = (-1, \quad 0)$		$6/50$	$6/25$
$m_7 = (-1, \quad 1)$		$9/50$	
$m_8 = (-1, \quad -1)$			$9/25$
$m_9 = (1, \quad 1)$	$9/25$		

Table 2.26. *Conditional probabilities:* $p_{(m_l/i)} = Pr\{m_l/a_k = i\}$

14) Expression of the probability p_{m_l}:

$$p_{m_l} = \sum_{i=-1}^{1} p(i, m_l) = \sum_{i=-1}^{1} p_i \times p_{(m_l/i)} \quad \text{and} \quad \sum_{l=1}^{9} p_{m_l} = 1$$

$\boldsymbol{m_l = [a_{k-1}, a_{k+1}]}$	$\boldsymbol{p_{m_l}}$	$\boldsymbol{I_{m_l}}$
$m_1 = (0, \quad 0)$	$3/10 \times 4/25 + 2/5 \times 4/25 + 3/10 \times 4/25 = 4/25$	0
$m_2 = (0, \quad 1)$	$3/10 \times 6/25 + 2/5 \times 6/50 = 3/25$	$V/10$
$m_3 = (0, \quad -1)$	$2/5 \times 6/50 + 3/10 \times 6/25 = 3/25$	$-V/10$
$m_4 = (1, \quad 0)$	$3/10 \times 6/25 + 2/5 \times 6/50 = 3/25$	$V/10$
$m_5 = (1, \quad -1)$	$2/5 \times 9/50 = 1.8/25$	0
$m_6 = (-1, \quad 0)$	$2/5 \times 6/50 + 3/10 \times 6/25 = 3/25$	$-V/10$
$m_7 = (-1, \quad 1)$	$2/5 \times 9/50 = 1.8/25$	0
$m_8 = (-1, \quad -1)$	$3/10 \times 9/25 = 2.7/25$	$-V/5$
$m_9 = (1, \quad 1)$	$3/10 \times 9/25 = 2.7/25$	$V/5$

Table 2.27. *Intersymbol interference amplitudes and associated probabilities with each possible interfering message*

with:

$$p_0 = Pr\{a_k = 0\} = Pr\{b_k = 0\} = 2/5$$

$$p_1 = Pr\{a_k = 1\} = p_{-1} = Pr\{a_k = -1\} = 1/2 \times Pr\{b_k = 1\} = 3/10$$

and we actually have:

$$\sum_{l=1}^{9} p_{m_l} = 1$$

15) Possible values of intersymbol interference and associated probabilities.

I_{m_l}	I_{-2}	I_{-1}	I_0	I_1	I_2
Value of I_{m_l}	$-V/5$	$-V/10$	0	$V/10$	$V/5$
$Pr\{I_{m_l}\}$	1/9	2/9	3/9	2/9	1/9

Table 2.28. *Inter-symbol interference values and associated probabilities*

And we have:

$$\sum Pr\{I_{m_l}\} = 1$$

16) Expression of the probability of error on the symbols a:

$$P_{e,a} = \sum_{i=-1}^{1} p_i \left[\sum_{l=1}^{9} p_{(m_l/i)} \sum_{m=-1}^{1} P_{e(m/i,m_l)_{m \neq i}} \right]$$

17) Expression of each of the four conditional probabilities of possible error.

The corrected and sampled signal is:

$$s_c(kT) = a_k p(0) + I_{m_l}(kT) + b_1(kT)$$

The decision thresholds are such that:

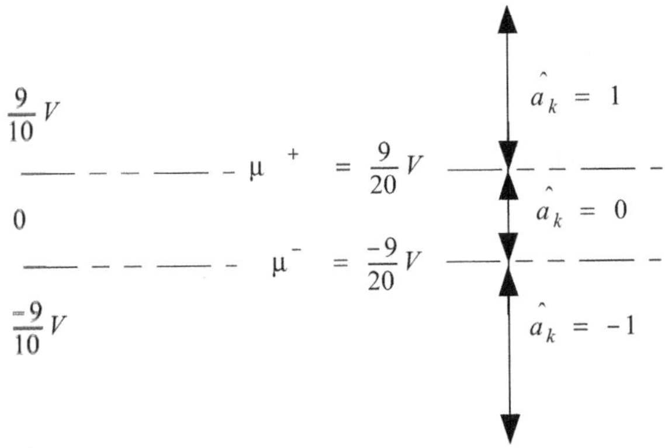

Figure 2.41. *Sample values without ISI and noise, optimal thresholds and decision classes of a_k: $\hat{a}_k$*

The noise $b_1(kT)$ is written:

$$b_1(kT) = s_c(kT) - a_k p(0) - I_{m_l}(kT)$$

If:

$$s_c(kT) \in \left]c, d\right]$$

Then, we have:

$$b_1(kT) \in \left]c - a_k p(0) - I_{m_l}(kT), d - a_k p(0) - I_{m_l}(kT)\right]$$

– *Case of transmission* $a_k = 0$.

If we decide:

$$\hat{a}_k = -1 \rightarrow s_c(kT) \in \left]-\infty, \mu^-\right] \rightarrow b_1(kT) \in \left]-\infty, \mu^- - I_{m_l}(kT)\right]$$

$$\rightarrow P_{e(-1/0,m_l)} = \frac{1}{\sigma_{b_1}\sqrt{2\pi}} \int_{-\infty}^{\mu^- - I_{m_l}(kT)} \exp\left[\frac{-1}{2}\frac{b_1^2}{\sigma_{b_1}^2}\right] db_1$$

If we decide:

$$\hat{a}_k = 1 \rightarrow s_c(kT) \in \left]\mu^+, \infty\right[\rightarrow b_1(kT) \in \left]\mu^+ - I_{m_l}(kT), \infty\right[$$

$$\rightarrow P_{e(1/0,m_l)} = \frac{1}{\sigma_{b_1}\sqrt{2\pi}} \int_{\mu^+ - I_{m_l}(kT)}^{\infty} \exp\left[\frac{-1}{2}\frac{b_1^2}{\sigma_{b_1}^2}\right] db_1$$

– *Case of transmission* $a_k = -1$.

If we decide:

$$\hat{a}_k = 0 \rightarrow s_c(kT) \in \left]\mu^-, \mu^+\right]$$

$$\rightarrow b_1(kT) \in \left]\mu^- + p(0) - I_{m_l}(kT),\ \mu^+ + p(0) - I_{m_l}(kT)\right]$$

$$\rightarrow P_{e(0/-1,m_l)} = \frac{1}{\sigma_{b_1}\sqrt{2\pi}} \int_{\mu^- + p(0) - I_{m_l}(kT)}^{\mu^+ + p(0) - I_{m_l}(kT)} \exp\left[\frac{-1}{2}\frac{b_1^2}{\sigma_{b_1}^2}\right] db_1$$

– *Case of transmission* $a_k = 1$.

If we decide:

$$\hat{a}_k = 0 \rightarrow s_c(kT) \in \left]\mu^-, \mu^+\right]$$

$$\rightarrow b_1(kT) \in \left]\mu^- - p(0) - I_{m_l}(kT),\ \mu^+ - p(0) - I_{m_l}(kT)\right]$$

$$\rightarrow P_{e(0/1,m_l)} = \frac{1}{\sigma_{b_1}\sqrt{2\pi}} \int_{\mu^- - p(0) - I_{m_l}(kT)}^{\mu^+ - p(0) - I_{m_l}(kT)} \exp\left[\frac{-1}{2}\frac{b_1^2}{\sigma_{b_1}^2}\right] db_1$$

18) For each message m_l, calculation of intervals of noise dynamics and values of the conditional probabilities of error, are given in Table 2.29.

The noise amplitude intervals should be expressed as a function of σ_{b_1}. We have:

$$\left[\frac{S}{b}\right]_{c,dB} = 20 \times \log_{10}\left[\frac{p(0)}{\sigma_{b_1}}\right] = 20 \times \log_{10}\left[\frac{9V/10}{\sigma_{b_1}}\right] = 13.064 \text{ dB}$$

$$\rightarrow \frac{9V}{10\sigma_{b_1}} = 4.5 \rightarrow V = 5\sigma_{b_1} \text{ and } \mu^+ = \frac{9}{4}\sigma_{b_1};\ \mu^- = \frac{-9}{4}\sigma_{b_1};\ p(0) = \frac{9}{2}\sigma_{b_1}$$

19) Calculation of the probability of error $P_{e,a}$:

$$P_{e,a} = p_{-1}\left[\frac{4}{25} \times 0.0122 + \frac{6}{25} \times 0.0401 + \frac{6}{25} \times 0.0401 + \frac{9}{25} \times 0.1056\right]$$

$$+p_0\left[\frac{4}{25}(0.0122 + 0.0122) + \frac{6}{50}(0.0030 + 0.0401)\right.$$

$$+\frac{6}{50}(0.0401 + 0.0030) + \frac{6}{50}(0.0030 + 0.0401)$$

$$+\frac{9}{50}(0.0122 + 0.0122) + \frac{6}{50}(0.0401 + 0.0030)$$

$$\left.+\frac{9}{50}(0.0122 + 0.0122)\right]$$

$$+p_1\left[\frac{4}{25} \times 0.0122 + \frac{6}{25} \times 0.0401 + \frac{6}{25} \times 0.0401 + \frac{9}{25}\right.$$

$$\left.\times 0.1056\right] = 0.0490336$$

20) We keep only the most unfavorable case of intersymbol interference according to the table giving the values of the conditional probabilities for each message:

$$P_{e(-1/0,m_3)} = 0.0401 \leftrightarrow m_3 = (0, \quad -1); \qquad I_{-1} = -0.5\,\sigma_{b_1}$$

$$P_{e(1/0,m_2)} = 0.0401 \leftrightarrow m_2 = (0, \quad 1); \qquad I_1 = 0.5\,\sigma_{b_1}$$

$$P_{e(0/-1,m_9)} = 0.1056 \leftrightarrow m_9 = (1, \quad 1); \qquad I_2 = \sigma_{b_1}$$

$$P_{e(0/1,m_8)} = 0.1056 \leftrightarrow m_8 = (-1, \quad -1); \qquad I_{-2} = \sigma_{b_1}$$

hence:

$$P_{e1,a} = p_{-1} \times P_{e(0/-1,m_9)} + p_0 \times \left[P_{e(-1/0,m_3)} + P_{e(1/0,m_2)}\right] + p_1$$

$$\times P_{e(0/1,m_8)}$$

$$P_{e1,a} = \frac{3}{10} \times 0.1056 + \frac{2}{5} \times [0.0401 + 0.0401] + \frac{3}{10} \times 0.1056 = 0.038416$$

m_l	m_1	m_2	m_3	m_4	m_5	m_6	m_7	m_8	m_9
$ul = \mu^- - I_{m_l}$ $ll = -\infty$	$-2.25\ \sigma_{b_1}$ $-\infty$	$-2.75\ \sigma_{b_1}$ $-\infty$	$-1.75\ \sigma_{b_1}$ $-\infty$	$-2.75\ \sigma_{b_1}$ $-\infty$	$-2.25\ \sigma_{b_1}$ $-\infty$	$-1.75\ \sigma_{b_1}$ $-\infty$	$-2.25\ \sigma_{b_1}$ $-\infty$	X	X
$P_{e(-1/0,m_l)}$	0.0122	0.0030	0.0401	0.0030	0.0122	0.0401	0.0122	X	X
$ul = \infty$ $ll = \mu^+ - I_{m_l}$	∞ $2.25\ \sigma_{b_1}$	∞ $1.75\ \sigma_{b_1}$	∞ $2.75\ \sigma_{b_1}$	∞ $1.75\ \sigma_{b_1}$	∞ $2.25\ \sigma_{b_1}$	∞ $2.75\ \sigma_{b_1}$	∞ $2.25\ \sigma_{b_1}$	X	X
$P_{e(1/0,m_l)}$	0.0122	0.0401	0.0030	0.0401	0.0122	0.0030	0.0122	X	X
$ul = \mu^+ + p(0) - I_{m_l}$ $ll = \mu^- + p(0) - I_{m_l}$	$6.75\ \sigma_{b_1}$ $2.25\ \sigma_{b_1}$	$6.25\ \sigma_{b_1}$ $1.75\ \sigma_{b_1}$	X	$6.25\ \sigma_{b_1}$ $1.75\ \sigma_{b_1}$	X	X	X	X	$5.75\ \sigma_{b_1}$ $1.25\ \sigma_{b_1}$
$P_{e(0/-1,m_l)}$	0.0122	0.0401	X	0.0401	X	X	X	X	0.1056
$ul = \mu^+ - p(0) - I_{m_l}$ $ll = \mu^- - p(0) - I_{m_l}$	$-2.25\ \sigma_{b_1}$ $-6.75\ \sigma_{b_1}$	X	$-1.75\ \sigma_{b_1}$ $-6.25\ \sigma_{b_1}$	X	X	$-1.75\ \sigma_{b_1}$ $-6.25\ \sigma_{b_1}$	X	$-1.25\ \sigma_{b_1}$ $-5.75\ \sigma_{b_1}$	X
$P_{e(0/1,m_l)}$	0.0122	X	0.0401	X	X	0.0401	X	0.1056	X

Table 2.29. *Noise interval dynamics and associated probabilities according to the intersymbol interference amplitude*

21) Expression and value of the probability of error $P_{e,b}$:

$$P_{e,b} = Pr\{b = 0\} \times Pr\{\hat{b} = 1/b = 0\} + Pr\{b = 1\} \times Pr\{\hat{b} = 0/b = 1\}$$

$$P_{e,b} = p_0 \times \left[P_{e(-1/0,m_3)} + P_{e(1/0,m_2)}\right] + p_{-1} \times P_{e(0/-1,m_9)} + p_1 \times P_{e(0/1,m_8)}$$

therefore: $P_{e,b} = P_{e1,a}$.

This result was predictable with the simplification of the statement because:

$$Pr\{\hat{a}_k = 1/a_k = -1\} = 0$$

$$Pr\{\hat{a}_k = -1/a_k = 1\} = 0$$

But these errors on a_k do not introduce errors on b_k, hence: $P_{e,b} = P_{e1,a}$.

22) Probability of error $P_{e2,a}$ in the absence of ISI. We have:

$$P_{e2,a} = \sum_{i=-1}^{1} p_i \times \left[\sum_{m=-1}^{1} P_{e(m/i)_{m \neq i}}\right]$$

hence:

$$P_{e2,a} = p_{-1} \times P_{e(0/-1)} + p_0 \times \left[P_{e(-1/0)} + P_{e(1/0)}\right] + p_1 \times P_{e(0/1)}$$

$$P_{e2,a} = \frac{3}{10} \times 0.0122 + \frac{2}{5} \times [0.0122 + 0.0122] + \frac{3}{10} \times 0.0122 = 0.01708$$

We have, from the normalized Gaussian law table:

$$\int_{2.25}^{\infty} p(x)dx = 0.5 - \int_{0}^{2.25} p(x)dx = 0.5 - 0.4878 = 0.0122$$

$$\int_{-\infty}^{-2.75} = 0.5 - \int_{-2.75}^{0} = 0.5 - \int_{0}^{2.75} = 0.5 - 0.4970 = 0.0030$$

$$\int_{-6.75}^{-2.25} = 0.5 - \left[\int_{-\infty}^{-6.75} + \int_{-2.25}^{0}\right] = 0.5 - \left[\int_{6.75}^{\infty} + \int_{0}^{2.25}\right]$$

$$= 0.5 - [0 + 0.4878] = 0.0122$$

$$\int_{1.75}^{6.25} = \int_{0}^{6.25} - \int_{0}^{1.75} = 0.5 - 0.4599 = 0.0401$$

$$\int_{1.25}^{5.75} = \int_{0}^{5.75} - \int_{0}^{1.25} = 0.5 - 0.3944 = 0.1056$$

2.10. Problem 25 – Baseband transmission and reception using a partial response linear coding (1)

The problem of baseband transmitting and receiving independent binary information on a reduced capacity channel is considered.

The transmission and reception system in question uses partial response linear coding according to the Figure 2.42.

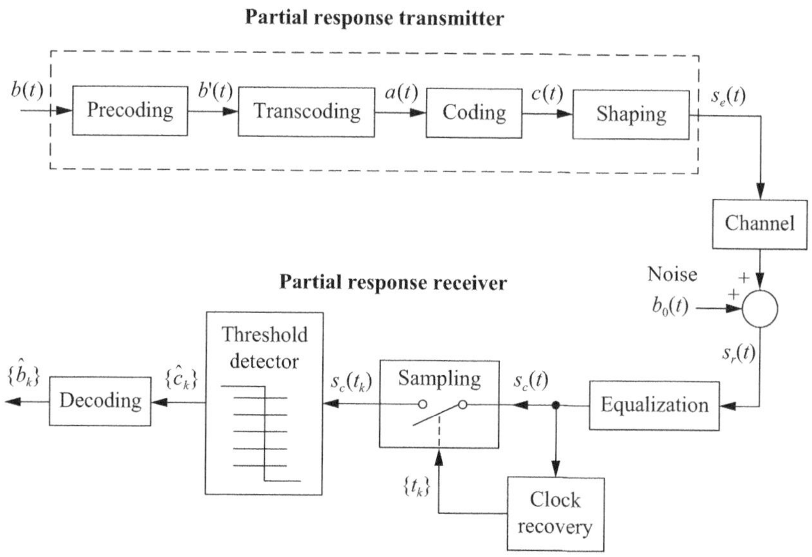

Figure 2.42. *Baseband transmission and reception chain with partial response linear coder*

Where:

$$b(t) = \sum_{n} b_n \delta(t - nT_b) \qquad b_n \in \{0, 1\}$$

$$b'(t) = \sum_n b'_n \delta(t - nT_b) \qquad b'_n \in \{0, 1\}$$

$$a(t) = \sum_n a_n \delta(t - nT_b) \qquad a_n = 2b'_n - 1 \qquad a_n \in \{-1, 1\}$$

$$c(t) = \sum_n c_n \delta(t - nT_b) \qquad c_n \text{ positive, negative or null integer}$$

The partial response linear coding used in this problem is the NRZ duobinary coding, characterized by its transfer function:

$$H(z) = 1 + z^{-1} = 1 + D$$

Where D is the delay operator T (time slot allocated to the transmission of a symbol c_k).

1) Give the transfer function $P(z)$ of the precoder filter as well as its equation providing b'_k as a function of b_k.

2) Give the equation of the encoder generating c_k from a_k.

3) Give the relationships allowing us to estimate the binary symbols emitted b_k : $\hat{b}_k$ from the symbols received c_k : $\hat{c}_k$, with and without precoding. Comment on each case.

4) Give the block diagram of the precoder, transcoder and duobinary combined encoder.

The shaping filter has an impulse response $x(t)$ considered as an NRZ signal of duration T and amplitude V.

Let us take the 14-bit $\{b_k\}$ time sequence shown in Figure 2.43 (time running from left to right).

5) Determine (temporal representations will be plotted directly in Figure 2.43):

– the time sequence $\{b'_k\}$ associated with the output of the precoder (the latter is considered initialized to zero);

– the corresponding temporal sequences $\{a_k\}$, $\{c_k\}$, $\{\hat{b}_k\}$ and signal $s_e(t)$.

The transmission channel is modeled by a linear filtering and additive noise at the output of the channel. The latter is a stationary second-order Gaussian noise, with a zero mean and a broad spectrum.

We assume that at the output of the equalizer, the signal-to-noise ratio obtained is:

$$\left[\frac{S}{b}\right]_{c,dB} = 20 \times \log_{10}\left[\frac{p(t_0)}{\sigma_{b_1}}\right] = 7.96 \text{ dB}$$

First, we consider the classical baseband transmission system (without precoding and coding).

6) Give the expression of the signal $s_c(kT + t_0)$ at the input of the decision unit according to the symbols a.

Take the case of the duobinary partial response transmission and reception system.

7) Particularize the expression of the signal $s_c(kT + t_0)$.

We assume for the rest of the problem that: $p(T + t_0) = p(t_0) = V$.

8) Deduce the new expression of $s_c(kT + t_0)$ according to the symbols c_k.

9) Calculate the conditional probabilities of error:

$$P_{e_k} = Pr\{\hat{c}_k \neq k / c_k = k\} \text{ with } k = \{2, -2, 0\}$$

10) Deduce the total probability of error: $P_e = Pr\{\hat{c}_k \neq c_k\}$ with: $c_k = \{2, -2, 0\}$.

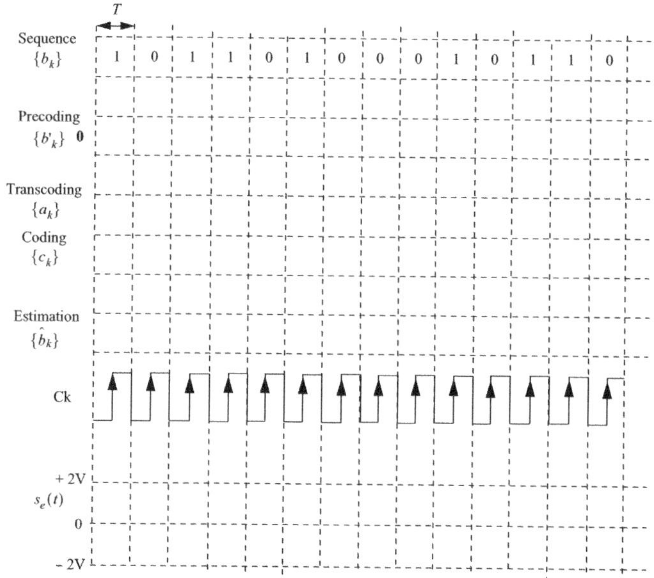

Figure 2.43. *Temporal diagrams of duobinary coding and decoding*

NOTE.– If X is a Gaussian random process, with mean value m and standard deviation σ, you will take:

$$Pr\{|X - m| > 7.5\,\sigma\} = 2 \times 10^{-8}$$

$$Pr\{|X - m| > 2.5\,\sigma\} = 2 \times 10^{-4}$$

Solution of problem 25

1) Transfer function $P(z)$ of the precoder and equation giving b'_k as a function of b_k:

$$P(z) = \frac{1}{H(z)} = \frac{1}{1 + z^{-1}} = \frac{B'(z)}{B(z)} \rightarrow B'(z) = B(z) - B'(z) \times z^{-1}$$

$$\rightarrow b'_k = b_k \oplus b'_{k-1}$$

2) Equation of the coder giving c_k as a function of a_k:

$$H(z) = \frac{C(z)}{A(z)} = 1 + z^{-1} \rightarrow C(z) = A(z) + A(z) \times z^{-1} \rightarrow c_k = a_k + a_{k-1}$$

3) Equation allowing the estimation of the emitted symbols b_k : $\hat{b}_k$ from the received symbols c_k : $\hat{c}_k$.

With precoding, the transcoder provides:

$$a_k = 2b'_k - 1$$

hence:

$$c_k = a_k + a_{k-1} = [2b'_k - 1 + 2b'_{k-1} - 1\,] = 2[b'_k + b'_{k-1} - 1]$$

$$\rightarrow \frac{1}{2}c_k + 1 = b'_k + b'_{k-1} = b'_k \oplus b'_{k-1} = b_k \text{ from 1)}$$

Thus, we get a direct estimation of the emitted sequence $\{b_k\}$ from the received sequence $\{\hat{c}_k\}$.

Without precoding:

$$b'_k = b_k \rightarrow \hat{b}_k = \frac{1}{2}\hat{c}_k + 1 - \hat{b}_{k-1} \bmod 2$$

This leads to a propagation of decision errors. Indeed, if b_{k-1} is badly decoded, it will also be the case for b_k.

4) Block diagram of the precoder, transcoder and duobinary coder.

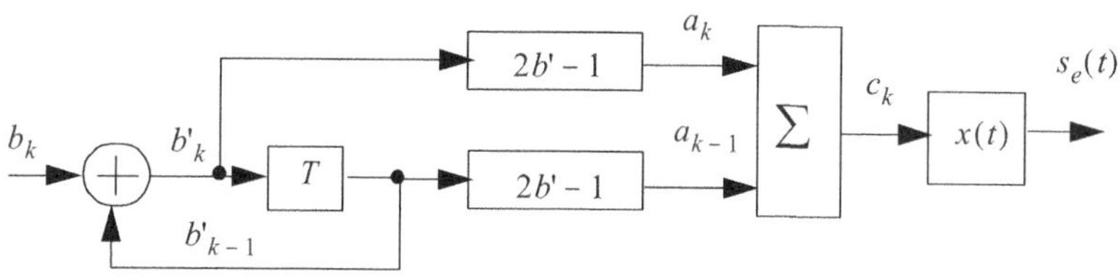

Figure 2.44. *Duobinary precoder, transcoder and coder scheme*

5) Chronograms of duobinary coding and decoding.

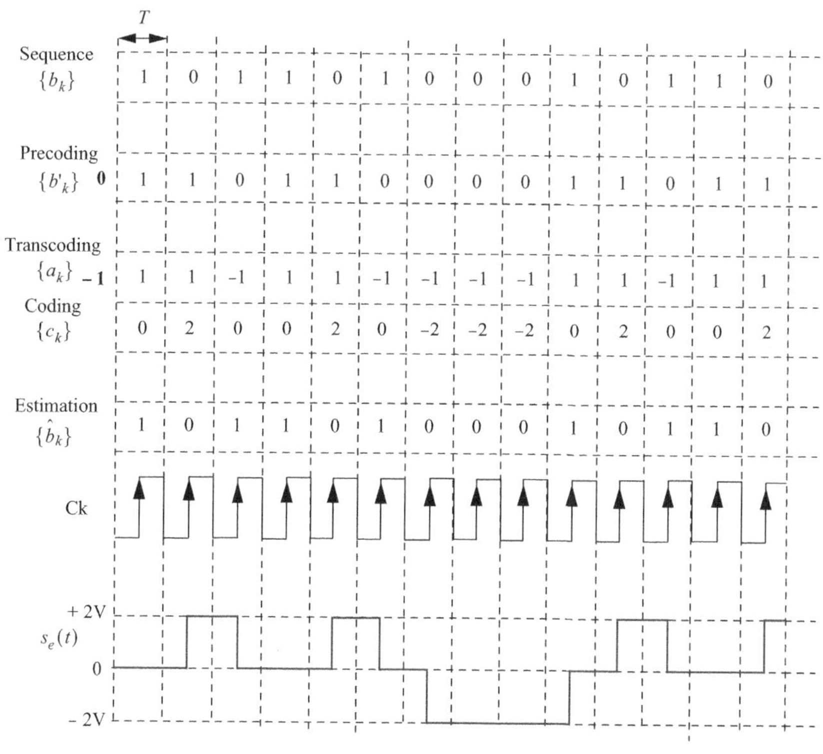

Figure 2.45. *Chronograms of duobinary coding and decoding. For a color version of this figure, see www.iste.co.uk/assad/digital2.zip*

6) Expression of the signal $s_c(kT + t_0)$:

$$s_c(kT + t_0) = \boldsymbol{a_k p}(t_0) + \sum_{\boldsymbol{n \neq k}} a_n p[(k - n)T + t_0] + b_1(kT + t_0)$$

7) Duobinary partial response transmission and reception system: expression of $s_c(kT + t_0)$.

With $p(T + t_0) = p(t_0) = V$, the intersymbol interference is now given by:

$$I_{m_l}(kT) = \sum_{n \neq k} a_n p[(k - n)T + t_0] = a_{k-1} p(T + t_0)$$

$$\rightarrow s_c(kT + t_0) = a_k p(t_0) + a_{k-1} p(T + t_0) + b_1(kT + t_0)$$

8) New expression of $s_c(kT + t_0)$:

$$s_c(kT + t_0) = [a_k + a_{k-1}] p(t_0) + b_1(kT + t_0)$$

$$s_c(kT + t_0) = c_k p(t_0) + b_1(kT + t_0) = V c_k + b_1(kT + t_0)$$

9) Conditional probabilities of error:

$$P_{e_k} = Pr\{\hat{c}_k \neq k / c_k = k\} \text{ with } k = \{2, -2, 0\}$$

From the result obtained in response 8, we have:

$$b_1(kT + t_0) = s_c(kT + t_0) - V c_k$$

If $s_c(kT + t_0) \in [c, d[$ then $b_1(kT + t_0) \in [c - V c_k, d - V c_k[$

and $Pr\{\hat{c}_k \neq k / c_k = k\} = Pr\{c - V c_k \leq b_1(kT + t_0) < d - V c_k\}$

The decision thresholds are located in the middle of two adjacent levels obtained without noise:

$$V c_k = \begin{cases} 2V \\ 0 \\ -2V \end{cases} \rightarrow \mu_1 = V \text{ and } \mu_{-1} = -V$$

Let's express the decision thresholds according to σ_{b_1}:

$$\left[\frac{S}{b}\right]_{c,dB} = 20 \times \log_{10}\left[\frac{V}{\sigma_{b_1}}\right] = 7.96 \text{ dB} \rightarrow V = 2.5\,\sigma_{b_1} \rightarrow \begin{cases} \mu_1 = 2.5\,\sigma_{b_1} \\ \mu_{-1} = -2.5\,\sigma_{b_1} \end{cases}$$

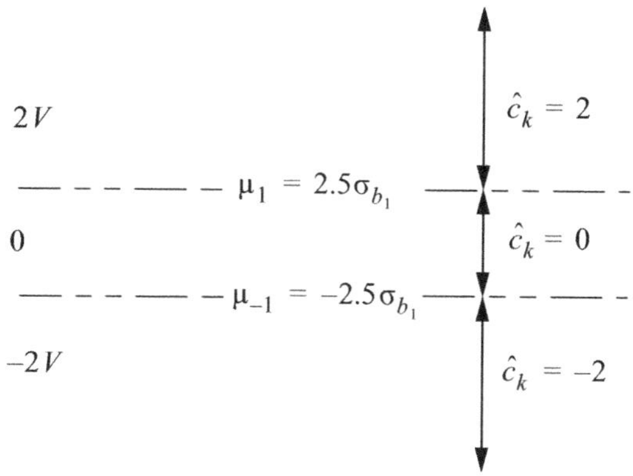

Figure 2.46. *Values of sample $c_k V$, optimum thresholds and decision classes of c_k: $\hat{c}_k$*

We have three possible decision values of symbol $\hat{c}_k$, one without errors and two with errors.

Transmission of $c_k = 2$, we decide on reception:

$$\hat{c}_k = 0 \ if \ \ s_c(kT + t_0) \in [\mu_{-1}, \mu_1[\ \rightarrow b_1(kT + t_0) \in [\mu_{-1} - Vc_k, \mu_1 - Vc_k[$$

$$\rightarrow b_1(kT + t_0) \in \left[-7.5\ \sigma_{b_1}, -2.5\ \sigma_{b_1}\right[$$

$$\hat{c}_k = -2 \ if \ \ s_c(kT + t_0) \in [-\infty, \mu_{-1}[\ \rightarrow b_1(kT + t_0) \in [-\infty, \mu_{-1} - Vc_k[$$

$$\rightarrow b_1(kT + t_0) \in \left[-\infty, -7.5\ \sigma_{b_1}\right[$$

$$\rightarrow P_{e_2} = Pr\{\hat{c}_k = 0/c_k \ = 2\} + Pr\{\hat{c}_k = -2/c_k \ = 2\} = 10^{-4} + 10^{-8} \cong 10^{-4}$$

Transmission of $c_k = 0$, we decide on reception:

$$\hat{c}_k = -2 \ if \ \ s_c(kT + t_0) \in [-\infty, \mu_{-1}[\ \rightarrow b_1(kT + t_0) \in [-\infty, \mu_{-1} - Vc_k[$$

$$\rightarrow b_1(kT + t_0) \in \left[-\infty, -2.5\ \sigma_{b_1}\right[$$

$$\hat{c}_k = 2 \text{ if } s_c(kT + t_0) \in [\mu_1, \infty[\ \rightarrow b_1(kT + t_0) \in [\mu_1 - Vc_k, \infty[$$

$$\rightarrow b_1(kT + t_0) \in \left[2.5\ \sigma_{b_1}, \infty\right[$$

$$\rightarrow P_{e_0} = Pr\{\hat{c}_k = -2/c_k = 0\} + Pr\{\hat{c}_k = 2/c_k = 0\} = 10^{-4} + 10^{-4}$$
$$= 2 \times 10^{-4}$$

Transmission of $c_k = -2$, we decide on reception:

$$\hat{c}_k = 0 \text{ if } s_c(kT + t_0) \in [\mu_{-1}, \mu_1[\rightarrow b_1(kT + t_0) \in [\mu_{-1} - Vc_k, \mu_1 - Vc_k[$$

$$\rightarrow b_1(kT + t_0) \in \left[2.5\,\sigma_{b_1}, 7.5\,\sigma_{b_1}\right[$$

$$\hat{c}_k = 2 \text{ if } s_c(kT + t_0) \in [\mu_1, \infty[\rightarrow b_1(kT + t_0) \in [\mu_1 - Vc_k, \infty[$$

$$\rightarrow b_1(kT + t_0) \in \left[7.5\,\sigma_{b_1}, \infty\right[$$

$$\rightarrow P_{e_{-2}} = Pr\{\hat{c}_k = 0/c_k = -2\} + Pr\{\hat{c}_k = 2/c_k = -2\} = 10^{-4} + 10^{-8}$$
$$\cong 10^{-4}$$

10) Calculation of the total probability of error:

$$P_e = Pr\{\hat{c}_k \neq c_k\}$$

The total probability of error is then given by:

$$P_e = \sum_k p_k \times P_{e_k} \text{ with } p_k = Pr\{c_k = k\} \text{ and } k = \{-2, 0, 2,\}$$

The binary symbols b_k are independent and identically distributed on the alphabet $\{0, 1\}$, hence:

$$Pr\{b_k = 0\} = Pr\{b_k = 1\} = 1/2$$

The probabilities of transmitting the symbols c_k are respectively:

$$p_0 = Pr\{c_k = 0\} = \frac{1}{2};\quad p_2 = Pr\{c_k = 2\} = \frac{1}{4};\quad p_{-2} = Pr\{c_k = -2\} = \frac{1}{4}$$

Hence, the total probability of error:

$$P_e = p_0 \times P_{e_0} + p_2 \times P_{e_2} + p_{-2} \times P_{e_{-2}}$$

$$P_e = \frac{1}{2} \times 2 \times 10^{-4} + \frac{1}{4} \times 10^{-4} + \frac{1}{4} \times 10^{-4} = 1.5 \times 10^{-4}$$

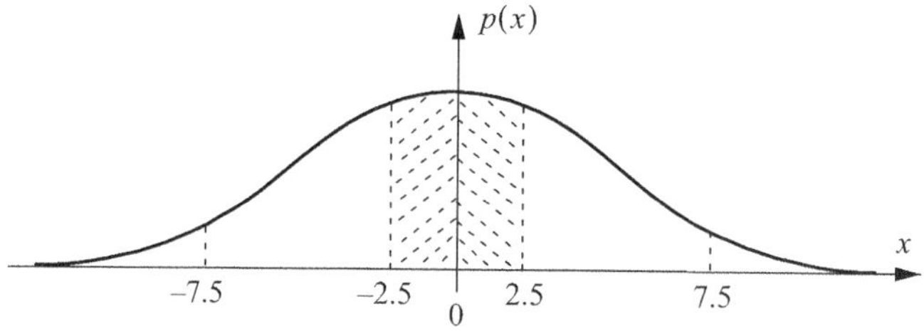

Figure 2.47. *Gaussian probability law and distribution intervals*

Calculation of integrals:

$$\int_{-\infty}^{\infty} p(x)dx = 1 = \int_{-\infty}^{-7.5} + \int_{-7.5}^{-2.5} + \int_{-2.5}^{2.5} + \int_{2.5}^{7.5} + \int_{7.5}^{\infty}$$

$$\int_{-\infty}^{-7.5} = \int_{7.5}^{\infty} \quad \text{and} \quad \int_{-7.5}^{-2.5} = \int_{2.5}^{7.5} \quad \text{and} \quad \int_{-\infty}^{-2.5} = \int_{2.5}^{\infty}$$

$$\int_{-\infty}^{-2.5} + \int_{-2.5}^{2.5} + \int_{2.5}^{\infty} = 1 \rightarrow \int_{-2.5}^{2.5} = 1 - 2\int_{2.5}^{\infty} = 1 - 2 \times 10^{-4}$$

$$\rightarrow 2\int_{2.5}^{7.5} = 1 - 2\int_{7.5}^{\infty} - \int_{-2.5}^{2.5} = 1 - 2 \times 10^{-8} - 1 + 2 \times 10^{-4} \cong 2 \times 10^{-4}$$

$$\rightarrow \int_{2.5}^{7.5} = \int_{-7.5}^{-2.5} \cong 10^{-4}$$

2.11. Problem 26 – Baseband transmission and reception using a partial response linear coding (2)

The problem of transmitting and receiving independent binary information on a reduced capacity channel is considered. The baseband transmission and reception system in question uses the partial response linear coding according to the following Figures 2.48 and 2.49.

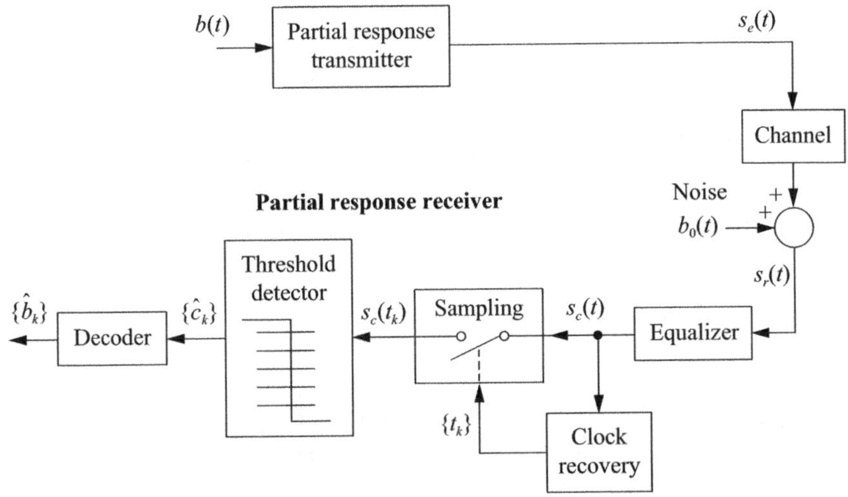

Figure 2.48. *Partial response transmitter and receiver block diagram*

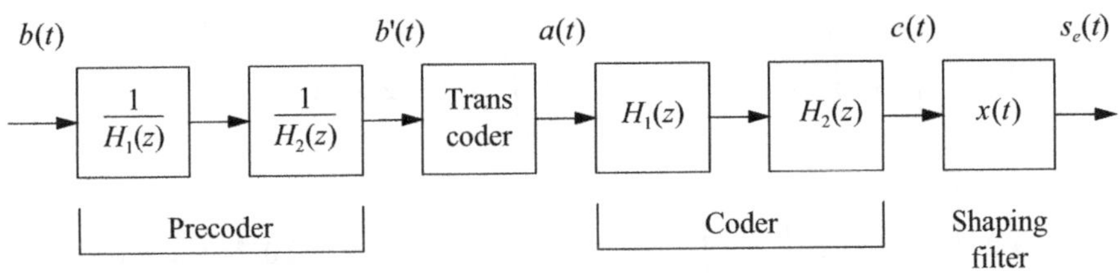

Figure 2.49. *Details of the partial response transmitter block diagram*

with:

$$b(t) = \sum_n b_n \delta(t - nT_b) \qquad b_n \in \{0, 1\}$$

$$b'(t) = \sum_n b'_n \delta(t - nT_b) \qquad b'_n \in \{0, 1\}$$

$$a(t) = \sum_n a_n \delta(t - nT_b) \qquad a_n = 2b'_n - 1 \qquad a_n \in \{-1, 1\}$$

$$c(t) = \sum_n c_n \delta(t - nT_b) \quad c_n \text{ positive, negative or null integer}$$

The partial response linear coding used in this problem is defined by the two following transfer functions:

$$H_1(z) = 1 - z^{-3/2} \text{ and } H_2(z) = 1 + z^{-3/2}$$

The transmission channel is modeled by a linear filtering and additive noise at the output of the channel. Noise is considered as a second-order stationary, Gaussian random process, with zero mean and broad spectrum.

1) Give the transfer function of the encoder filter $H(z)$.

2) Give the transfer function of the precoder filter $P(z)$ as well as its equation providing b'_k as a function of b_k.

3) Give the equation of the encoder providing c_k as a function of a_k.

4) Give the relationships providing the estimation of the transmitting symbols b_k from the received symbols $\hat{c}_k$, with and without precoding. Comment on each case.

5) Give the implementation scheme of the precoder, transcoder and combined encoder.

The shaping filter has an impulse response $x(t)$ considered as an NRZ signal of duration T and amplitude $V/2$.

Let us take the 14-bit sequence $\{b_k\}$ shown in Table 2.30 (time running from left to right).

6) Determine (the temporal diagrams will be given in Table 2.30):

– the time sequence $\{b'_k\}$ associated with the output of the precoder (the latter is considered initialized to zero);

– the corresponding time sequences $\{a_k\}$, $\{c_k\}$, $\{\hat{b}_k\}$ and plot the signal $s_e(t)$.

We assume that at the output of the equalizer, the signal-to-noise ratio obtained is:

$$\left[\frac{S}{b}\right]_{c,dB} = 20 \times \log_{10}\left[\frac{p(0)}{\sigma_{b_1}}\right] = 10.88$$

We first consider the classical baseband transmission system (without precoding and coding).

7) Give the expression of the signal $s_c(kT)$ at the input of the decision unit.

In the case of the partial response transmission and reception system defined in Figure 2.48.

8) Particularize the expression of $s_c(kT)$.

Assuming that: $p(3T) = p(0) = V$.

9) Deduce the new expression of $s_c(kT)$ according to, in particular, the symbols c_k.

10) Calculate the conditional probabilities of error:

$$P_{e_k} = Pr\{\hat{c}_k \neq k / c_k = k\} \text{ with } k = \{-2, 0, 2\}$$

11) Deduce the total probability of error:

$$P_e = Pr\{\hat{c}_k \neq c_k\} \text{ with } c_k = \{2, -2, 0\}$$

NOTE.– If X is a Gaussian random process, with mean value m and standard deviation σ, you will take:

$$Pr\{|X - m| > 10.5\,\sigma\} = 4 \times 10^{-9}$$

$$Pr\{|X - m| > 3.5\,\sigma\} = 6 \times 10^{-5}$$

$\{b_k\}$				1	0	0	1	0	1	1	1	0	1	0	0	1	0
b'_k	0	0	0														
$\{a_k\}$																	
$\{c_k\}$																	
$\{\ddot{b}_k\}$																	
$s_e(t)$																	

Table 2.30. *Temporal representation of the proposed partial response coding and decoding system*

Solution of problem 26

1) Transfer function of the encoding filter:

$$H(z) = H_1(z) \times H_2(z) = \left[1 - z^{-3/2}\right] \times \left[1 + z^{-3/2}\right] = 1 - z^{-3}$$

2) Transfer function of the precoder:

$$P(z) = \frac{1}{H(z)} = \frac{1}{1 - z^{-3}} = \frac{B'(z)}{B(z)} \rightarrow B'(z) = B(z) + B'(z) \times z^{-3}$$

$$\rightarrow b'_k = b_k \oplus b'_{k-3}$$

3) Equation defining the encoder:

$$H(z) = \frac{C(z)}{A(z)} = 1 - z^{-3} \rightarrow C(z) = A(z) - A(z) \times z^{-3} \rightarrow c_k = a_k - a_{k-3}$$

4) Equation giving the estimate of b_k.

With precoding, the transcoder gives: $a_k = 2b'_k - 1$, hence:

$$c_k = a_k - a_{k-3} = [2b'_k - 1 - 2b'_{k-3} + 1] = 2[b'_k - b'_{k-3}] = 2b_k$$

$$\rightarrow \hat{b}_k = \frac{1}{2}|\hat{c}_k| \text{ mod } 2 \rightarrow \begin{cases} \hat{c}_k = 0 \rightarrow \hat{b}_k = 0 \\ \hat{c}_k = \pm 2 \rightarrow \hat{b}_k = 1 \end{cases}$$

So a direct estimate of the sequence $\{b_k\}$ issued from the sequence $\{c_k\}$ received.

Without precoding:

$$b'_k = b_k \rightarrow c_k = 2[b_k - b_{k-3}] \rightarrow \hat{b}_k = \frac{1}{2}|\hat{c}_k| \oplus \hat{b}_{k-3} \text{ mod } 2$$

This leads to a propagation of decision errors. Indeed, if b_{k-3} is badly decoded, it will be also the same case for b_k.

5) Block diagram of the combined precoder, transcoder and encoder:

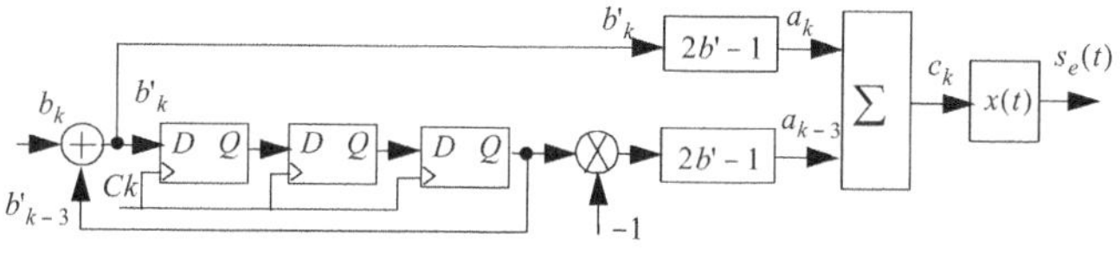

Figure 2.50. *Combined scheme of precoder, transcoder and encoder*

6) Temporal sequences:

$\{b_k\}$				1	0	0	1	0	1	1	1	0	1	0	0	1	0
b'_k	**0**	**0**	**0**	1	0	0	0	0	1	1	1	1	0	1	1	1	1
$\{a_k\}$	-1	-1	-1	1	-1	-1	-1	-1	1	1	1	1	-1	1	1	1	1
$\{c_k\}$				2	0	0	-2	0	2	2	2	0	-2	0	0	2	0
$\{\hat{b}_k\}$				1	0	0	1	0	1	1	1	0	1	0	0	1	0
$s_e(t)$			V 0 $-V$	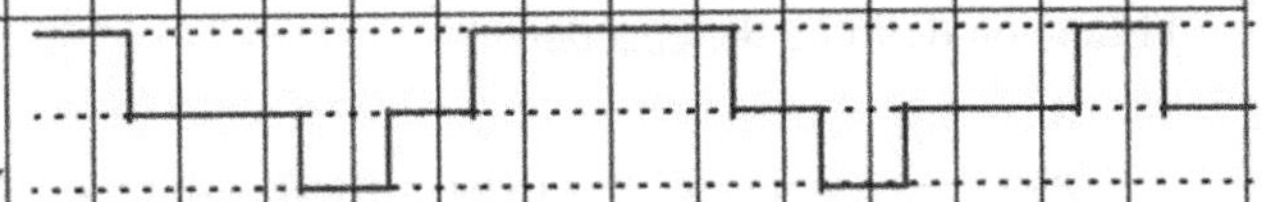													

Table 2.31. *Temporal sequences. For a color version of this table, see www.iste.co.uk/assad/digital2.zip*

7) Expression of the signal $s_c(kT)$:

$$s_c(kT) = a_k p(0) + \sum_{n \neq k} a_n p[(k-n)T] + b_1(kT)$$

8) In this case, the intersymbol interference is:

$$I_{m_l}(kT) = \sum_{n \neq k} a_n p[(k-n)T] = -a_{k-3} p(3T)$$

$$\rightarrow s_c(kT) = a_k p(0) - a_{k-3} p(3T) + b_1(kT)$$

9) In this particular case, we have:

$$s_c(kT) = [a_k - a_{k-3}] p(0) + b_1(kT) = c_k p(0) + b_1(kT) = V c_k + b_1(kT)$$

10) The different conditional probabilities of errors are as follows:

$$P_{e_2} = Pr\{\hat{c}_k \neq 2 / c_k = 2\}$$

$$\rightarrow P_{e_2} = Pr\{\hat{c}_k = 0 / c_k = 2\} + Pr\{\hat{c}_k = -2 / c_k = 2\}$$

$$P_{e_{-2}} = Pr\{\hat{c}_k \neq -2 / c_k = -2\}$$

$$\rightarrow P_{e_{-2}} = Pr\{\hat{c}_k = 0/c_k = -2\} + Pr\{\hat{c}_k = 2/c_k = -2\}$$

$$P_{e_0} = Pr\{\hat{c}_k \neq 0/c_k = 0\}$$

$$\rightarrow P_{e_0} = Pr\{\hat{c}_k = 2/c_k = 0\} + Pr\{\hat{c}_k = -2/c_k = 0\}$$

a) Calculation of conditional probabilities of errors. According to the result obtained in question 9, we have:

$$b_1(kT) = s_c(kT) - Vc_k$$

If $s_c(kT) \in [c, d[$ then $b_1(kT) \in [c - Vc_k, d - Vc_k[$

and $Pr\{\hat{c}_k \neq k/c_k = k\} = Pr\{c - Vc_k \leq b_1(kT) < d - Vc_k\}$

The decision thresholds $[c, d[$ are such that:

$$c, d \in \{\mu_m, \pm\infty\}; \quad \mu_m = mp(0) = mV; \quad m = \{-1, 1\}$$

$$Vc_k = \begin{cases} 2V \\ 0 \\ -2V \end{cases} \rightarrow \mu_1 = V \text{ et } \mu_{-1} = -V$$

Let's express the decision thresholds according to σ_{b_1}:

$$\left[\frac{S}{b}\right]_{c,dB} = 20 \times \log_{10}\left[\frac{V}{\sigma_{b_1}}\right] = 10.88 \text{ dB} \rightarrow V = 3.5\,\sigma_{b_1} \rightarrow \begin{cases} \mu_1 = 3.5\,\sigma_{b_1} \\ \mu_{-1} = -3.5\,\sigma_{b_1} \end{cases}$$

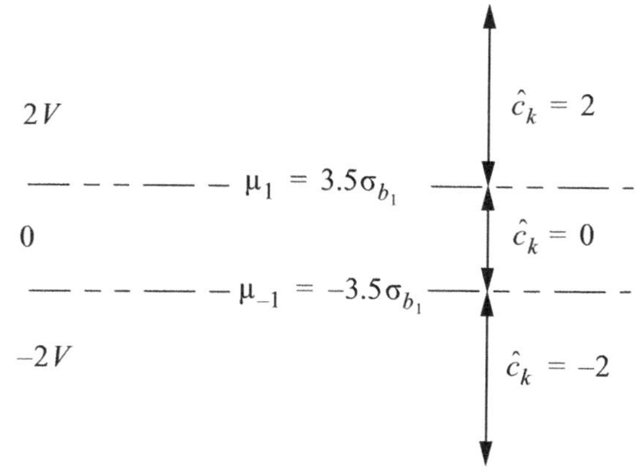

Figure 2.51. *Values of sample $c_k V$, optimum thresholds and decision classes of c_k: $\hat{c}_k$*

b) Application of the decision rules. We have three possible decision values on c_k, one without errors and two with errors.

c) Transmission of symbol $c_k = 2$, we decide on reception:

$$\hat{c}_k = 0 \text{ if } s_c(kT) \in [\mu_{-1}, \mu_1[\rightarrow b_1(kT) \in [\mu_{-1} - Vc_k, \mu_1 - Vc_k[$$

$$Pr\{\hat{c}_k = 0/c_k = 2\} = Pr\{-V - 2V \leq b_1(kT) < V - 2V\}$$

$$Pr\{\hat{c}_k = 0/c_k = 2\} = Pr\{-10.5\,\sigma_{b_1} \leq b_1(kT) < -3.5\,\sigma_{b_1}\} \cong 3 \times 10^{-5}$$

$$\hat{c}_k = -2 \text{ if } s_c(kT) \in [-\infty, \mu_{-1}[\rightarrow b_1(kT) \in [-\infty, \mu_{-1} - Vc_k[$$

$$Pr\{\hat{c}_k = -2/c_k = 2\} = Pr\{-\infty \leq b_1(kT) < -V - 2V\}$$

$$Pr\{\hat{c}_k = -2/c_k = 2\} = Pr\{-\infty \leq b_1(kT) < -10.5\,\sigma_{b_1}\} = 2 \times 10^{-9}$$

$$\rightarrow P_{e_2} = Pr\{\hat{c}_k = 0/c_k = 2\} + Pr\{\hat{c}_k = -2/c_k = 2\}$$
$$\cong 3 \times 10^{-5} + 2 \times 10^{-9} \cong 3 \times 10^{-5}$$

d) Transmission of symbol $c_k = 0$, we decide on reception:

$$\hat{c}_k = -2 \text{ if } s_c(kT) \in [-\infty, \mu_{-1}[\rightarrow b_1(kT) \in [-\infty, \mu_{-1} - Vc_k[$$

$$Pr\{\hat{c}_k = -2/c_k = 0\} = Pr\{-\infty \leq b_1(kT) < -V\}$$

$$Pr\{\hat{c}_k = -2/c_k = 0\} = Pr\{-\infty \leq b_1(kT) < -3.5\,\sigma_{b_1}\} \cong 3 \times 10^{-5}$$

$$\hat{c}_k = 2 \text{ if } s_c(kT) \in [\mu_1, \infty[\rightarrow b_1(kT) \in [\mu_1 - Vc_k, \infty[$$

$$Pr\{\hat{c}_k = 2/c_k = 0\} = Pr\{V \leq b_1(kT) < \infty\}$$

$$Pr\{\hat{c}_k = 2/c_k = 0\} = Pr\{3.5\,\sigma_{b_1} \leq b_1(kT) < \infty\} \cong 3 \times 10^{-5}$$

$$P_{e_0} = Pr\{\hat{c}_k = -2/c_k = 0\} + Pr\{\hat{c}_k = 2/c_k = 0\} \cong 3 \times 10^{-5} + 3 \times 10^{-5}$$
$$\cong 6 \times 10^{-5}$$

e) Transmission of symbol $c_k = -2$, we decide on reception:

$$\hat{c}_k = 0 \text{ if } s_c(kT) \in [\mu_{-1}, \mu_1[\rightarrow b_1(kT) \in [\mu_{-1} - Vc_k, \mu_1 - Vc_k[$$

$$Pr\{\hat{c}_k = 0/c_k = -2\} = Pr\{-V + 2V \leq b_1(kT) < V + 2V\}$$

$$Pr\{\hat{c}_k = 0/c_k = -2\} = Pr\{3.5\,\sigma_{b_1} \leq b_1(kT) < 10.5\,\sigma_{b_1}\} \cong 3 \times 10^{-5}$$

$\hat{c}_k = 2 \text{ if } s_c(kT) \in [\mu_1, \infty[\ \rightarrow b_1(kT) \in [\mu_1 - Vc_k, \infty[$

$Pr\{\hat{c}_k = 2/c_k = -2\} = Pr\{3V \leq b_1(kT) < \infty\}$

$Pr\{\hat{c}_k = 2/c_k = -2\} = Pr\{10.5\, \sigma_{b_1} \leq b_1(kT) < \infty\} = 2 \times 10^{-9}$

$\rightarrow P_{e_{-2}} = Pr\{\hat{c}_k = 0/c_k = -2\} + Pr\{\hat{c}_k = 2/c_k = -2\}$

$\cong 3 \times 10^{-5} + 2 \times 10^{-9} \cong 3 \times 10^{-5}$

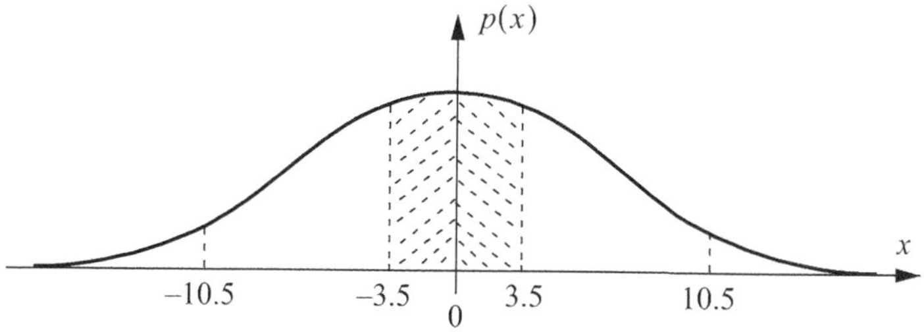

Figure 2.52. *Gaussian probability law with zero mean and unit standard deviation, and distribution intervals*

Calculation of integrals:

$$\int_{-\infty}^{\infty} p(x)dx = 1 = \int_{-\infty}^{-10.5} + \int_{-10.5}^{-3.5} + \int_{-3.5}^{3.5} + \int_{3.5}^{10.5} + \int_{10.5}^{\infty}$$

$$\int_{-\infty}^{-10.5} = \int_{10.5}^{\infty} \quad \text{and} \quad \int_{-10.5}^{-3.5} = \int_{3.5}^{10.5} \quad \text{and} \quad \int_{-\infty}^{-3.5} = \int_{3.5}^{\infty}$$

$$\int_{-\infty}^{-3.5} + \int_{-3.5}^{3.5} + \int_{3.5}^{\infty} = 1 \rightarrow \int_{-3.5}^{3.5} = 1 - 2\int_{3.5}^{\infty} = 1 - 2 \times 3 \times 10^{-5}$$

$$= 1 - 6 \times 10^{-5}$$

$$\rightarrow 2\int_{3.5}^{10.5} = 1 - 2\int_{10.5}^{\infty} - \int_{-3.5}^{3.5} = 1 - 2 \times 2 \times 10^{-9} - 1 + 6 \times 10^{-5}$$

$$\cong 6 \times 10^{-5}$$

$$\rightarrow \int_{3.5}^{10.5} = \int_{-10.5}^{-3.5} \cong 3 \times 10^{-5}$$

11) The total probability of error is then:

$$P_e = \sum_k p_k \times P_{e_k} \quad \text{with} \quad p_k = Pr\{c_k = k\} \quad \text{and} \quad k = \{-2, 0, 2,\}$$

The symbols b_k are independent and identically distributed on the alphabet $\{0, 1\}$, hence:

$$Pr\{b_k = 0\} = Pr\{b_k = 1\} = 1/2$$

The probabilities of emission of the symbols c_k are:

$$p_0 = Pr\{c_k = 0\} = \frac{1}{2}; \quad p_2 = Pr\{c_k = 2\} = \frac{1}{4}; \quad p_{-2} = Pr\{c_k = -2\} = \frac{1}{4}$$

Hence finally:

$$P_e = p_0 \times P_{e_0} + p_2 \times P_{e_2} + p_{-2} \times P_{e_{-2}}$$

$$P_e = \frac{1}{2} \times 6 \times 10^{-5} + \frac{1}{4} \times 3 \times 10^{-5} + \frac{1}{4} \times 3 \times 10^{-5} = 4.5 \times 10^{-5}$$

3

Digital Transmissions with Carrier Modulation: Problems 27 to 33

3.1. Problem 27 – Digital transmissions with carrier modulation

We consider the general system of transmission of digital information with modulation of a carrier represented in the block diagram of Figure 3.1.

The symbols b_n delivered by the binary source are emitted every T_b seconds. The baseband encoder of the transmission system generates two baseband signals $I(t)$ and $Q(t)$ as follows:

– separation of the binary sequence $\{b_n\}$ into two binary sequences $\{b_{2n}\}$ and $\{b_{2n+1}\}$;

– transcoding of the sequences $\{b_{2n}\}$ and $\{b_{2n+1}\}$ into two sequences of symbols $\{a_n\}$ and $\{a'_n\} \in M_1 -$ ary and $M_2 -$ ary respectively;

– pulse amplitude modulation using the basic pulse $x(t)$:

$$x(t) = \begin{cases} 1 & \text{for } t \in [-T/2, T/2[\\ 0 & \text{elsewhere} \end{cases}$$

The two baseband signals $I(t)$ and $Q(t)$ are expressed by:

$$I(t) = \sum_n a_n x(t - nT) \quad \text{and} \quad Q(t) = \sum_n a'_n x(t - nT)$$

and indicate that symbols $\{a_n\}$ and $\{a'_n\}$ are emitted every T seconds.

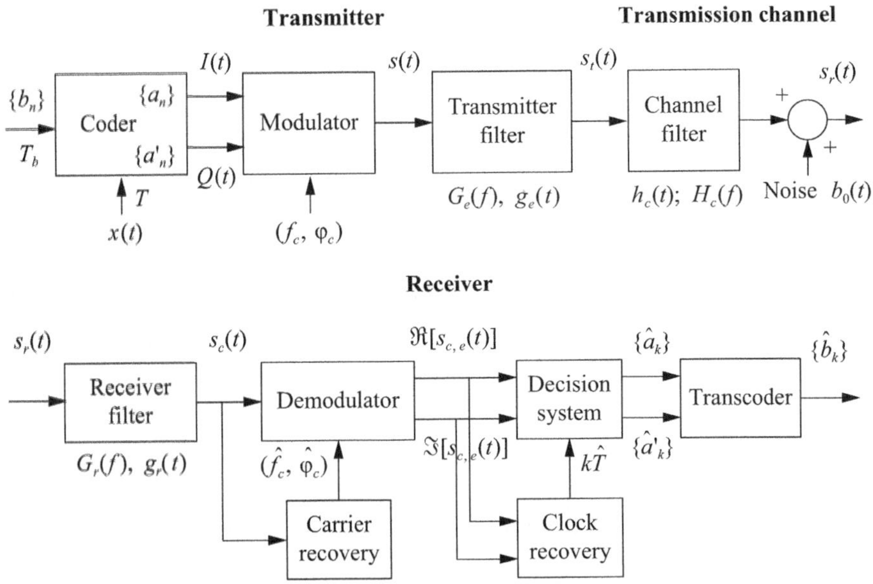

Figure 3.1. *General block diagram of a digital transmission system with carrier modulation and demodulation (Transmitter – Channel – Receiver)*

The modulator is defined by the carrier signal:

$$p_c(t) = V\exp[j(2\pi f_c t + \varphi_c)]$$

It constructs a real signal $s(t)$ by an adequate linear modulation of the digital signal to be modulated:

$$c(t) = I(t) + Q(t)$$

The linear transmission filter (of complex gain $G_e(f)$) and reception filter (of complex gain $G_r(f)$) are of the band-pass type around the frequencies f_c and $-f_c$. The transmission channel is supposed to be modeled by a linear filter (of complex gain $H_c(f)$) to which is added an observation noise $b_0(t)$.

This latter is assumed to be a Gaussian white random process, with a power spectral density equal to $\Gamma_0/2$ in the frequency band covered by the receiving band-pass filter.

Moreover, let: $H(f) = G_e(f) \times H_c(f) \times G_r(f)$

and: $h(t) = g_e(t) \otimes h_c(t) \otimes g_r(t)$ be assumed real.

1) What is the real signal $s(t)$ at the output of the modulator in the general case of amplitude and phase shift keying modulation (APSK) at $M_1 \times M_2$ states? Then determine its complex envelope, denoted $s_e(t)$. Finally, in the expression of the real signal $s(t)$, show explicitly the two components modulating in quadrature the carrier $p_c(t)$.

2) From the previous results, determine the real signal $s(t)$, its complex envelope $s_e(t)$ and the components of the signal $s(t)$ modulating in quadrature the carrier for the following modulations:

a) QAM: quadrature amplitude modulation at $M_1 \times M_2$ states;

b) PSK: phase shift keying modulation at M states;

c) ASK: amplitude shift keying modulation at M states.

3) Considering that the bitrate is set to D, and that $M = M_1 = M_2 = 2^k$, what is the symbol rate D_s of the signal $s(t)$ for each of the four preceding digital modulations (that of question 1 plus the three of question 2)?

From now on, it is assumed that the amplitude spectrum of the transmitted digital modulation signal has a frequency band limited to:

$$|f| \in \left[f_c - \frac{\Delta f}{2}, f_c + \frac{\Delta f}{2}\right] \quad \text{with } \Delta f \ll f_c$$

We call $H_e(f)$ the frequency gain of the low-pass filter equivalent to the total filter of frequency gain $H(f)$ (which is a band-pass filter around frequencies f_c and $-f_c$) and $y_e(t)$ the signal at the output of the equivalent low-pass filter (frequency gain $H_e(f)$) when the input signal is the previous signal $x(t)$ (noise excepted).

We denote:

$$y_e(t) = p(t) + jq(t)$$

In the same way, by denoting $b(t)$, the noise $b_0(t)$ filtered by the receiving band-pass filter, we note:

$b_e(t) = b_{e,p}(t) + jb_{e,q}(t)$, the complex envelope of the filtered noise $b(t)$.

4) What is the response of the global filter $H(f)$ to the signal $x(t) \times p_c(t)$?

5) If the impulse response of the equivalent low-pass filter $H_e(f)$ to the total band-pass filter $H(f)$ is: $h_e(t) = h_{e,p}(t) + jh_{e,q}(t)$, then demonstrate that the impulse response $h(t)$ of the total band-pass filter is:

$$h(t) = 2\left[h_{e,p}(t)\cos(2\pi f_c t + \varphi_c) - h_{e,q}(t)\sin(2\pi f_c t + \varphi_c)\right]$$

In the following, we consider a 4-QAM digital modulation for which the symbols a_n and a'_n take the values 1 and –1. Decisions are made at instants of the form: $t_k = kT$.

6) a) Considering the equivalent baseband transmission and reception system: the digital modulated signals are replaced by their complex envelope, the band-pass filters by their equivalent low-pass filter, determine the complex envelope, denoted $s_{c,e}(t)$, of signal $s_c(t)$ at the output of filter of frequency gain $H_e(f)$. Particularize this one at the decision instants $t_k = kT$: i.e. $s_{c,e}(kT)$.

b) By separating the real and imaginary parts of the complex envelope $s_{c,e}(kT)$, determine the intersymbol interference on the symbol a_k, denoted $I_{m_l}(kT)$, on the one hand, and on the symbol a'_k, denoted $I'_{m'_l}(kT)$, on the other hand.

7) It is assumed in the following, again in the context of a 4-QAM modulation that the characteristics of the band-pass filters and of the transmission channel are such that the equivalent low-pass filter $H_e(f)$ satisfies the Hermitian symmetry.

a) Show that $y_e(t)$ is then a real signal ($q(t) = 0$).

b) Show that the intersymbol interference is of a purely intra-channel type.

Solution of problem 27

1) The real signal $s(t)$ at the output of the modulator is written:

$$s(t) = \Re\left\{V\sum_n \rho_n x(t - nT)\exp[j(2\pi f_c t + \varphi_c + \psi_n)]\right\}$$

or:

$$s(t) = \Re\left\{V\sum_n \rho_n \exp(j\psi_n)x(t - nT)\exp[j(2\pi f_c t + \varphi_c)]\right\}$$

with:

$$\rho_n\exp(j\psi_n) = c_n = a_n + ja'_n\,;\quad \rho_n = \left[a_n^2 + a_n'^2\right]^{1/2}\,;\quad \psi_n = \tan^{-1}\left[\frac{a'_n}{a_n}\right]$$

The pair (ρ_n, ψ_n) belongs to a set of $M_1 \times M_2$ possible amplitude-phase pairs: ρ_n being $M_1 -$ ary and ψ_n being $M_2 -$ ary.

The signal $s(t)$ can also be written:

$$s(t) = \Re\{s_e(t) \times \exp[j(2\pi f_c t + \varphi_c)]\}$$

hence its complex envelope $s_e(t)$ is:

$$s_e(t) = V \sum_n \rho_n \exp(j\psi_n) x(t - nT)$$

$$= V \left[\sum_n \rho_n \cos(\psi_n) x(t - nT) + j \sum_n \rho_n \sin(\psi_n) x(t - nT) \right]$$

The signal $s(t)$ is also written:

$$s(t) = V \sum_n \rho_n \cos(\psi_n) x(t - nT) \cos(2\pi f_c t + \varphi_c)$$

$$-V \sum_n \rho_n \sin(\psi_n) x(t - nT) \sin(2\pi f_c t + \varphi_c)$$

or:

$$s(t) = V[I(t) \cos(2\pi f_c t + \varphi_c) - Q(t) \sin(2\pi f_c t + \varphi_c)]$$

The signals $I(t)$ and $Q(t)$ are linear combinations of the baseband digital signals.

2) From the preceding results, we have:

a) QAM: $\rho_n \exp(j\psi_n) = c_n = a_n + ja'_n$.

Hence, from response 1, we can write:

$$s(t) = \Re\left\{ V \sum_n [a_n + ja'_n]\, x(t - nT) \exp[j(2\pi f_c t + \varphi_c)] \right\}$$

$$s_e(t) = V \sum_n c_n\, x(t - nT) = V \left[\sum_n a_n\, x(t - nT) + j \sum_n a'_n\, x(t - nT) \right]$$

$$s(t) = V\sum_n a_n\, x(t-nT)\cos(2\pi f_c t + \varphi_c)$$

$$-V\sum_n a'_n\, x(t-nT)\sin(2\pi f_c t + \varphi_c)$$

a_n being $M_1 -$ ary and a'_n being $M_2 -$ ary.

Or:

$$s(t) = V[I(t)\cos(2\pi f_c t + \varphi_c) - Q(t)\sin(2\pi f_c t + \varphi_c)]$$

b) In PSK modulation, we have: $\rho_n = 1$.

$$s(t) = \Re\left\{V\sum_n x(t-nT)\exp[j(2\pi f_c t + \varphi_c + \psi_n)]\right\}$$

with:

$$\psi_i = \frac{\pi}{M} + 2i\frac{\pi}{M}; \quad i = \{0, 1, \cdots M-1\} \text{ and } M > 2$$

Also:

$$s(t) = \Re\left\{V\sum_n \exp(j\psi_n)\, x(t-nT)\exp[j(2\pi f_c t + \varphi_c)]\right\}$$

$$\rightarrow s_e(t) = V\sum_n \exp(j\psi_n)\, x(t-nT)$$

$$= V\left[\sum_n \cos(\psi_n)\, x(t-nT) + j\sum_n \sin(\psi_n)\, x(t-nT)\right]$$

The signal $s(t)$ can also be written:

$$s(t) = V\sum_n \cos(\psi_n)\, x(t-nT)\cos(2\pi f_c t + \varphi_c)$$

$$-V\sum_n \sin(\psi_n)\, x(t-nT)\sin(2\pi f_c t + \varphi_c)$$

$$s(t) = V[I(t)\cos(2\pi f_c t + \varphi_c) - Q(t)\sin(2\pi f_c t + \varphi_c)]$$

By setting:

$$a_n = \cos(\psi_n)\,; \quad a'_n = \sin(\psi_n)$$

We note that the PSK modulation is identical to the QAM modulation.

c) ASK: $a'_n = 0$.

From the QAM modulation expression in which a'_n is set to zero, we have:

$$s(t) = \Re\left\{V\sum_n a_n\, x(t-nT)\exp[j(2\pi f_c t + \varphi_c)]\right\}$$

with:

$$a_i = \frac{2i}{M-1} - 1\,; \quad i = \{0, 1, \cdots M-1\} \text{ and } M = 2^k$$

$$s_e(t) = V\sum_n a_n\, x(t-nT)$$

$$s(t) = V\sum_n a_n\, x(t-nT)\cos(2\pi f_c t + \varphi_c)$$

a_n being $M-$ary.

Or also:

$$s(t) = V[I(t)\cos(2\pi f_c t + \varphi_c)]$$

3) One has:

$$D = \frac{1}{T_b};\ D_s = \frac{1}{T_s};\quad T_s = kT_b\,;\quad M = 2^k \rightarrow k = \log_2 M$$

$$\rightarrow D_s = \frac{1}{T_s} = \frac{1}{kT_b} = \frac{D}{k} = \frac{D}{\log_2 M}$$

with D: bitrate; D_s: symbol rate.

– APSK modulation:

$$\text{at } M_1 \times M_2 = 2^k \times 2^k = 2^{2k} \text{ states} \rightarrow D_s = \frac{D}{2k}$$

– QAM modulation:

$$\text{at } M_1 \times M_2 = 2^k \times 2^k = 2^{2k} \text{ states } \rightarrow D_s = \frac{D}{2k}$$

– PSK modulation:

$$\text{at } M = 2^k \text{ states } \rightarrow D_s = \frac{D}{k}$$

– ASK modulation:

$$\text{at } M = 2^k \text{ states } \rightarrow D_s = \frac{D}{k}$$

4) We have:

$$y_e(t) = p(t) + jq(t)$$

$$b_e(t) = b_{e,p}(t) + jb_{e,q}(t)$$

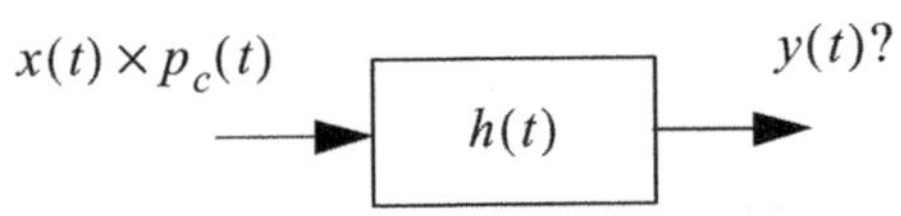

We can write:

$$y(t) = [x(t) \times p_c(t)] \otimes h(t)$$

But this relation is not very useful here.

To find the expression of $y(t)$, one has just to apply the definition of the complex envelope:

$$y(t) = \Re\{y_e(t) \times \exp[j(2\pi f_c t + \varphi_c)]\}$$

$$y(t) = \Re\{[p(t) + jq(t)] \times [\cos(2\pi f_c t + \varphi_c) + j\sin(2\pi f_c t + \varphi_c)]\}$$

So:

$$y(t) = p(t) \times \cos(2\pi f_c t + \varphi_c) - q(t) \times \sin(2\pi f_c t + \varphi_c)$$

5) Demonstration of:

$$h(t) = 2\left[h_{e,p}(t)\cos(2\pi f_c t + \varphi_c) - h_{e,q}(t)\sin(2\pi f_c t + \varphi_c)\right]$$

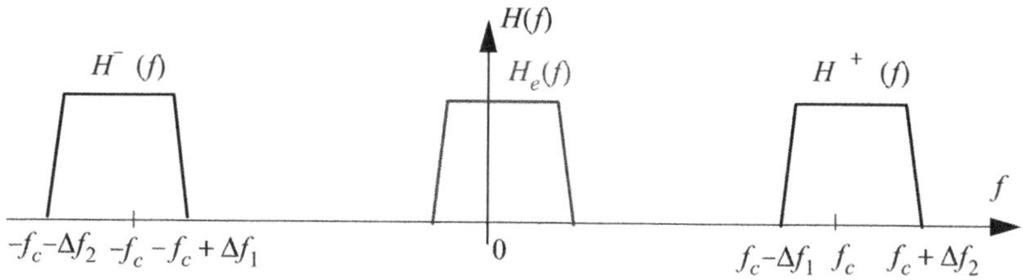

Figure 3.2. *Supports of transfer function $H(f)$ and $H_e(f)$. For a color version of this figure, see www.iste.co.uk/assad/digital2.zip*

$H(f)$ breaks down into:

$$H(f) = H^{+}(f) + H^{-}(f)$$

Yet $H(f)$ is such that: $H(-f) = H^{*}(f)$ because $h(t)$ is a real function (* denotes the conjugate complex).

By definition:

$$H_e(f) = H^{+}(f + f_c)$$

$$\rightarrow \begin{cases} H^{+}(f) = H_e(f - f_c) = H_e(f) \otimes \delta(f - f_c) \\ H^{-}(f) = [H^{+}(-f)]^{*} = [H_e(-f) \otimes \delta(-f - f_c)]^{*} \end{cases}$$

but:

$$[\delta(\ \ )]^{*} = \delta(\ \ ); \quad \delta(-f - f_c) = \delta(f + f_c)$$

$$\rightarrow H^{-}(f) = H_e^{*}(-f) \otimes \delta(f + f_c)$$

Finally, we get:

$$H(f) = H^{+}(f) + H^{-}(f) = H_e(f) \otimes \delta(f - f_c) + H_e^{*}(-f) \otimes \delta(f + f_c)$$

and as:

$$F\{z^{*}(t)\} = Z^{*}(-f)$$

hence:

$$h(t) = h_e(t) \exp[j(2\pi f_c t + \varphi_c)] + h_e^{*}(t) \exp[-j(2\pi f_c t + \varphi_c)]$$

so:

$$h(t) = \left[h_{e,p}(t) + jh_{e,q}(t)\right][\cos(2\pi f_c t + \varphi_c) + j\sin(2\pi f_c t + \varphi_c)]$$
$$+\left[h_{e,p}(t) - jh_{e,q}(t)\right][\cos(2\pi f_c t + \varphi_c) - j\sin(2\pi f_c t + \varphi_c)]$$

Finally, we get:

$$h(t) = 2\left[h_{e,p}(t)\cos(2\pi f_c t + \varphi_c) - h_{e,q}(t)\sin(2\pi f_c t + \varphi_c)\right]$$

6) 4-QAM modulation: a_n and $a'_n \in \{-1, 1\}$.

a) Determination of the complex envelope, denoted $s_{c,e}(t)$, of the signal $s_c(t)$ at the output of the filter $H_e(f)$?

We have:

$$s_e(t) = V\sum_n [a_n + ja'_n]\,x(t - nT)$$

As the action of $x(t)$ on the input of the equivalent baseband system is $y_e(t) = p(t) + jq(t)$ at the output of filter $H_e(f)$ (noise-free), and taking into account the response of filtered noise turned into its equivalent baseband noise, that is $b_e(t)$, the response $s_{c,e}(t)$ is then:

$$s_{c,e}(t) = V\sum_n [a_n + ja'_n]\,[p(t - nT) + jq(t - nT)] + b_e(t)$$

With $t = t_k = kT$, then we have:

$$s_{c,e}(kT) = V\sum_n [a_n + ja'_n]\,\{p[(k-n)T] + jq[(k-n)T]\} + b_e(kT)$$

b) Intersymbol interference on the symbols a_k and a'_k, respectively:

$$\Re\left[s_{c,e}(kT)\right] = V\left\{a_k p(0) + \sum_{n\neq k} a_n p[(k-n)T] - \sum_n a'_n q[(k-n)T]\right\}$$
$$+b_{e,p}(kT)$$

$$\Im\left[s_{c,e}(kT)\right] = V\left\{a'_k p(0) + \sum_{n\neq k} a'_n p[(k-n)T] + \sum_{n} a_n q[(k-n)T]\right\}$$
$$+b_{e,q}(kT)$$

Hence:

$$I_{m_l}(kT) = V\left[\underbrace{\sum_{n\neq k} a_n p[(k-n)T]}_{\text{ISI intra-channel}} - \underbrace{\sum_{n} a'_n q[(k-n)T]}_{\text{ISI inter-channel}}\right]$$

$$I'_{m'_l}(kT) = V\left[\underbrace{\sum_{n\neq k} a'_n p[(k-n)T]}_{\text{ISI intra-channel}} + \underbrace{\sum_{n} a_n q[(k-n)T]}_{\text{ISI inter-channel}}\right]$$

7) Show that $y_e(t)$ is a real signal and that the intersymbol interference is purely an intra-channel interference.

a) As: $H_e(f) = G_{e,e}(f) \times H_{c,e}(f) \times G_{r,e}(f)$ satisfies the Hermitian symmetry, then $h_e(t)$ is a real function, hence $h_e(t) = h_{e,p}(t)$ and $y_e(t) = x(t)\otimes h_e(t)$ is a real function. Therefore, we have: $q(t) = 0$ and $y_e(t) = p(t)$.

b) From question 6 (b), we have:

$$I_{m_l}(kT) = V\sum_{n\neq k} a_n p[(k-n)T]\,; \qquad I'_{m'_l}(kT) = V\sum_{n\neq k} a'_n p[(k-n)T]$$

therefore, the intersymbol interference is of a purely intra-channel type.

3.2. Problem 28 – 4-QAM digital modulation transmission (1)

The transmission of binary information based on a 4-QAM digital modulation is considered. The block diagram of this transmission system (transmitter, transmission channel, receiver) is given in Figure 3.3. In this type of modulation, symbols $\{a_n\}$ and $\{a'_n\}$ take the values on the set $\{1, - - 1\}$.

The symbols b_n delivered by the binary source are emitted every T_b seconds. The baseband encoder of the transmission system generates two baseband signals $I(t)$ and $Q(t)$ as follows:

– separation of the binary sequence $\{b_n\}$ into two binary sequences $\{b_{2n}\}$ and $\{b_{2n+1}\}$;

– transcoding the sequences $\{b_{2n}\}$ and $\{b_{2n+1}\}$ into two sequences of symbols $\{a_n\}$ and $\{a'_n\} \in \{1, --1\}$;

– pulse amplitude modulation using the basic pulse $x(t)$:

$$x(t) = \begin{cases} 1 & \text{for } t \in [-T/2, T/2[\\ 0 & \text{elsewhere} \end{cases}$$

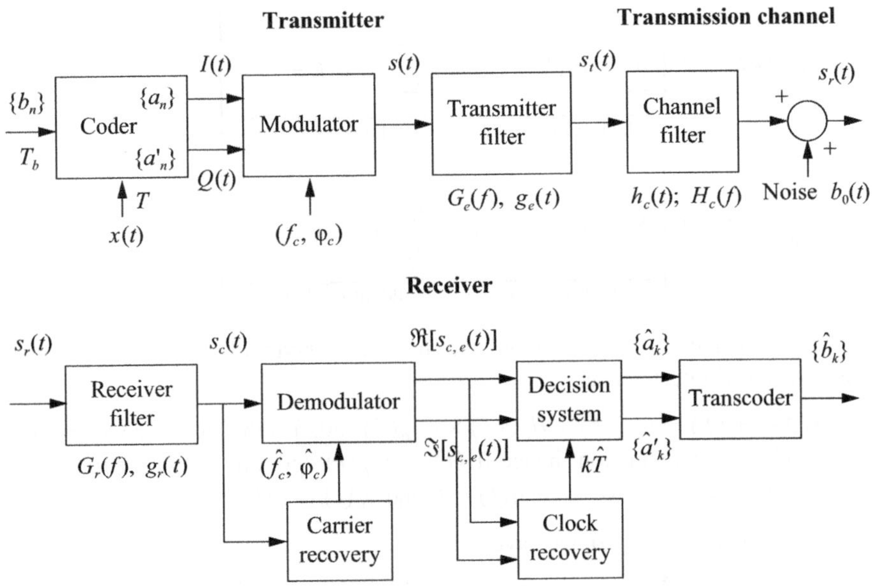

Figure 3.3. *General block diagram of a digital transmission system with quadrature amplitude modulation*

The two baseband signals $I(t)$ and $Q(t)$ are expressed by:

$$I(t) = \sum_n a_n x(t - nT)$$

$$Q(t) = \sum_n a'_n x(t - nT)$$

and indicate that symbols $\{a_n\}$ and $\{a'_n\}$ are emitted every T seconds.

The modulator is defined by the carrier signal:

$$p_c(t) = V\exp[j(2\pi f_c t + \varphi_c)]$$

It constructs a real signal $s(t)$ by an adequate linear modulation of the digital signal to be modulated:

$$c(t) = I(t) + jQ(t)$$

The linear transmission filter (of complex gain $G_e(f)$) and reception filter (of complex gain $G_r(f)$) are of the band-pass type around the frequencies f_c and $-f_c$. The transmission channel is supposed to be modeled by a linear filter (of complex gain $H_c(f)$) to which is added an observation noise $b_0(t)$. This latter is assumed to be a Gaussian white random process, with a power spectral density equal to $\Gamma_0/2$ in the frequency band covered by the receiving band-pass filter.

Moreover, let:

$$H(f) = G_e(f) \times H_c(f) \times G_r(f)$$

and:

$$h(t) = g_e(t) \otimes h_c(t) \otimes g_r(t)$$

be assumed real.

1) a) Write the real signal $s(t)$ at the output of the 4-QAM modulator at $M_1 \times M_2$ states. Show explicitly in the expression of the real signal $s(t)$ the two components modulating in quadrature the carrier $p_c(t)$.

b) Then determine its complex envelope, noted $s_e(t)$.

From now on, it is assumed that the amplitude spectrum of the transmitted digital modulation signal has a frequency band limited to:

$$|f| \in \left[f_c - \frac{\Delta f}{2}, f_c + \frac{\Delta f}{2}\right] \text{ with } \Delta f \ll f_c$$

We call $H_e(f)$ the frequency gain of the low-pass filter equivalent to the total filter of frequency gain $H(f)$ (which is a band-pass filter around the frequencies f_c and $-f_c$) and $y_e(t)$ the signal at the output of the equivalent low-pass filter (frequency gain $H_e(f)$) when the input signal is the previous signal $x(t)$ (noise excepted).

We denote:

$$y_e(t) = p(t) + jq(t)$$

In the same way, by denoting $b(t)$, the noise $b_0(t)$ filtered by the receiving band-pass filter, we denote by $b_e(t)$ the complex envelope of the filtered noise $b(t)$:

$$b_e(t) = b_{e,p}(t) + jb_{e,q}(t)$$

2) a) Determine the complex envelope, denoted $s_{c,e}(t)$, of signal $s_c(t)$ at the output of the filter of frequency gain $H_e(f)$, and taking into account the noise $b_e(t)$.

b) Particularize this one at the decision instants $t_k = kT$, that is $s_{c,e}(kT)$.

3) By separating the real and imaginary parts of the complex envelope $s_{c,e}(kT)$, determine the intersymbol interference on the symbol a_k, denoted $I_{m_l}(kT)$, on the one hand, and on the symbol a'_k, denoted $I'_{m'_l}(kT)$, on the other hand.

It is considered that:

$$P(f) = F\{p(t)\} = \begin{cases} T \text{ for } f \in \left[\frac{-(1+\alpha)}{2T}, \frac{(1+\alpha)}{2T}\right] \\ 0 \text{ otherwise} \end{cases} \quad \text{with } \alpha = \frac{1}{6}$$

$$Q(f) = F\{q(t)\} = \begin{cases} \alpha T \text{ for } f \in \left[\frac{-1}{2T}, \frac{1}{2T}\right] \\ 0 \text{ otherwise} \end{cases}$$

4) Determine the expressions of $p(t)$ and $q(t)$. Then deduce:

$p(0), p(\pm T), p(\pm 2T), q(0), q(\pm iT)$ with i integer $\neq 0$

5) Give the two simple expressions of the intersymbol interference $I_{m_l}(kT)$ and $I'_{m'_l}(kT)$ in the case where we consider $p(iT) = 0 \ \forall\, i \neq \{0, -1, 1\}$, then the values of these intersymbol interferences for the messages m_l interfering with a_k and for the messages m'_l interfering with a'_k (you will take for simplification $\pi \cong 3$).

After the quadrature demodulation, both noises $b_{e,p}(kT)$ and $b_{e,q}(kT)$ are supposed to have the same power σ_e^2. Let the signal-to-noise ratio defined as:

$$\left[\frac{S}{B}\right] = 20 \times \log_{10}\left[\frac{Vp(0)}{\sigma_e}\right]$$

be equal to 14.4 dB after demodulation for each signal $I(t)$ and $Q(t)$.

From now on, we assume that symbols b_n are independent and equiprobable.

6) Give the expression of the probability $P_{e,a}$ of error on symbol a_k at the output of the decision block, and the probability $P_{e,a'}$ of error on symbol a'_k.

7) a) Calculate the conditional probabilities of error on symbol a_k:

$$Pr\{\hat{a}_k = a_j/a_k = a_i, m_l\}\,;\; j \neq i\,;\; a_j = \{-1,1\}\,;\; a_i = \{-1,1\}$$

for each of the messages m_l interfering with a_k .

b) Deduce the probability of error $P_{e,a}$ on symbol a_k.

8) a) Calculate the conditional probabilities of error on symbol a'_k:

$$Pr\{\hat{a}'_k = a'_j/a'_k = a'_i, m'_l\}\,;\; j \neq i\,;\; a'_j = \{-1,1\}\,;\; a'_i = \{-1,1\}$$

for each of the messages m'_l interfering with a'_k .

b) Deduce the probability of error $P_{e,a'}$ on symbol a'_k .

9) By considering that symbols a_k and a'_k are independent and equiprobable, what is the total probability of error $P_{e,c}$ on the quaternary symbol $\{c_k = a_k, a'_k\}$ transmitted?

NOTE.– If X is a Gaussian random process, of mean value m and standard deviation σ, you will take:

$$Pr\{|X - m| > 3\,\sigma\} = 2.6 \times 10^{-3}$$

$$Pr\{|X - m| > 4.5\,\sigma\} = 6 \times 10^{-6}$$

$$Pr\{|X - m| > 6\,\sigma\} \cong 0$$

Solution of problem 28

1) a) The real signal $s(t)$ at the output of the modulator (with 4-QAM modulation) is written:

$$s(t) = \Re\left\{V\sum_n [a_n + ja'_n]\,x(t - nT)\exp[j(2\pi f_c t + \varphi_c)]\right\}$$

$$s(t) = V\sum_n a_n\,x(t - nT)\cos(2\pi f_c t + \varphi_c)$$

$$-V\sum_n a'_n\,x(t - nT)\sin(2\pi f_c t + \varphi_c)$$

b) By definition, $s(t)$ is also written:

$$s(t) = \Re\{s_e(t) \times \exp[j(2\pi f_c t + \varphi_c)]\}$$

Hence the complex envelope of signal $s(t)$ is:

$$s_e(t) = V \sum_n [a_n + ja'_n]\, x(t - nT)$$

2) a) As the action of $x(t)$ at the input of the equivalent baseband system is $y_e(t) = p(t) + jq(t)$ at the output of the filter of frequency gain $H_e(f)$ (noise-free), and taking into account the response of filtered noise turned into its equivalent baseband noise, that is $b_e(t)$, the response $s_{c,e}(t)$ is then:

$$s_{c,e}(t) = V \sum_n [a_n + ja'_n]\, [p(t - nT) + jq(t - nT)] + b_e(t)$$

b) With: $t = t_k = kT$, $s_{c,e}(kT)$ is written:

$$s_{c,e}(kT) = V \sum_n [a_n + ja'_n]\, \{p[(k - n)T] + jq[(k - n)T]\} + b_e(kT)$$

3) Determination of the intersymbol interference on symbols a_k and a'_k, respectively.

We have:

$$\Re[s_{c,e}(kT)] = V \left\{a_k p(0) + \sum_{n \neq k} a_n p[(k - n)T] - \sum_n a'_n q[(k - n)T]\right\}$$

$$+ b_{e,p}(kT)$$

$$\Im[s_{c,e}(kT)] = V \left\{a'_k p(0) + \sum_{n \neq k} a'_n p[(k - n)T] + \sum_n a_n q[(k - n)T]\right\}$$

$$+ b_{e,q}(kT)$$

hence:

$$I_{m_l}(kT) = V \left[\underbrace{\sum_{n \neq k} a_n p[(k - n)T]}_{\text{ISI intra-channel}} - \underbrace{\sum_n a'_n q[(k - n)T]}_{\text{ISI inter-channel}}\right]$$

$$I'_{m'_l}(kT) = V\left[\underbrace{\sum_{n\neq k} a'_n p[(k-n)T]}_{\text{ISI intra-channel}} + \underbrace{\sum_{n} a_n q[(k-n)T]}_{\text{ISI inter-channel}}\right]$$

4) We have successively:

$$p(t) = F^{-1}\{P(f)\} = T \times \frac{(1+\alpha)}{T} \times \frac{\sin[2\pi\,(1+\alpha)t/2T]}{2\pi(1+\alpha)t/2T}$$

$$= (1+\alpha) \times \frac{\sin[\pi\,(1+\alpha)t/T]}{\pi(1+\alpha)t/T}$$

$$q(t) = F^{-1}\{Q(f)\} = \frac{\alpha T}{T} \times \frac{\sin[2\pi t/2T]}{2\pi t/2T} = \alpha \times \frac{\sin[\pi\,t/T]}{\pi\,t/T}$$

$$p(0) = (1+\alpha) = 1 + 1/6 = 7/6$$

$$p(\pm T) = \frac{7}{6} \times \frac{\sin\left[\pi + \frac{\pi}{6}\right]}{\pi \times \frac{7}{6}} = \frac{-1}{2\pi} \cong \frac{-1}{6}$$

because: $\sin\left[\pi + \frac{\pi}{6}\right] = -\sin\left[\frac{\pi}{6}\right] = \frac{-1}{2}$

$$p(\pm 2T) = \frac{7}{6} \times \frac{\sin\left[2\pi + \frac{\pi}{3}\right]}{2\pi \times \frac{7}{6}} = \frac{\sqrt{3}/2}{2\pi} = \frac{\sqrt{3}}{4\pi} \cong \frac{\sqrt{3}}{12}$$

$$q(0) = \alpha = 1/6\,;\ q(\pm iT) = 0\ \forall\, i \neq 0\,;\ i\text{: integer}$$

5) One has successively:

$$I_{m_l}(kT) = V\left[\sum_{n=k-1;\,n\neq k}^{k+1} a_n p[(k-n)T] - \sum_{n} a'_n q[(k-n)T]\right]$$

$$I_{m_l}(kT) = V[a_{k-1}p(T) + a_{k+1}p(-T) - a'_k q(0)] = \frac{-V}{6}[a_{k-1} + a_{k+1} + a'_k]$$

$$I'_{m'_l}(kT) = V\left[\sum_{n=k-1;\,n\neq k}^{k+1} a'_n p[(k-n)T] + \sum_{n} a_n q[(k-n)T]\right]$$

$$I'_{m'_l}(kT) = V[a'_{k-1}p(T) + a'_{k+1}p(-T) + a_k q(0)] = \frac{-V}{6}[a'_{k-1} + a'_{k+1} - a_k]$$

Hence, the values of the intersymbol interference for decision on the symbols a_k and a'_k are given in Table 3.1.

$m_l = \{a_{k-1}, a'_k, a_{k+1}\}$	$I_{m_l}(kT)$	$m'_l = \{a'_{k-1}, a_k, a'_{k+1}\}$	$I'_{m'_l}(kT)$
-1 -1 -1	$V/2$	-1 -1 -1	$V/6$
-1 -1 1	$V/6$	-1 -1 1	$-V/6$
-1 1 -1	$V/6$	-1 1 -1	$V/2$
-1 1 1	$-V/6$	-1 1 1	$V/6$
1 -1 -1	$V/6$	1 -1 -1	$-V/6$
1 -1 1	$-V/6$	1 -1 1	$-V/2$
1 1 -1	$-V/6$	1 1 -1	$V/6$
1 1 1	$-V/2$	1 1 1	$-V/6$

Table 3.1. *Amplitudes of intersymbol interference for decision on a_k and a'_k*

6) The probability of error on symbols a_k and a'_k are, respectively:

$$P_{e,a} = p_{-1}\left[\sum_{l=1}^{8} p_{m_l} \times P_{e_{-1}}\right] + p_1\left[\sum_{l=1}^{8} p_{m_l} \times P_{e_1}\right]$$

$$P_{e,a'} = p'_{-1}\left[\sum_{l=1}^{8} p_{m'_l} \times P_{e'_{-1}}\right] + p'_1\left[\sum_{l=1}^{8} p_{m'_l} \times P_{e'_1}\right]$$

where:

$$p_i = Pr\{a_k = i\} = \frac{1}{M} = \frac{1}{2}\ ;\ p'_i = Pr\{a'_k = i\} = \frac{1}{M} = \frac{1}{2}\ ;\ i = \{-1, 1\}$$

since the symbols are equiprobable. M is the number of levels a symbol can take.

There are $M^L = 2^3 = 8$ interfering messages m_l with symbol a_k. Likewise, there are $M^{L'} = 2^3 = 8$ interfering messages m'_l with symbol a'_k. Moreover:

$$p_{m_l} = \frac{1}{M^L} = \frac{1}{8}\ ;\ p_{m'_l} = \frac{1}{M^{L'}} = \frac{1}{8}$$

from the fact that the binary symbols are equiprobable, then the messages are equiprobable.

Finally, note that:

$$P_{e_{-1}} = P_{e(1/-1,\, m_l)} = Pr\{\hat{a}_k = 1/a_k = -1, m_l\}$$

$$P_{e_1} = P_{e(-1/1,\, m_l)} = Pr\{\hat{a}_k = -1/a_k = 1, m_l\}$$

$$P_{e'_{-1}} = P_{e'(1/-1, m'_l)} = Pr\{\hat{a}'_k = 1/a'_k = -1, m'_l\}$$

$$P_{e'_1} = P_{e'(-1/1, m'_l)} = Pr\{\hat{a}'_k = -1/a'_k = 1, m'_l\}$$

7) a) Calculation of the conditional probabilities of error on symbols a_k:

$$P_{e_{-1}} = P_{e(1/-1, m_l)} = \frac{1}{\sigma_e\sqrt{2\pi}} \int_{Vp(0)-I_{m_l}(kT)}^{\infty} \exp\left[-\frac{1}{2}\frac{b_{e,p}^2}{\sigma_e^2}\right] db_{e,p}$$

$$P_{e_1} = P_{e(-1/1, m_l)} = \frac{1}{\sigma_e\sqrt{2\pi}} \int_{-\infty}^{-\left[Vp(0)+I_{m_l}(kT)\right]} \exp\left[-\frac{1}{2}\frac{b_{e,p}^2}{\sigma_e^2}\right] db_{e,p}$$

The signal-to-noise ratio is:

$$\left[\frac{S}{b}\right] = 20 \times \log_{10}\left[\frac{Vp(0)}{\sigma_e}\right] = 20 \times \log_{10}\left[\frac{V \times 7/6}{\sigma_e}\right] = 14.4 \text{ dB}$$

$$\rightarrow \frac{7V}{6\sigma_e} \cong 5.25 \rightarrow V = 4.5\,\sigma_e \rightarrow \frac{5V}{3} = 7.5\,\sigma_e \rightarrow \frac{4V}{3} = 6\,\sigma_e \rightarrow \frac{2V}{3} = 3\,\sigma_e$$

Hence, Table 3.2 gives the probabilities of error conditional to the interfering messages m_l.

$\boldsymbol{m_l = \{a_{k-1}, a'_k, a_{k+1}\}}$	$\boldsymbol{Vp(0) - I_{m_l}}$	$\boldsymbol{P_{e_{-1}}}$	$\boldsymbol{-[Vp(0) + I_{m_l}]}$	$\boldsymbol{P_{e_1}}$
-1 -1 -1	$3\,\sigma_e$	1.3×10^{-3}	$-7.5\,\sigma_e$	0
-1 -1 1	$4.5\,\sigma_e$	3×10^{-6}	$-6\,\sigma_e$	0
-1 1 -1	$4.5\,\sigma_e$	3×10^{-6}	$-6\,\sigma_e$	0
-1 1 1	$6\,\sigma_e$	0	$-4.5\,\sigma_e$	3×10^{-6}
1 -1 -1	$4.5\,\sigma_e$	3×10^{-6}	$-6\,\sigma_e$	0
1 -1 1	$6\,\sigma_e$	0	$-4.5\,\sigma_e$	3×10^{-6}
1 1 -1	$6\,\sigma_e$	0	$-4.5\,\sigma_e$	3×10^{-6}
1 1 1	$7.5\,\sigma_e$	0	$-3\,\sigma_e$	1.3×10^{-3}

Table 3.2. *Probabilities of error on a_k conditional on possible m_l interfering messages*

b) Probability of error $P_{e,a}$:

$$P_{e,a} = \frac{1}{2} \times \frac{1}{8} \times \sum_{l=1}^{8} [P_{e_{-1}} + P_{e_1}]$$

$$P_{e,a} = \frac{1}{16} [1.3 \times 10^{-3} + 3 \times 3 \times 10^{-6} + 3 \times 3 \times 10^{-6} + 1.3 \times 10^{-3}]$$

$= 1.636 \times 10^{-4}$

8) a) Calculation of the conditional probabilities of error on symbols a'_k:

$$P_{e'_{-1}} = P_{e'(1/-1,m'_l)} = \frac{1}{\sigma_e \sqrt{2\pi}} \int_{Vp(0)-I'_{m'_l}(kT)}^{\infty} \exp\left[-\frac{1}{2}\frac{b_{e,q}^2}{\sigma_e^2}\right] db_{e,q}$$

$$P_{e'_1} = P_{e'(-1/1,m'_l)} = \frac{1}{\sigma_e \sqrt{2\pi}} \int_{-\infty}^{-\left[Vp(0)+I'_{m'_l}(kT)\right]} \exp\left[-\frac{1}{2}\frac{b_{e,q}^2}{\sigma_e^2}\right] db_{e,q}$$

Hence, Table 3.3 gives the probabilities of error on a'_k, conditional on the interfering messages m'_l.

$m'_l = \{a'_{k-1}, a_k, a'_{k+1}\}$	$Vp(0) - I'_{m'_l}$	$P_{e'_{-1}}$	$-\left[Vp(0) + I'_{m'_l}\right]$	$P_{e'_1}$
-1 -1 -1	$4.5\,\sigma_e$	3×10^{-6}	$-6\,\sigma_e$	0
-1 -1 1	$6\,\sigma_e$	0	$-4.5\,\sigma_e$	3×10^{-6}
-1 1 -1	$3\,\sigma_e$	1.3×10^{-3}	$-7.5\,\sigma_e$	0
-1 1 1	$4.5\,\sigma_e$	3×10^{-6}	$-6\,\sigma_e$	0
1 -1 -1	$6\,\sigma_e$	0	$-4.5\,\sigma_e$	3×10^{-6}
1 -1 1	$7.5\,\sigma_e$	0	$-3\,\sigma_e$	1.3×10^{-3}
1 1 -1	$4.5\,\sigma_e$	3×10^{-6}	$-6\,\sigma_e$	0
1 1 1	$6\,\sigma_e$	0	$-4.5\,\sigma_e$	3×10^{-6}

Table 3.3. *Probabilities of error on a'_k conditional on possible m'_l interfering messages*

b) Probability of error $P_{e,a'}$:

$$P_{e,a'} = \frac{1}{2} \times \frac{1}{8} \times \sum_{l=1}^{8} \left[P_{e'_{-1}} + P_{e'_1}\right]$$

$$P_{e,a'} = \frac{1}{16}[1.3 \times 10^{-3} + 3 \times 3 \times 10^{-6} + 3 \times 3 \times 10^{-6} + 1.3 \times 10^{-3}]$$

$$= 1.636 \times 10^{-4} = P_{e,a}$$

9) The probability of error per quaternary symbol emitted $c_k = \{a_k, a'_k\}$ is:

$$P_{e,c} = Pr\{\hat{c}_k \neq c_k\} = Pr\{\hat{a}_k \neq a_k \text{ or } \hat{a}'_k \neq a'_k\}$$

$$P_{e,c} = Pr\{\hat{a}_k \neq a_k\} + Pr\{\hat{a}'_k \neq a'_k\} - Pr\{\hat{a}_k \neq a_k \text{ and } \hat{a}'_k \neq a'_k\}$$

Since a_k and a'_k are independent symbols, then we have:

$$P_{e,c} = Pr\{\hat{a}_k \neq a_k\} + Pr\{\hat{a}'_k \neq a'_k\} - Pr\{\hat{a}_k \neq a_k\} \times Pr\{\hat{a}'_k \neq a'_k\}$$

$$\rightarrow P_{e,c} = P_{e,a} + P_{e,a'} - P_{e,a} \times P_{e,a'} \cong 3.27 \times 10^{-4}$$

3.3. Problem 29 – Digital transmissions with 2-ASK modulation

We consider the transmission system with 2-state ASK digital amplitude modulation given by the block diagram of Figure 3.4.

The symbols b_n delivered by the binary source are emitted every T_b seconds with the following probability law:

$$Pr\{b = 0\} = 1/4\,; \quad Pr\{b = 1\} = 3/4$$

and the binary symbols b_n are considered independent.

The transcoding of binary information $\{b_n\}$ into symbols $\{a_n\}$ corresponds to the following rule:

$$\text{if: } b = 1 \text{ then: } a = 1\,; \quad \text{if: } b = 0 \text{ then: } a = -1$$

The baseband encoder of the transmission system generates the baseband signal $I(t)$ by pulse amplitude modulation using the basic pulse $x(t)$:

$$x(t) = \begin{cases} 1 & \text{for } t \in [-T/2, T/2[\\ 0 & \text{otherwise} \end{cases}$$

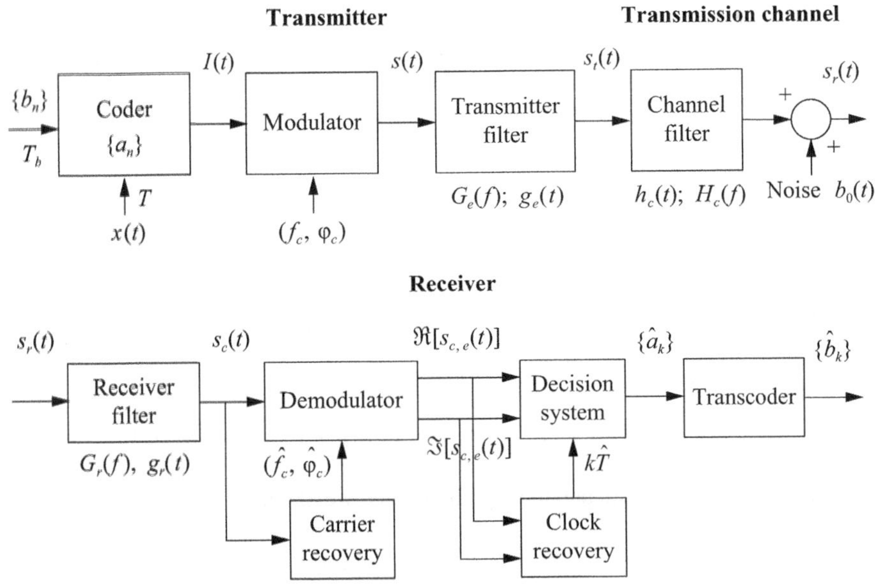

Figure 3.4. *Block diagram of a digital transmission system with 2-ASK modulation*

The baseband signal $I(t)$ is expressed as:

$$I(t) = \sum_n a_n x(t - nT)$$

The modulator is defined by the carrier signal:

$$p_c(t) = V\exp[j(2\pi f_c t + \varphi_c)]$$

It constructs a real signal $s(t)$ by an adequate linear modulation of the baseband signal $I(t)$ to be modulated.

The linear transmission filter (of complex gain $G_e(f)$) and reception filter (of complex gain $G_r(f)$) are of the band-pass type around frequencies f_c and $-f_c$. The transmission channel is assumed to be modeled by a linear filter (of complex frequency gain $H_c(f)$) to which is added an observation noise $b_0(t)$. This latter is assumed to be a Gaussian white random process, with a power spectral density equal to $\Gamma_0/2$ in the frequency band covered by the receiving band-pass filter.

Moreover, let:

$$H(f) = G_e(f) \times H_c(f) \times G_r(f)$$

and:

$$h(t) = g_e(t) \otimes h_c(t) \otimes g_r(t)$$

be assumed real.

1) Write explicitly the real signal $s(t)$ at the output of the modulator then determine its complex envelope: $s_e(t)$.

From now on, it is assumed that the amplitude spectrum of the transmitted digital modulation signal has a frequency band limited to:

$$|f| \in \left[f_c - \frac{\Delta f}{2}, f_c + \frac{\Delta f}{2}\right] \text{ with } \Delta f \ll f_c$$

We call $H_e(f)$ the frequency gain of the low-pass filter equivalent to the total filter of frequency gain $H(f)$ (which is a band-pass filter around frequencies f_c and $-f_c$) and $y_e(t)$ the signal at the output of the equivalent low-pass filter (frequency gain $H_e(f)$) when the input signal is the previous signal $x(t)$ (noise excepted).

We denote:

$$y_e(t) = p(t) + jq(t)$$

In the same way, by denoting $b(t)$, the noise $b_0(t)$ filtered by the receiving band-pass filter, we note $b_e(t)$ the complex envelope of the filtered noise $b(t)$:

$$b_e(t) = b_{e,p}(t) + jb_{e,q}(t)$$

2) a) Determine the complex envelope, denoted $s_{c,e}(t)$, of the signal $s_c(t)$ at the output of filter $H_e(f)$ and in taking into account the noise $b_e(t)$. Particularize this one at the instant $t_k = kT$ of decision, that is $s_{c,e}(kT)$.

b) By separating real and imaginary parts of the complex envelope $s_{c,e}(kT)$, determine the intersymbol interference on symbol a_k, denoted $I_{m_l}(kT)$, on the one hand, and on symbol a'_k, denoted $I'_{m'_l}(kT)$, on the other hand.

3) Is it interesting to consider thereafter that the filter $H_e(f)$ verifies the Hermitian symmetry in the case of a 2-ASK modulation? Justify your answer.

After quadrature demodulation, both noises $b_{e,p}(kT)$ and $b_{e,q}(kT)$ are supposed to have the same power σ_e^2. The signal-to-noise ratio given by:

$$\left[\frac{S}{b}\right]_{dB} = 20 \times \log_{10}\left[\frac{Vp(0)}{\sigma_e}\right]$$

is equal to 12 dB for signal $\Re\left[s_{c,e}(kT)\right]$ after demodulation.

It is considered that the amplitude spectrum $P(f)$ of the real part is given by:

$$P(f) = F\{p(t)\} = \begin{cases} T \text{ for } f \in \left[\frac{-(1+\alpha)}{2T}, \frac{(1+\alpha)}{2T}\right] \\ 0 \text{ otherwise} \end{cases} \quad \text{with } \alpha = \frac{5}{6}$$

4) a) Determine the expression of the signal $p(t)$ and give the values of $p(0)$ and $p(\pm T)$.

b) Give the relation of the optimal threshold $\mu_0 = \mu_{0opt}$ of the decision block and calculate its value according to σ_e (to simplify, you will take later in the problem $\pi \cong 3$).

c) In the case where we consider that only the symbols adjacent to a given symbol a_k interfere with it, so the interfering messages are of the form: $m_l = \{a_{k-1}, a_{k+1}\}$, calculate the intersymbol interference $I_{m_l}(kT)$ and the probability $p_{m_l} = Pr\{m_l\}$ for each message m_l interfering with a_k.

5) a) Determine the expression of the probability of error $P_{e,a}$ on symbol a_k.

b) For each message m_l interfering with symbol a_k, calculate the conditional probabilities of erroneous decisions:

$$P_{e_{-1}} = P_{e(1/-1,\, m_l)} = Pr\{\hat{a}_k = 1/a_k = -1, m_l\}$$

$$P_{e_1} = P_{e(-1/1,\, m_l)} = Pr\{\hat{a}_k = -1/a_k = 1, m_l\}$$

c) Deduce the probability of error $P_{e,a}$ on symbol a_k. What is the probability of error $P_{e,b}$ on the binary information b_k transmitted?

NOTE.– If X is a Gaussian random process with mean value m and standard deviation σ, you will take:

$$Pr\{|X - m| > 3.119\, \sigma\} = 1.92 \times 10^{-3}$$

$$Pr\{|X - m| > 3.395\, \sigma\} = 6.8 \times 10^{-4}$$

$$Pr\{|X-m|>3.842\,\sigma\}=1.4\times 10^{-4}$$

$$Pr\{|X-m|>4.118\,\sigma\}=6\times 10^{-5}$$

$$Pr\{|X-m|>4.5\,\sigma\}\cong 0$$

Solution of problem 29

1) Signals $s(t)$ and $s_e(t)$ are given by:

$$s(t)=\Re\left\{V\sum_n a_n\,x(t-nT)\exp[j(2\pi f_c t+\varphi_c)]\right\}$$

$$=V\sum_n a_n\,x(t-nT)\cos(2\pi f_c t+\varphi_c)$$

hence its real envelope $s_e(t)$:

$$s_e(t)=V\sum_n a_n\,x(t-nT)$$

2) a) One has: $s_e(t)=V\sum_n a_n\,x(t-nT)$ and since the action of $x(t)$ at the input of the equivalent baseband system is $y_e(t)=p(t)+jq(t)$ at the output of filter $H_e(f)$ (noise-free), and taking into account the response of filtered noise turned into its equivalent baseband noise, that is $b_e(t)$, the response $s_{c,e}(t)$ is then:

$$s_{c,e}(t)=V\sum_n a_n\,[p(t-nT)+jq(t-nT)]+b_e(t)$$

For $t=t_k=kT$, $s_{c,e}(kT)$ is written:

$$s_{c,e}(kT)=V\sum_n a_n\,\{p[(k-n)T]+jq[(k-n)T]\}+b_e(kT)$$

So:

$$\Re\left[s_{c,e}(kT)\right]=V\left\{a_k p(0)+\sum_{n\neq k} a_n p[(k-n)T]\right\}+b_{e,p}(kT)$$

$$\Im\left[s_{c,e}(kT)\right]=V\left\{\sum_n a_n q[(k-n)T]\right\}+b_{e,q}(kT)$$

b) Hence:

$$I_{m_l}(kT) = V\left[\sum_{n\neq k} a_n p[(k-n)T]\right];\ I'_{m'_l}(kT) = V\left[\sum_{n} a_n q[(k-n)T]\right]$$

Since the useful signal is only carried by the real component in phase (because in 2-ASK, $a'_n = 0\ \forall\, n$), only the real part of $s_{c,e}(kT)$ is interesting.

3) Considering that $H_e(f)$ verifies Hermitian symmetry does not matter, because, $a'_n = 0\ \forall\ n$, the real part of $s_{c,e}(kT)$ does not change. After demodulation, only the channel resulting from the demodulation with the carrier in phase is reconstructed and thus, even if the imaginary part of $s_{c,e}(kT)$ is non-zero, this does not matter.

4) a) We have:

$$p(t) = F^{-1}\{P(f)\} = (1+\alpha) \times \frac{\sin[\pi\,(1+\alpha)t/T]}{\pi(1+\alpha)t/T}$$

$$p(0) = (1+\alpha) = 1 + 5/6 = 1.833$$

$$p(\pm T) = \frac{\sin\left[\pi + \frac{5\pi}{6}\right]}{\pi} = \frac{-\sin\left[\frac{5\pi}{6}\right]}{\pi} = \frac{-\sin\left[\frac{\pi}{6}\right]}{\pi} = \frac{-1}{2\pi} \cong \frac{-1}{6}$$

b) The optimal threshold is given by:

$$\mu_{0,e} = \mu_{0opt,e} = \frac{\sigma_e^2}{2Vp(0)}\log_e\left[\frac{p_{-1}}{p_1}\right]$$

$$p_{-1} = 1/4\ ;\ p_1 = 3/4\ \rightarrow\ \mu_{0,e} = -0.138\ \sigma_e$$

The signal-to-noise ratio is such that:

$$\left[\frac{S}{b}\right]_{dB} = 20 \times \log_{10}\left[\frac{Vp(0)}{\sigma_e}\right] = 12\text{ dB}$$

hence:

$$Vp(0) = 3.98\,\sigma_e\ \rightarrow V = 2.17\,\sigma_e$$

c) Amplitude $I_{m_l}(kT)$ and probability p_{m_l} of the intersymbol interference:

$$I_{m_l}(kT) = V\left[\sum_{n=k-1\,;\,n\neq k}^{k+1} a_n p[(k-n)T]\right]$$

$$I_{m_l}(kT) = V[a_{k-1}p(T) + a_{k+1}p(-T)] = \frac{-V}{6}[a_{k-1} + a_{k+1}]$$

Hence the amplitude and probabilities of the intersymbol interference for decision on symbol a_k given in Table 3.4.

l	$m_l = \{a_{k-1}, a_{k+1}\}$	$\frac{-V}{6}[a_{k-1} + a_{k+1}]$	p_{m_l}
1	-1 -1	$V/3 = 0.723\,\sigma_e$	1/16
2	-1 1	0	3/16
3	1 -1	0	3/16
4	1 1	$-V/3 = -0.723\,\sigma_e$	9/16

Table 3.4. *Amplitude $I_{m_l}(kT)$ and probability p_{m_l} of the intersymbol interference*

5) a) Expression of the probability of error $P_{e,a}$ on symbol a_k:

$$P_{e,a} = p_{-1}\left[\sum_{l=1}^{4} p_{m_l} \times P_{e_{-1}}\right] + p_1\left[\sum_{l=1}^{4} p_{m_l} \times P_{e_1}\right]$$

b) Calculation of the conditional probabilities of error on symbol a_k:

$$P_{e_{-1}} = P_{e(1/-1,m_l)} = \frac{1}{\sigma_e\sqrt{2\pi}} \int_{\mu_{0,e}+Vp(0)-I_{m_l}(kT)}^{\infty} \exp\left[-\frac{1}{2}\frac{b_{e,p}^2}{\sigma_e^2}\right] db_{e,p}$$

$$P_{e_1} = P_{e(-1/1,m_l)} = \frac{1}{\sigma_e\sqrt{2\pi}} \int_{-\infty}^{\mu_{0,e}-\left[Vp(0)+I_{m_l}(kT)\right]} \exp\left[-\frac{1}{2}\frac{b_{e,p}^2}{\sigma_e^2}\right] db_{e,p}$$

Hence, Table 3.5 gives the conditional probabilities of erroneous decisions based on interfering messages.

$m_l = \{a_{k-1}, a_{k+1}\}$	$\mu_{0,e} + Vp(0) - I_{m_l}$	$P_{e_{-1}}$	$\mu_{0,e} - [Vp(0) + I_{m_l}]$	P_{e_1}
-1 -1	$3.119\ \sigma_e$	9.6×10^{-4}	$-4.841\ \sigma_e$	0
-1 1	$3.842\ \sigma_e$	7×10^{-5}	$-4.118\ \sigma_e$	3×10^{-5}
1 -1	$3.842\ \sigma_e$	7×10^{-5}	$-4.118\ \sigma_e$	3×10^{-5}
1 1	$4.565\ \sigma_e$	0	$-3.395\ \sigma_e$	3.4×10^{-4}

Table 3.5. *Conditional probabilities of erroneous decisions based on interfering messages*

c) Calculation of the probability of error:

$$P_{e,a} = p_{-1}\left[\sum_{l=1}^{4} p_{m_l} \times P_{e_{-1}}\right] + p_1\left[\sum_{l=1}^{4} p_{m_l} \times P_{e_1}\right]$$

$$P_{e,a} = \frac{1}{4}\left[\frac{1}{16} \times 9.6 \times 10^{-4} + 2 \times \frac{3}{16} \times 7 \times 10^{-5}\right]$$

$$+\frac{3}{4}\left[2 \times \frac{3}{16} \times 3 \times 10^{-5} + \frac{9}{16} \times 3.4 \times 10^{-4}\right]$$

$$\cong 1.73 \times 10^{-4}$$

We have: $P_{e,b} = P_{e,a}$ since a symbol a depends only on a single symbol b.

3.4. Problem 30 – 4-QAM digital modulation transmission (2)

We consider the digital transmission system with 4-QAM modulation given by the block diagram of Figure 3.5. For this type of modulation, symbols a_n and a'_n take the values $\{1, -1\}$ with the following probability law:

$$Pr\{a_n = 1\} = Pr\{a'_n = 1\} = p_1 = 0.65$$

$$Pr\{a_n = -1\} = Pr\{a'_n = -1\} = p_{-1} = 0.35$$

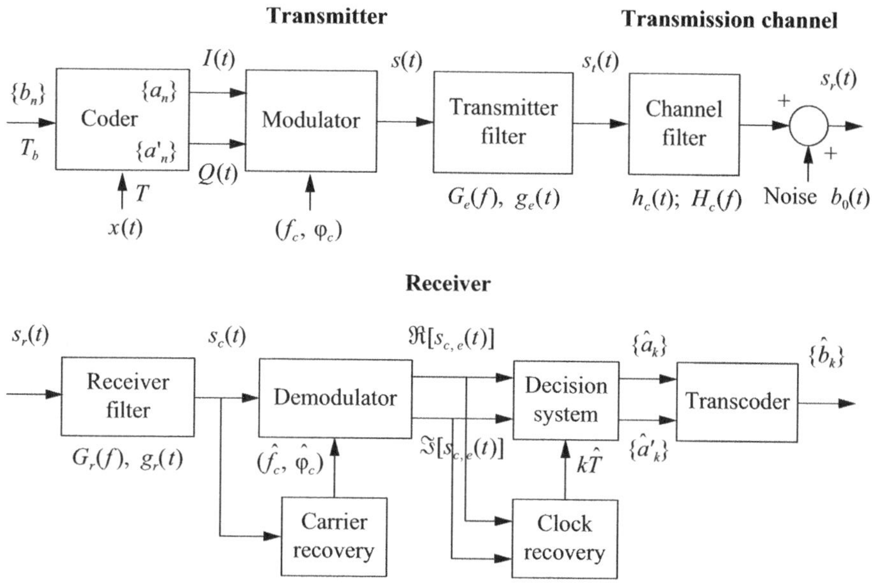

Figure 3.5. *Block diagram of a digital transmission system with quadrature amplitude modulation*

We call $H_e(f)$ the frequency gain of the low-pass filter equivalent to the total filter $H(f)$ (which is band-pass around frequencies f_c and $-f_c$) and we denote $y_e(t)$ the signal at the output of the equivalent low-pass filter (frequency gain $H_e(f)$) when the signal input is the previous signal $x(t)$ (noise excepted):

$$y_e(t) = p(t) + jq(t)$$

In the same way, we call $b(t)$, the noise $b_0(t)$ filtered by the receiving band-pass filter and we denote $b_e(t)$ its complex envelope:

$$b_e(t) = b_{e,p}(t) + jb_{e,q}(t)$$

The two components (real and imaginary parts) of the noise $b_e(t)$ are assumed to be Gaussian, zero mean, of the same variance σ_e^2 and decorrelated. Moreover, after demodulation, the signal-to-noise ratio on each channel is the following:

$$\left[\frac{S}{b}\right]_{dB} = 20 \times \log_{10}\left[\frac{Vp(0)}{\sigma_e}\right] = 9.6 \text{ dB}$$

1) Give the expression of the complex envelope $s_e(t)$ of the 4-QAM modulation signal.

2) Determine the complex envelope, denoted $s_{c,e}(t)$, of signal $s_c(t)$ at the output of the filter of frequency gain $H_e(f)$, and taking into account the noise $b_e(t)$. Particularize this one at the decision instants $t_k = kT$, that is $s_{c,e}(kT)$.

3) Write the real part $\Re[s_{c,e}(kT)]$, and imaginary part $\Im[s_{c,e}(kT)]$ of the signal $s_{c,e}(kT)$, showing explicitly the useful signal, intersymbol interference (intra and inter-channel), and noise on each of these two parts.

It is considered that the amplitude spectrum $P(f)$ of signal $p(t)$ is a constant function, equal to T on the frequency domain:

$$\left[\frac{-(1+\alpha)}{2T}, \frac{(1+\alpha)}{2T}\right]$$

with $\alpha = 1/4$, and zero elsewhere.

It is also considered that the amplitude spectrum $Q(f)$ of signal $q(t)$ is a constant function, equal to $[1+\alpha]T$ on the frequency domain:

$$\left[\frac{-1}{2T}, \frac{1}{2T}\right]$$

and zero elsewhere.

4) Determine the expressions of signals $p(t)$ and $q(t)$ and give the values of:

$$p(0)\ ;\ p(\pm T)\ ;\ q(0)\ ;\ q(\pm T)$$

(in the rest of the problem you will take $\pi \cong 3$).

5) Give the expressions of the intersymbol interferences $I_{m_l}(kT)$ and $I'_{m'_l}(kT)$ in the case where we have $p(iT) = 0\ \forall\ i \neq \{0, -1, 1\}$.

6) Calculate the values of the intersymbol interference $I'_{m'_l}(kT)$ for the different possible combinations of the message m'_l interfering with symbol a'_k.

7) Give the expression of the optimal threshold $\mu'_{0,e} = \mu_{0,e} = \mu_{0opt,e}$ and calculate its value according to σ_e.

8) Show that, even without noise, the probability of error is very high.

Also, it is decided to more completely correct the equalization filter on the in-phase channel so that the amplitude spectrum $P(f)$ of signal $p(t)$ is constant, equal to T, on the frequency domain:

$$\left[\frac{-1}{2T}, \frac{1}{2T}\right]$$

and zero elsewhere.

Moreover, the amplitude spectrum $Q(f)$ of signal $q(t)$ is adjusted to the value $\alpha T/2$ over the frequency domain:

$$\left[\frac{-1}{2T}, \frac{1}{2T}\right]$$

and zero elsewhere.

9) Under these conditions, give the new expression of the intersymbol interference $I'_{m'_l}(kT)$ and calculate its value for the different messages m'_l.

10) Give the expressions of the conditional probabilities of errors:

$$P_{e'_{-1}} = P_{e'(1/-1,m'_l)} = Pr\{\hat{a}'_k = 1/a'_k = -1, m'_l\}$$

$$P_{e'_1} = P_{e'(-1/1,m'_l)} = Pr\{\hat{a}'_k = -1/a'_k = 1, m'_l\}$$

and calculate their values.

11) Give the expression of the probability of error $P_{e,a'} = Pr\{\hat{a}'_k \neq a'_k\}$

and calculate its value.

NOTE.– If X is a Gaussian random process, with zero mean $(m = 0)$ and reduced standard deviation $(\sigma = 1)$, we will assume we have approximately:

$$Pr\{|X| > 3.2\} = 1.4 \times 10^{-3}\ ; \quad Pr\{|X| > 2.6\} = 9.4 \times 10^{-3}$$

$$Pr\{|X| > 2.8\} = 5.2 \times 10^{-3}\ ; \quad Pr\{|X| > 3.4\} = 6 \times 10^{-4}$$

Solution of problem 30

1) The complex envelope is given by:

$$s_e(t) = V \sum_n [a_n + ja'_n]\, x(t - nT)$$

2) Determination of the complex envelope $s_{c,e}(t)$.

Since the action of $x(t)$ at the input of the equivalent baseband system is $y_e(t) = p(t) + jq(t)$ at the output of filter of frequency gain $H_e(f)$ (noise-free), and taking into account the response of filtered noise turned into its equivalent baseband noise, that is $b_e(t)$, the response $s_{c,e}(t)$ is then:

$$s_{c,e}(t) = V \sum_n [a_n + ja'_n]\,[p(t - nT) + jq(t - nT)] + b_e(t)$$

For $t = t_k = kT$, $s_{c,e}(kT)$ is written:

$$s_{c,e}(kT) = V \sum_n [a_n + ja'_n]\,\{p[(k-n)T] + jq[(k-n)T]\} + b_e(kT)$$

3) Expressions of the real $\Re[s_{c,e}(kT)]$ and imaginary $\Im[s_{c,e}(kT)]$ parts of the signal $s_{c,e}(kT)$:

$$\Re[s_{c,e}(kT)] = V a_k p(0) + I_{m_l}(kT) + b_{e,p}(kT)$$

$$\Im[s_{c,e}(kT)] = V a'_k p(0) + I'_{m'_l}(kT) + b_{e,q}(kT)$$

with:

$$I_{m_l}(kT) = V\left[\underbrace{\sum_{n\neq k} a_n p[(k-n)T]}_{\text{ISI intra-channel}} - \underbrace{\sum_n a'_n q[(k-n)T]}_{\text{ISI inter-channel}}\right]$$

$$I'_{m'_l}(kT) = V\left[\underbrace{\sum_{n\neq k} a'_n p[(k-n)T]}_{\text{ISI intra-channel}} + \underbrace{\sum_n a_n q[(k-n)T]}_{\text{ISI inter-channel}}\right]$$

4) We have:

$$p(t) = F^{-1}\{P(f)\} = (1+\alpha) \times \frac{\sin[\pi\,(1+\alpha)t/T]}{\pi(1+\alpha)t/T}$$

$$q(t) = F^{-1}\{Q(f)\} = (1+\alpha) \times \frac{\sin[\pi\, t/T]}{\pi\, t/T}$$

$$p(0) = (1+\alpha) = 1 + 1/4 = 5/4 = 1.25$$

$$p(\pm T) = \frac{\sin[5\pi/4]}{\pi} = \frac{-\sqrt{2}}{2\pi} \cong -0.235 \quad (\pi \cong 3)$$

$$q(0) = (1+\alpha) = 1.25;\ q(\pm iT) = 0\ \forall\, i \neq 0;\ i\text{: integer}$$

5) Expressions of the intersymbol interference:

$$I_{m_l}(kT) = V\left[\sum_{n=k-1;\, n\neq k}^{k+1} a_n p[(k-n)T] - \sum_n a'_n q[(k-n)T]\right]$$

$$I_{m_l}(kT) = V[a_{k-1}p(T) + a_{k+1}p(-T) - a'_k q(0)]$$

$$\cong \frac{-V}{3\sqrt{2}}\left[a_{k-1} + a_{k+1} + \frac{5}{4} \times 3\sqrt{2} \times a'_k\right]$$

$$I'_{m'_l}(kT) = V\left[\sum_{n=k-1;\, n\neq k}^{k+1} a'_n p[(k-n)T] + \sum_n a_n q[(k-n)T]\right]$$

$$I'_{m'_l}(kT) = V[a'_{k-1}p(T) + a'_{k+1}p(-T) + a_k q(0)]$$

$$\cong \frac{-V}{3\sqrt{2}}\left[a'_{k-1} + a'_{k+1} - \frac{5}{4} \times 3\sqrt{2} \times a_k\right]$$

$$I'_{m'_l}(kT) \cong \frac{-V}{3\sqrt{2}}\left[a'_{k-1} + a'_{k+1} - \frac{15\sqrt{2}}{4} \times a_k\right]$$

6) Amplitudes of the intersymbol interference.

These are given in Table 3.6 for the different possible intersymbol interferences.

$m'_l = \{a'_{k-1}, a_k, a'_{k+1}\}$	$I'_{m'_l}(kT)$
-1 -1 -1	$-0.788\,V$
-1 -1 1	$-1.25\,V$
-1 1 -1	$1.72\,V$
-1 1 1	$1.25\,V$
1 -1 -1	$-1.25\,V$
1 -1 1	$-1.72\,V$
1 1 -1	$1.25\,V$
1 1 1	$0.778\,V$

Table 3.6. *Amplitudes of intersymbol interference $I'_{m'_l}(kT)$ for the different possible interfering messages m'_l*

7) Calculation of the optimal threshold. We have:

$$\mu'_{0,e} = \mu_{0,e} = \mu_{0opt,e} = \frac{\sigma_e^2}{2Vp(0)} \log_e \left[\frac{p_{-1}}{p_1}\right] = \frac{\sigma_e^2}{2V \times 1.25} \log_e \left[\frac{0.35}{0.65}\right]$$
$$\cong -\frac{0.247}{V} \sigma_e^2$$

$$\left[\frac{S}{b}\right]_{dB} = 20 \times \log_{10} \left[\frac{Vp(0)}{\sigma_e}\right] = 9.6 \text{ dB} \rightarrow \frac{Vp(0)}{\sigma_e} \cong 3 \rightarrow V \cong \frac{3\sigma_e}{1.25}$$

hence:

$$\mu_{0,e} \cong -0.103\, \sigma_e$$

8) Without noise, $\Im\left[s_{c,e}(kT)\right]$ is written:

$$\Im\left[s_{c,e}(kT)\right] = Va'_k p(0) + I'_{m'_l}(kT) = 1.25\, Va'_k + I'_{m'_l}(kT)$$
$$= 3\sigma_e \times a'_k + I'_{m'_l}(kT)$$

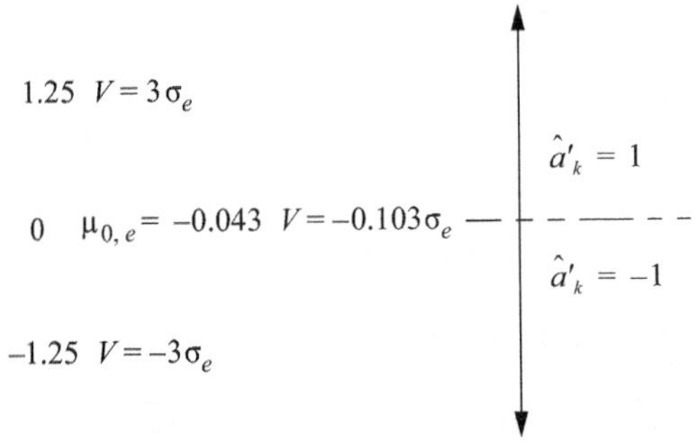

Figure 3.6. *Sample value $a'_k V$, optimal threshold and estimation classes*

From the graph represented in Figure 3.6 and previous results on the values of $I'_{m'_l}(kT)$, we find that for taking a wrong decision, it suffices that:

$$\left|I'_{m'_l}(kT)\right| > 1.25\, V$$

Specifically, the probability of taking an erroneous decision on a'_k is (see Table 3.6):

$$Pr\{\text{of taking a wrong decision on } [a'_k = \pm 1]\} = 1/8 = 0.125$$

This is a very large value compared to a usual probability of error.

9) New expression of the intersymbol interference $I'_{m'_l}(kT)$. We now have:

$$p(t) = \frac{\sin[\pi\, t/T]}{\pi\, t/T}$$

thus:

$$p(0) = 1\,;\ p(\pm iT) = 0\ \ \forall\ \ i \neq 0\,;\ \ i\text{: integer}$$

And:

$$q(t) = \frac{\alpha}{2} \times \frac{\sin[\pi\, t/T]}{\pi\, t/T}$$

thus:

$$q(0) = \frac{\alpha}{2} = 0.125;\ \ q(\pm iT) = 0\ \ \forall\ \ i \neq 0\,;\ \ i\text{: integer}$$

hence:

$$I'_{m'_l}(kT) = Va_k \times q(0) = 0.125 \times Va_k = 0.3\,\sigma_e\, a_k$$

The amplitudes of the intersymbol interference for the different possible interfering messages are given in Table 3.7.

l	$m'_l = a_k$	$I'_{m'_l}(kT) = 0.3\,\sigma_e\, a_k$
1	-1	$-0.3\,\sigma_e$
2	1	$0.3\,\sigma_e$

Table 3.7. *Amplitude of intersymbol interference* $I'_{m'_l}(kT)$ *for the different possible interfering messages* m'_l

10) Expressions of the conditional probabilities of error:

$$P_{e'_{-1}} = P_{e'(1/-1,m'_l)} = \frac{1}{\sigma_e\sqrt{2\pi}} \int\limits_{\mu_{0,e}+Vp(0)-I'_{m'_l}(kT)}^{\infty} \exp\left[-\frac{1}{2}\frac{b_{e,q}^2}{\sigma_e^2}\right] db_{e,q}$$

$$P_{e'_1} = P_{e'(-1/1,m'_l)} = \frac{1}{\sigma_e\sqrt{2\pi}} \int_{-\infty}^{\mu_{0,e}-\left[Vp(0)+I'_{m'_l}(kT)\right]} \exp\left[-\frac{1}{2}\frac{b_{e,q}^2}{\sigma_e^2}\right] db_{e,q}$$

Hence the probabilities of error on a'_k conditional on the interfering messages given in Table 3.8.

$m'_l = a_k$	$\mu_{0,e} + Vp(0) - I'_{m'_l}$	$P_{e'_{-1}}$	$\mu_{0,e} - \left[Vp(0) + I'_{m'_l}\right]$	$P_{e'_1}$
-1	$[-0.103 + 3 + 0.3]\sigma_e$ $\cong 3.2\,\sigma_e$	7×10^{-4}	$[-0.103 - 3 + 0.3]\sigma_e$ $\cong -2.8\,\sigma_e$	2.6×10^{-3}
1	$[-0.103 + 3 - 0.3]\sigma_e$ $\cong 2.6\,\sigma_e$	4.7×10^{-3}	$[-0.103 - 3 - 0.3]\sigma_e$ $\cong -3.4\,\sigma_e$	3×10^{-4}

Table 3.8. *Probabilities of error on a'_k conditional on interfering messages m'_l*

11) Probability of error on symbol a'_k:

$$P_{e,a'} = p'_{-1}\left[\sum_{l=1}^{2} p_{m'_l} \times P_{e'_{-1}}\right] + p'_1\left[\sum_{l=1}^{2} p_{m'_l} \times P_{e'_1}\right]$$

The number of interfering messages on a'_k is M^L, with:

$$M = 2;\ L = 1 \rightarrow M^L = 2^1 = 2$$

$$p_{m'_l} = p_{a_k} \rightarrow \begin{cases} p_{m'_1} = p_{-1} = 0.35 \ \text{for} \ a_k = -1 \\ p_{m'_2} = p_1 = 0.65 \ \text{for} \ a_k = 1 \end{cases}$$

$$P_{e,a'} = 0.35[0.35 \times 7 \times 10^{-4} + 0.65 \times 4.7 \times 10^{-3}]$$

$$+0.65[0.35 \times 2.6 \times 10^{-3} + 0.65 \times 3 \times 10^{-4}]$$

$$= 1.873 \times 10^{-3}$$

3.5. Problem 31 – Digital transmissions with 4-QAM digital modulation: case of single and double paths propagation

The 4-QAM digital modulation (also called QPSK modulation) transmission system given by the block diagram of Figure 3.7 is considered. In this kind of modulation symbols a_n and a'_n take their values on $\{1, -1\}$.

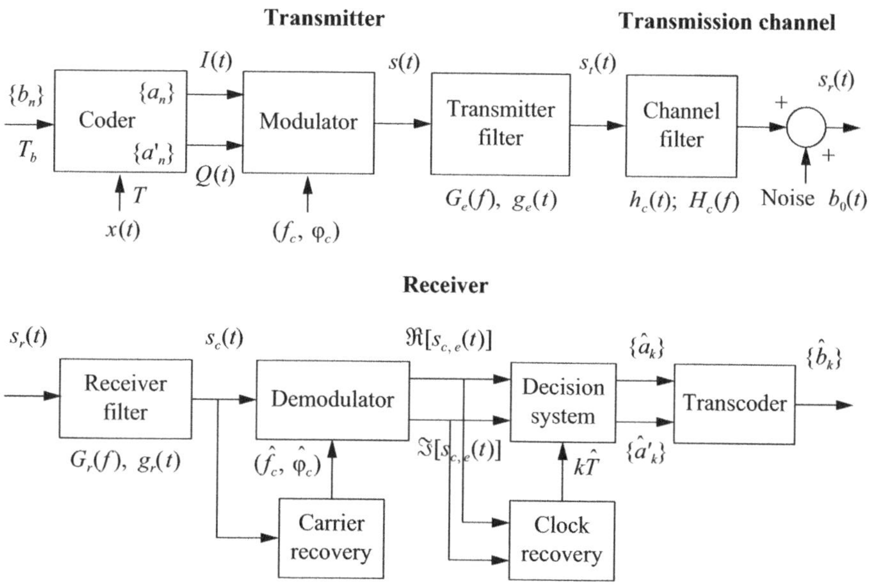

Figure 3.7. *Block diagram of a digital transmission system with QAM modulation*

A. *Case of a single path propagation*

The symbols b_n delivered by the binary source are emitted every T_b seconds. The baseband encoder of the transmission system generates two baseband signals $I(t)$ and $Q(t)$ as follows:

– separation of the binary sequence $\{b_n\}$ into two binary sequences $\{b_{2n}\}$ and $\{b_{2n+1}\}$;

– transcoding the sequences $\{b_{2n}\}$ and $\{b_{2n+1}\}$ into two sequences of symbols $\{a_n\}$ and $\{a'_n\} \in \{1, -1\}$;

– pulse amplitude modulation using the basic pulse $x(t)$:

$$x(t) = \begin{cases} 1 & \text{for } t \in [-T/2, T/2[\\ 0 & \text{elsewhere} \end{cases}$$

The two baseband signals $I(t)$ and $Q(t)$ are expressed by:

$$I(t) = \sum_n a_n x(t - nT) \quad \text{and} \quad Q(t) = \sum_n a'_n x(t - nT)$$

and indicate that the symbols $\{a_n\}$ and $\{a'_n\}$ are emitted every T seconds.

The modulator is defined by the carrier signal:

$$p_c(t) = V\exp[j(2\pi f_c t + \varphi_c)]$$

It constructs a real signal $s(t)$ by adequate linear modulation of the digital signal to be modulated:

$$c(t) = I(t) + jQ(t)$$

The linear transmission filter (of complex gain $G_e(f)$) and reception filter (of complex gain $G_r(f)$) are of the band-pass type around frequencies f_c and $-f_c$. The transmission channel is assumed to be modeled by a linear filter (of complex gain $H_c(f)$) to which is added an observation noise $b_0(t)$. This latter is assumed to be a Gaussian white random process, with power spectral density equal to $\Gamma_0/2$ in the frequency band covered by the receiving band-pass filter.

Moreover, let:

$$H(f) = G_e(f) \times H_c(f) \times G_r(f)$$

and:

$$h(t) = g_e(t) \otimes h_c(t) \otimes g_r(t)$$

be assumed real.

1) Write the real signal $s(t)$ obtained at the output of the modulator in quadrature at $M_1 \times M_2$ states. Show explicitly in the expression of the real signal $s(t)$ the two components modulating in quadrature the carrier $p_c(t)$.

2) Determine its complex envelope, denoted $s_e(t)$.

From now on, it is assumed that the amplitude spectrum of the transmitted digital modulation signal has a frequency band limited to:

$$|f| \in \left[f_c - \frac{\Delta f}{2}, f_c + \frac{\Delta f}{2}\right] \text{ with } \Delta f \ll f_c$$

We call $H_e(f)$ the frequency gain of the low-pass filter equivalent to the total filter of frequency gain $H(f)$ (which is a band-pass filter around the frequencies f_c and $-f_c$) and $y_e(t)$ the signal at the output of the equivalent low-pass filter (frequency gain $H_e(f)$) when the input signal is the previous signal $x(t)$ (noise excepted). We denote:

$$y_e(t) = p(t) + jq(t)$$

In the same way, by denoting $b(t)$, the noise $b_0(t)$ filtered by the receiving band-pass filter, we note:

$$b_e(t) = b_{e,p}(t) + jb_{e,q}(t)$$

the complex envelope of the filtered noise $b(t)$. The two components of the noise $b_e(t)$ (real and imaginary parts) are assumed to be Gaussian, zero mean, of the same variance σ_e^2 and decorrelated.

3) Determine the complex envelope, denoted $s_{c,e}(t)$, of the signal $s_c(t)$ at the output of the equivalent low-pass filter of frequency gain $H_e(f)$, and in taking the noise $b_e(t)$ into account.

4) Particularize this one at the decision instants $t_k = kT$, that is $s_{c,e}(kT)$.

5) By separating the real and imaginary parts of the complex envelope $s_{c,e}(kT)$, determine the intersymbol interference on the symbol a_k, denoted $I_{m_l}(kT)$, on the one hand, and on the symbol a'_k, denoted $I'_{m'_l}(kT)$, on the other hand.

6) In the following, we assume that the characteristics of the band-pass filters and of the channel are such that the equivalent low-pass filter $H_e(f)$ satisfies the Hermitian symmetry. Show that $y_e(t)$ is then a real signal ($q(t) = 0$).

7) Then give the new expression of the signal $s_{c,e}(kT)$.

Moreover, the amplitude spectrum $P(f)$ is:

$$P(f) = F\{p(t)\} = \begin{cases} \alpha T \text{ for } f \in \left[\frac{-1}{2T}, \frac{1}{2T}\right] \\ 0 \text{ elsewhere} \end{cases}$$

8) Show that the intersymbol interference $I_{m_l}(kT)$ is null.

The signal-to-noise ratio:

$$\left[\frac{S}{b}\right]_{db} = 20 \times \log_{10}\left[\frac{Vp(0)}{\sigma_e}\right]$$

after demodulation of each of the $I(t)$ and $Q(t)$ signals is equal to 14 dB.

9) For independent and equiprobable binary symbols b_n, calculate the probability of error $P_{e,a}$ on symbol a_k.

B. *Case of a double path propagation (a direct and a delayed transmission path)*

In this section, we will keep the hypothesis that the equivalent low-pass filter has an even frequency gain $H_e(f)$ (hence $q(t) = 0$). The intersymbol interference $I_{m_l}(kT)$ *due to the delayed path* is assumed to be limited only to symbols adjacent to a given symbol a_k (so symbols a_{k-1} and a_{k+1}).

Similarly, the inter-symbol interference $I'_{m'_l}(kT)$ *also due to the delayed path* is assumed to be limited only to symbols adjacent to a given symbol a'_k (so symbols a'_{k-1} and a'_{k+1}).

The equivalent baseband signal $s_{c,e}(t)$ is now considered to be the sum of two signals:

– the first, denoted $s^D_{c,e}(t)$, corresponds to a transmission through the direct path;

– the second signal corresponds to a delayed transmission (with a delay τ) due to a complex reflection.

Under these conditions, the complex envelope of the resulting signal received can be written in the form:

$$s_{c,e}(t) = s^D_{c,e}(t) + \rho \times \exp[-j2\pi f_c \tau] \times s^D_{c,e}(t-\tau)$$

with: $\rho = |\rho|\exp[j\varphi]$ a complex reflection coefficient.

A record of the signal $s_{c,e}(kT)$ shows that the amplitude of the latter oscillates between two levels:

– a maximum level (summation in phase of the two components);

– a minimum level (summation in opposite phase of the two components)

and that the maximum dynamics between these two levels is given by the following expression (standing wave ratio):

$$\frac{1+|\rho|}{1-|\rho|} = 3$$

For the sake of simplification, we assume that the phase shift φ due to the complex reflection coefficient ρ compensates for that, due to the difference in the path lengths $[-2\pi f_c \tau]$ for given values of τ and f_c.

10) Determine the reflection coefficient modulus $|\rho|$.

11) Give the new expression of the signal $s_{c,e}(kT)$ (at the instant $t_k = kT$).

12) By separating real and imaginary parts of the complex envelope $s_{c,e}(kT)$, determine the intersymbol interference on symbol a_k, denoted $I_{m_l}(kT)$, on the one hand, and on symbol a'_k, denoted $I'_{m'_l}(kT)$, on the other hand.

13) Taking $\tau = T/2$, rewrite the new expression of intersymbol interference $I_{m_l}(kT)$ and calculate its value for the different messages m_l (to simplify, you will take $\pi \cong 3$).

14) Is the digital link affected compared to that of part A? If yes, why? Justify your answer.

15) Calculate the new probability of error on symbol a_k (equiprobable messages).

NOTE.– If X is a Gaussian random process, with mean value m and standard deviation σ, you will take:

$$Pr\{|X-m| > 5\,\sigma\} = 2 \times 10^{-6}\,; \quad Pr\{|X-m| > 5.55\,\sigma\} = 10^{-6}$$

$$Pr\{|X-m| > 4.44\,\sigma\} = 2 \times 10^{-5}\,; \quad Pr\{|X-m| > 6\,\sigma\} \cong 0$$

Solution of problem 31

A. Case of a single path propagation

1) The real signal $s(t)$ at the output of the modulator is written:

$$s(t) = \Re\left\{V\sum_n [a_n + ja'_n]\,x(t-nT)\exp[j(2\pi f_c t + \varphi_c)]\right\}$$

$$s(t) = V\sum_n a_n\,x(t-nT)\cos(2\pi f_c t + \varphi_c)$$

$$-V\sum_n a'_n\,x(t-nT)\sin(2\pi f_c t + \varphi_c)$$

2) By definition, $s(t)$ is also written:

$$s(t) = \Re\{s_e(t) \times \exp[j(2\pi f_c t + \varphi_c)]\}$$

Hence the complex envelope of signal $s(t)$ is:

$$s_e(t) = V\sum_n [a_n + ja'_n]\,x(t - nT)$$

3) Expression of the signal $s_{c,e}(t)$: since the action of $x(t)$ at the input of the equivalent baseband system is $y_e(t) = p(t) + jq(t)$ at the output of filter of frequency gain $H_e(f)$ (noise-free), and taking into account the response of filtered noise turned into its equivalent baseband noise, that is $b_e(t)$, the response $s_{c,e}(t)$ is then:

$$s_{c,e}(t) = V\sum_n [a_n + ja'_n]\,[p(t - nT) + jq(t - nT)] + b_e(t)$$

4) For $t = t_k = kT$, $s_{c,e}(kT)$ is written:

$$s_{c,e}(kT) = V\sum_n [a_n + ja'_n]\,\{p[(k - n)T] + jq[(k - n)T]\} + b_e(kT)$$

5) Intersymbol interference on a_k and a'_k respectively

$$\Re[s_{c,e}(kT)] = V\left\{a_k p(0) + \sum_{n\neq k} a_n p[(k - n)T] - \sum_n a'_n q[(k - n)T]\right\}$$
$$+ b_{e,p}(kT)$$
$$\Im[s_{c,e}(kT)] = V\left\{a'_k p(0) + \sum_{n\neq k} a'_n p[(k - n)T] + \sum_n a_n q[(k - n)T]\right\}$$
$$+ b_{e,q}(kT)$$

hence:

$$I_{m_l}(kT) = V\left[\underbrace{\sum_{n\neq k} a_n p[(k - n)T]}_{\text{ISI intra-channel}} - \underbrace{\sum_n a'_n q[(k - n)T]}_{\text{ISI inter-channel}}\right]$$

$$I'_{m'_l}(kT) = V\left[\underbrace{\sum_{n\neq k} a'_n p[(k - n)T]}_{\text{ISI intra-channel}} + \underbrace{\sum_n a_n q[(k - n)T]}_{\text{ISI inter-channel}}\right]$$

6) The filter of frequency gain $H_e(f)$ satisfies the Hermitian symmetry, then $h_e(t)$ is a real function. Therefore $y_e(t) = x(t) \otimes h_e(t) = p(t)$ is a real function ($q(t) = 0$).

7) In this case, the signal $s_{c,e}(kT)$ becomes:

$$s_{c,e}(kT) = V \sum_n [a_n + ja'_n] \{p[(k-n)T]\} + b_e(kT)$$

8) Show that the intersymbol interference $I_{m_l}(kT)$ is null.

We have:

$$q(t) = 0 \rightarrow I_{m_l}(kT) = V \left[\sum_{n \neq k} a_n p[(k-n)T] \right]$$

which depends on $p(t)$. And $p(t)$ is such that:

$$p(t) = F^{-1}\{P(f)\} = \alpha \times \frac{\sin[\pi t/T]}{\pi t/T} \rightarrow p(0) = \alpha;\ p(\pm iT) = 0\ \forall\ i \neq 0;\ i\text{: integer}$$

hence: $I_{m_l}(kT) = 0$.

9) Calculation of the probability of error $P_{e,a}$:

$$\Re\left[s_{c,e}(kT)\right] = V a_k p(0) + b_{e,p}(kT)$$

hence:

$$P_{e,a} = p_{-1} \times P_{e_{-1}} + p_1 \times P_{e_1}$$

Indeed, since the ISI is zero, the notion of interfering messages disappears and we have:

$$p_{-1} = p_1 = 1/M = 1/2;\quad \mu_{0,e} = 0$$

$$\rightarrow P_{e,a} = \frac{1}{2}\left[P_{e_{-1}} + P_{e_1}\right]$$

with:

$$P_{e_{-1}} = P_{e(1/-1,m_l)} = \frac{1}{\sigma_e \sqrt{2\pi}} \int_{Vp(0)}^{\infty} \exp\left[-\frac{1}{2}\frac{b_{e,p}^2}{\sigma_e^2}\right] db_{e,p}$$

$$P_{e_1} = P_{e(-1/1,m_l)} = \frac{1}{\sigma_e\sqrt{2\pi}} \int_{-\infty}^{-Vp(0)} \exp\left[-\frac{1}{2}\frac{b_{e,p}^2}{\sigma_e^2}\right] db_{e,p}$$

The signal-to-noise ratio is:

$$\left[\frac{S}{b}\right]_{dB} = 20 \times \log_{10}\left[\frac{Vp(0)}{\sigma_e}\right] = 14 \text{ dB} \rightarrow Vp(0) \cong 5\,\sigma_e$$

Hence:

$$P_{e,a} = \frac{1}{2}\left[P_{e_{-1}} + P_{e_1}\right] = \frac{1}{2}\left[10^{-6} + 10^{-6}\right] = 10^{-6}$$

B. Case of a double path propagation (a direct and a delayed transmission path)

10) Reflection coefficient modulus:

$$\frac{1+|\rho|}{1-|\rho|} = 3 \rightarrow |\rho| = 1/2$$

11) New expression of $s_{c,e}(kT)$:

$$s_{c,e}(kT) = V\sum_n [a_n + ja'_n]\{p[(k-n)T] + |\rho|p[(k-n)T-\tau]\} + b_e(kT)$$

12) Determination of the intersymbol interferences on a_k and a'_k . We have:

$$\Re[s_{c,e}(kT)] = V\left\{a_k[p(0) + |\rho|p(-\tau)] + \sum_{n=k-1;\ n\neq k}^{k+1} a_n\, p[(k-n)T]\right.$$

$$\left. +|\rho| \sum_{n=k-1;\ n\neq k}^{k+1} a_n\, p[(k-n)T-\tau]\right\} + b_{e,p}(kT)$$

The terms of the first summation are null because $p(\pm iT) = 0\ \forall\, i \neq 0; i$: integer, hence:

$$I_{m_l}(kT) = V|\rho|[a_{k-1}p(T-\tau) + a_{k+1}p(-T-\tau)]$$

And we have:

$$\Im[s_{c,e}(kT)] = V\left\{a'_k[p(0) + |\rho|p(-\tau)] + \sum_{n=k-1;\ n\neq k}^{k+1} a'_n\, p[(k-n)T]\right.$$

$$\left.+|\rho| \sum_{n=k-1;\ n\neq k}^{k+1} a'_n\, p[(k-n)T-\tau]\right\} + b_{e,q}(kT)$$

The terms of the first summation are null because $p(\pm iT) = 0\ \forall\, i \neq 0$; i: integer, hence:

$$I'_{m'_l}(kT) = V|\rho|[a'_{k-1}p(T-\tau) + a'_{k+1}p(-T-\tau)]$$

13) For $\tau = T/2$:

$$I_{m_l}(kT) = V|\rho|[a_{k-1}p(T/2) + a_{k+1}p(-3T/2)]$$

From the result of question 8:

$$p(\pm T/2) = \frac{2\alpha}{\pi} \cong \frac{2\alpha}{3};\ \ p(\pm 3T/2) = \frac{-2\alpha}{3\pi} \cong \frac{-2\alpha}{9}$$

And from the results of questions 9, 10 and 13, we have:

$$|\rho| = \frac{1}{2};\ \ \alpha V = p(0)V = 5\,\sigma_e$$

$$\rightarrow I_{m_l}(kT) = \frac{1}{2}V\left[\frac{2\alpha}{3}a_{k-1} - \frac{2\alpha}{9}a_{k+1}\right] = \frac{\alpha V}{9}[3a_{k-1} - a_{k+1}]$$

Hence, the amplitudes of the new intersymbol interference are given in Table 3.9.

$\boldsymbol{m_l = \{a_{k-1}, a_{k+1}\}}$	$\boldsymbol{I_{m_l}(kT) = \frac{\alpha V}{9}[3a_{k-1} - a_{k+1}]}$
-1 -1	$\frac{-2}{9}\alpha V = \frac{-10}{9}\sigma_e$
-1 1	$\frac{-4}{9}\alpha V = \frac{-20}{9}\sigma_e$
1 -1	$\frac{4}{9}\alpha V = \frac{20}{9}\sigma_e$
1 1	$\frac{2}{9}\alpha V = \frac{10}{9}\sigma_e$

Table 3.9. *Amplitudes of the new intersymbol interference* $I_{m_l}(kT)$

14) From the results of questions 12 and 13, the digital link is well affected because the ISI, caused by the delayed path via the function $p(t-\tau)$, is non-zero.

15) New probability of error on symbol a_k:

$$P_{e,a} = p_{-1}\left[\sum_{l=1}^{4} p_{m_l} \times P_{e_{-1}}\right] + p_1\left[\sum_{l=1}^{4} p_{m_l} \times P_{e_1}\right]$$

From the fact that the symbols are independent and equiprobable, we have:

$$p_{-1} = p_1 = 1/M = 1/2; \quad p_{m_l} = 1/M^L = 1/2^2 = 1/4$$

$$P_{e,a} = \frac{1}{8}\sum_{l=1}^{4}\left[P_{e_{-1}} + P_{e_1}\right]$$

with respectively:

$$P_{e_{-1}} = P_{e(1/-1,m_l)} = \frac{1}{\sigma_e\sqrt{2\pi}} \int_{V[p(0)+|\rho|p(-T/2)]-I_{m_l}(kT)}^{\infty} \exp\left[-\frac{1}{2}\frac{b_{e,p}^2}{\sigma_e^2}\right] db_{e,p}$$

$$P_{e_1} = P_{e(-1/1,m_l)} = \frac{1}{\sigma_e\sqrt{2\pi}} \int_{-\infty}^{-V[p(0)+|\rho|p(-T/2)]-I_{m_l}(kT)} \exp\left[-\frac{1}{2}\frac{b_{e,p}^2}{\sigma_e^2}\right] db_{e,p}$$

We have:

$$V[p(0) + |\rho|p(-T/2)] = V\left[\alpha + \frac{\alpha}{3}\right] = \frac{4}{3}\alpha V = \frac{4}{3} \times 5\,\sigma_e = \frac{20}{3}\,\sigma_e$$

hence the calculations of the conditional probability intervals and the associated conditional probabilities for symbol a_k presented in Table 3.10.

Hence, finally, the new probability of error on symbol a_k is:

$$P_{e,a} = \frac{1}{8} \times \sum_{l=1}^{4}\left[P_{e_{-1}} + P_{e_1}\right] = \left[10^{-5} + \frac{1}{2} \times 10^{-6} + \frac{1}{2} \times 10^{-6} + 10^{-5}\right]$$

$$= 2.625 \times 10^{-6}$$

$m_l = \{a_{k-1}, a_{k+1}\}$	$V[p(0) + \lvert\rho\rvert p(-T/2)] - I_{m_l}(kT)$	$P_{e_{-1}}$	$-V[p(0) + \lvert\rho\rvert p(-T/2)] - I_{m_l}(kT)$	P_{e_1}
-1 -1	$\frac{20}{3}\sigma_e + \frac{10}{9}\sigma_e = 7.77\,\sigma_e$	0	$-\left[\frac{20}{3}\sigma_e - \frac{10}{9}\sigma_e\right] = -5.55\,\sigma_e$	$\frac{10^{-6}}{2}$
-1 1	$\frac{20}{3}\sigma_e + \frac{20}{9}\sigma_e = 8.88\,\sigma_e$	0	$-\left[\frac{20}{3}\sigma_e - \frac{20}{9}\sigma_e\right] = -4.44\,\sigma_e$	10^{-5}
1 -1	$\frac{20}{3}\sigma_e - \frac{20}{9}\sigma_e = 4.44\,\sigma_e$	10^{-5}	$-\left[\frac{20}{3}\sigma_e + \frac{20}{9}\sigma_e\right] = -8.88\,\sigma_e$	0
1 1	$\frac{20}{3}\sigma_e - \frac{10}{9}\sigma_e = 5.55\,\sigma_e$	$\frac{10^{-6}}{2}$	$-\left[\frac{20}{3}\sigma_e + \frac{10}{9}\sigma_e\right] = -7.77\,\sigma_e$	0

Table 3.10. *Conditional probability intervals and the associated conditional probabilities for symbol a_k*

3.6. Problem 32 – Performance of digital modulations and 16-QAM digital modulation

For the digital radio modulations displayed in Table 3.11 hereinafter, operating at the same bitrate $D_b = 12$ Mbit/s and the same carrier frequency f_c (90 MHz), determine:

1) The fundamental frequency f_a (MHz), the necessary minimum frequency band of channel f_N (MHz) for transmitting, without intersymbol interference, the corresponding symbol rate D_s (Msymbol/s), the lower value of the frequency band (MHz) (minimum lower side frequency: LSF), and the upper value of the frequency band (MHz) (maximum upper side frequency: USF).

Modulation	f_a (MHz)	f_N (MHz) $= D_s$ (Msymbol/s)	LSF (MHz)	USF (MHz)
BPSK				
QPSK				
OQPSK				
8-PSK				
16-QAM				

Table 3.11. *Comparative characteristics of the minimum frequency bands required for different types of digital modulations*

The choice of a linear modulation method depends on several parameters (bitrate, channel bandwidth, noise, phase jitter, maximum cost). The constellation diagram provides an approximate but rapid comparative evaluation of the performance of various linear modulation methods with respect to noise and phase jitter.

The probability of error in the presence of noise depends on the type of modulation, the characteristics of the channel, those of the filters contained in the transmitter and the receiver and the nature of the noise. Only the influence of the type of modulation defined by its constellation diagram will be considered in the following, the noise being assumed to be a Gaussian white random process and the filters conforming to the first Nyquist criterion.

The comparison of different modulation methods requires the representation of their constellation diagrams in a scale ratio such that the average powers are equal. In all cases considered, the average power is equal to V^2 and they have the same symbol rate: $D_s = 1/T$.

On the other hand, if the diagram contains n symbols represented by points $M_1, M_2, \cdots, M_n$, if $V_1, V_2, \cdots, V_n$ are the corresponding signal amplitudes and if $p_1, p_2, \cdots, p_n$ are their associated probabilities, the average power P_m of the signal is given by:

$$P_m = \sum_{i=1}^{n} p_i \times V_i^2$$

In the case where all the symbols are equiprobable (case to be considered here), the mean power is the average of the square distances between the origin and each point of the constellation diagram: $P_m = E(d_i^2)$ where $E(\cdot)$ represents the statistical expectation.

Taking as a reference the following constellation diagram (Figure 3.8) of a 2-PSK modulation (also denoted BPSK modulation):

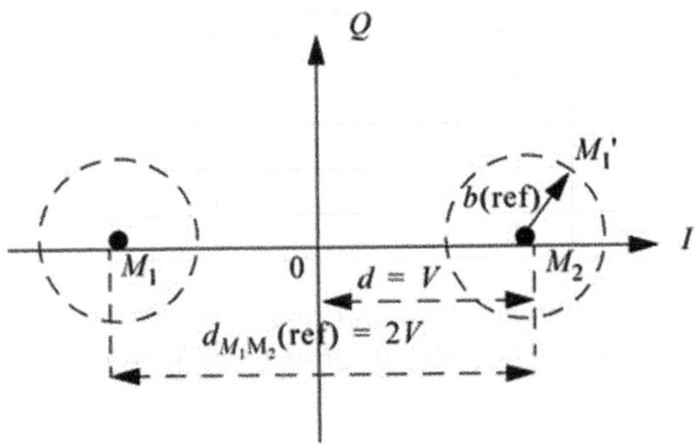

Figure 3.8. *Constellation diagram of a 2-PSK modulation (reference)*

2) Determine for each of the three constellation diagrams (4-ASK, 4-QAM (QPSK), 16-QAM) presented in Figure 3.9:

a) the distance d;

b) the distance between two adjacent states d_{M_i,M_j} ; $i \neq j$ expressed in d_{M_1,M_2}(ref);

c) the decrease δ in dB of the signal-to-noise ratio (S/B) with respect to the 2-PSK modulation.

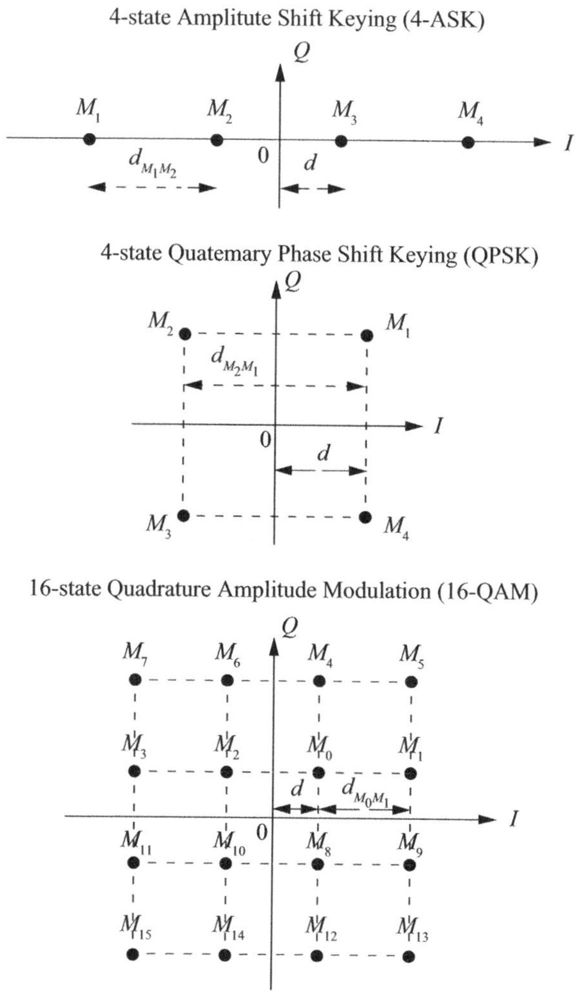

Figure 3.9. *Constellation diagrams of 4-ASK, 4-QAM (QPSK) and16-QAM digital modulations*

We now want to realize a 16-QAM modulation whose polar coordinates of the first state M_0 (state 0000) are:

$$\left[\rho_n = \sqrt{18}\ ;\ \psi_n = 45\,°\right]$$

3) Determine the pair of amplitudes $[a_n\ ;\ a'_n]$ allowing homogeneous distribution of the 16 states in the plane $\{I, Q\}$.

The carrier is defined by:

$$p_c(t) = V\exp[j(2\pi f_c t + \varphi_c)]$$

and the signal $x(t)$ is given by:

$$x(t) = \begin{cases} 1 & \text{for } t \in [-T/2, T/2[\\ 0 & \text{elsewhere} \end{cases}$$

4) What is the expression of the real signal $s(t)$ at the output of the modulator? Give its complex envelope $s_e(t)$.

We call $H_e(f)$ the frequency gain of the low-pass filter equivalent to the total filter of frequency gain $H(f)$ (which is a band-pass filter around the frequencies f_c and $-f_c$) and $y_e(t)$ the output of the low-pass filter of frequency gain $H_e(f)$ when the input signal is the previous signal $x(t)$ (noise excepted). We denote:

$$y_e(t) = p(t) + jq(t)$$

In the same way, by denoting $b(t)$, the noise $b_0(t)$ filtered by the receiving band-pass filter, we denote by $b_e(t)$ the complex envelope of $b(t)$:

$$b_e(t) = b_{e,p}(t) + jb_{e,q}(t)$$

5) Determine the complex envelope, denoted $s_{c,e}(t)$, of the signal $s_c(t)$ at the output of the filter of frequency gain $H_e(f)$, taking into account the noise $b_e(t)$.

6) Give the expression of the signal $s_c(t)$ and its amplitude spectrum $S_c(f)$.

7) Give the block diagram of the 16-QAM modulator built from two 4-QAM (QPSK) modulators.

8) We note $s_1(t)$ the output of the first modulator 4-QAM: M1(R, J), and by $s_2(t)$ the output of the second modulator 4-QAM: M2(Q, I). Give the expression of the signal $s(t)$ at the output of the 16-QAM modulator.

Solution of problem 32

1) We have:

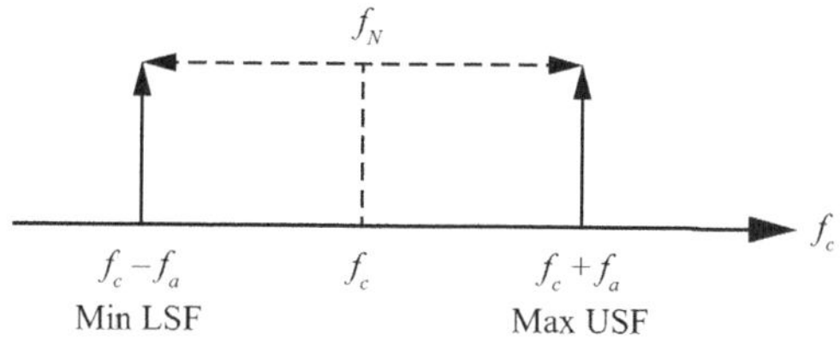

Figure 3.10. *Fundamental frequency characteristics of a digital modulation*

The minimum bandwidth f_N of a transmission channel in digital communication with carrier modulation is:

$$f_N = [f_c + f_a] - [f_c - f_a] = 2f_a = D_s \text{ and } D_s = \frac{D}{k} \text{ such that: } 2^k = M$$

The minimum lower side frequency (LSF) is:

$$\text{LSF} = f_c - f_a$$

The maximum upper side frequency (USF) is:

$$\text{USF} = f_c + f_a$$

According to the modulation used, we have the following results (Table 3.12):

Modulation	f_a **(MHz)**	f_N **(MHz)** $= D_s$ **(Msymbol/s)**	**LSF (MHz)**	**USF (MHz)**
BPSK	$\frac{D}{2} = 6$	$D = 12$	90 – 6 = 84	90 + 6 = 96
QPSK	$\frac{D}{4} = 3$	$\frac{D}{2} = 6$	90 – 3 = 87	90 + 3 = 93
OQPSK	$\frac{D}{2} = 6$	$D = 12$	90 – 6 = 84	90 + 6 = 96
8-PSK	$\frac{D}{6} = 2$	$\frac{D}{3} = 4$	90 – 2 = 88	90 + 2 = 92
16-QAM	$\frac{D}{8} = 1.5$	$\frac{D}{4} = 3$	90 – 1.5 = 88.5	90 + 1.5 = 91.5

Table 3.12. *Comparative characteristics of the minimum frequency bands required for different types of digital modulation*

2) We have for reference the constellation diagram (Figure 3.8) of a 2-PSK (BPSK) modulation, hence: $P_m = V^2$.

In the case of any digital modulation with the same average power V^2 and equiprobability of the constellation points (symbols), we have:

$$P_m = \sum_{i=1}^{n} p_i V_i^2 = \frac{1}{n}\sum_{i=1}^{n} V_i^2 = E(d_i^2) = V^2$$

– *4-ASK* modulation:

a) $\frac{1}{4} \times 2\left[d_{M_3}^2 + d_{M_4}^2\right] = \frac{1}{2}[d^2 + (3d)^2] = 5d^2 = V^2 \rightarrow d = \frac{V}{\sqrt{5}}$

b) $d_{M_1M_2} = 2d = \frac{2V}{\sqrt{5}} = \frac{d_{M_1M_2}(\text{ref})}{\sqrt{5}}$

c) $\delta = -20 \times \log_{10}\left(\frac{1}{\sqrt{5}}\right) \cong 7 \text{ dB}$

– *4-QAM (QPSK)* modulation:

a) $\frac{1}{4} \times 4\left[d_{M_1}^2\right] = 2d^2 = V^2 \rightarrow d = \frac{V}{\sqrt{2}}$

b) $d_{M_2M_1} = 2d = \sqrt{2}\,V = \frac{d_{M_1M_2}(\text{ref})}{\sqrt{2}}$

c) $\delta = -20 \times \log_{10}\left(\frac{1}{\sqrt{2}}\right) \cong 3 \text{ dB}$

– *16-QAM* modulation:

a) $d_{M_0}^2 = 2d^2\ ;\ d_{M_1}^2 = 10d^2\ ;\ d_{M_4}^2 = 10d^2\ ;\ d_{M_5}^2 = 18d^2$

$$\rightarrow \frac{1}{16} \times 4\left[d_{M_0}^2 + d_{M_1}^2 + d_{M_4}^2 + d_{M_5}^2\right] = \frac{1}{4}[40d^2] = 10d^2 = V^2 \rightarrow d = \frac{V}{\sqrt{10}}$$

b) $d_{M_0M_1} = 2d = \frac{2\,V}{\sqrt{10}} = \frac{d_{M_1M_2}(\text{ref})}{\sqrt{10}}$

c) $\delta = -20 \times \log_{10}\left(\frac{1}{\sqrt{10}}\right) \cong 10 \text{ dB}$

3) In the state M_0 (0 0 0 0), we have:

$$\left.\begin{matrix} \rho_n = \sqrt{18} \\ \psi_n = 45\,° \end{matrix}\right\} \rightarrow \begin{cases} a_n = \rho_n \cos(\psi_n) = 3 \\ a_n' = \rho_n \sin(\psi_n) = 3 \end{cases}$$

To have a homogeneous distribution of the 16 states in the (I, Q) plane, it is necessary to take:

$$\{a_n, a'_n\} = \{\pm 3, \pm 9\}$$

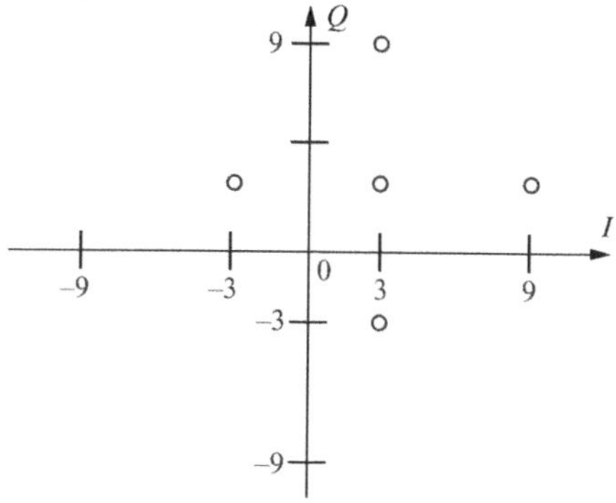

Figure 3.11. *Examples of positioning the points in the constellation diagram of the 16-QAM modulation*

4) The real signal $s(t)$ at the output of the modulator is written:

$$s(t) = \Re\left\{V \sum_n [a_n + ja'_n]\, x(t - nT) \exp[j(2\pi f_c t + \varphi_c)]\right\}$$

$$s(t) = \Re\{s_e(t) \times \exp[j(2\pi f_c t + \varphi_c)]\}$$

$$\rightarrow s_e(t) = V \sum_n [a_n + ja'_n]\, x(t - nT)$$

5) We have:

$$s_e(t) = V \sum_n [a_n + ja'_n]\, x(t - nT)$$

Since the action of $x(t)$ at the input of the equivalent baseband system is:

$$y_e(t) = p(t) + jq(t)$$

At the output of the filter of frequency gain $H_e(f)$ (noise-free), and taking into account the response of filtered noise turned into its equivalent baseband noise, i.e. $b_e(t)$, the response $s_{c,e}(t)$ is then:

$$s_{c,e}(t) = V\sum_n [a_n + ja'_n]\,[p(t-nT) + jq(t-nT)] + b_e(t)$$

6) Expression of signal $s_c(t)$ and its amplitude spectrum $S_c(f)$? We have:

$$s_c(t) = \Re\{s_{c,e}(t) \times \exp[j(2\pi f_c t + \varphi_c)]\}$$

$$s_c(t) = \Re\left\{\left[V\sum_n [a_n + ja'_n]\,[p(t-nT) + jq(t-nT)] + b_e(t)\right]\right.$$

$$\left.\times\,[\cos(2\pi f_c t + \varphi_c) + j\sin(2\pi f_c t + \varphi_c)]\right\}$$

$$\rightarrow s_c(t) = V\sum_n [a_n p(t-nT) - a'_n q(t-nT)] \times \cos(2\pi f_c t + \varphi_c)$$

$$-V\sum_n [a'_n p(t-nT) - a_n q(t-nT)] \times \sin(2\pi f_c t + \varphi_c) + b(t)$$

The signal $s_c(t)$ is also written:

$$s_c(t) = \frac{1}{2}\{s_{c,e}(t) \times \exp[j(2\pi f_c t + \varphi_c)] + s^*_{c,e}(t) \times \exp[-j(2\pi f_c t + \varphi_c)]\}$$

$$\rightarrow S_c(f) = \frac{1}{2}\{S_{c,e}(f) \otimes \delta(f - f_c) + S^*_{c,e}(-f) \otimes \delta(-f - f_c)\}$$

thus:

$$S_c(f) = \frac{1}{2}\{S_{c,e}(f - f_c) + S^*_{c,e}(-f - f_c)\}$$

7) See Figure 3.12. See also Volume 1, Chapter 7 for more details.

8) Expression of $s(t)$ at the output of the 16-QAM modulator based on two 4-QAM modulators:

$$s_1(t) = V[J(t)\cos(2\pi f_c t + \varphi_c) - R(t)\sin(2\pi f_c t + \varphi_c)]$$

$$s_2(t) = V[I(t)\cos(2\pi f_c t + \varphi_c) - Q(t)\sin(2\pi f_c t + \varphi_c)]$$

$$s(t) = s_1(t) + \frac{1}{2}s_2(t)$$

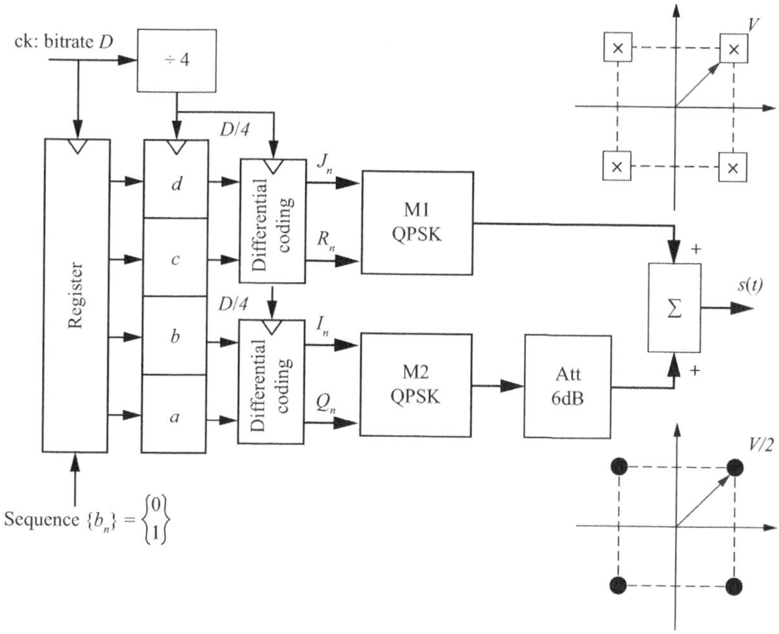

Figure 3.12. *Block diagram of the realization of the 16-QAM modulator constructed from two QPSK modulators*

3.7. Problem 33 – QAM encoding and transmission of motion information of digital video

The context is that of the transmission of coded video. Several categories of information are to be represented and coded in a compressed form. One of these categories is *motion information.*

Each frame I_t of a temporal sequence SI of digital frames $SI = \{\cdots, I_{t-1}, I_t, I_{t+1}, \cdots\}$ is divided into K macro-blocks MB_k, of size 16×16 pixels (we have $k = 1, 2, \cdots, K$). The SI sequence consists of L frames per second (typically in Europe $L = 25$). Each macro-block MB is associated with a motion vector (also called displacement vector) $\vec{D}$ which makes it possible to predict its content from the decoded preceding frame. $\vec{D}$ is a vector with two components d denoted (dx, dy), with values considered to be integers in this problem.

For simplicity, we suppose that in practice only seven values for dx, on the one hand, and for dy, on the other hand, are of significant probabilities. It is also assumed that dx and dy components have the same statistics (this is not true at all in practice). Each value of d (dx and dy) is associated with a symbol s. These values are given in Table 3.13 below with their associated probability.

Value	-3	-2	-1	0	1	2	3
s_i	s_1	s_2	s_3	s_4	s_5	s_6	s_7
Probability	0.0625	0.125	0.25	0.35	0.125	0.0625	0.025

Table 3.13. *Values and probabilities of a component d of the displacement vector* $\vec{D}$

1) a) Determine the entropy $H(d)$ of a component d (dx or dy) of the displacement vector $\vec{D}$ (we indicate: $\log_2(0.35) = -1.5146$ and $\log_2(0.025) = -5.322$).

b) Deduce the entropy $H(\vec{D})$ of the displacement vector $\vec{D}$ for a separate coding of dx and dy.

c) What would be the efficiency η_1 of a fixed-length code $C1$ (length L_1) encoding the components of displacement vector $\vec{D}$?

d) What is the bitrate per second D_{b_1} for a number of macro-blocks $K = 396$ per frame and with a frame rate per second $L = 25$ for encoding the displacement vectors $\vec{D}$?

2) Using in code $C1$ the natural binary coding in ascending order for encoding symbols s_i, determine the probability $p_{0,C1}$ of having a bit at zero in the bitstream of the displacement vectors $\vec{D}$. Deduce the probability $p_{1,C1}$ of having a bit at one.

3) Construct the Huffman code $C2$ giving the codeword S_i associated with each of the symbols s_i of a component d of vector $\vec{D}$.

NOTE.– In the construction of the code $C2$, the coding suffix associated with the element of greatest probability will systematically be set to 0.

4) Determine for code $C2$:

a) the average length $\bar{l}_2$ of the codewords of a displacement component;

b) the average length $\bar{L}_2$ of the codewords encoding vector $\vec{D}$ (again for a separate coding of dx and dy);

c) the efficiency η_2 of this code;

d) the average bitrate per second D_{b_2} for encoding vectors $\vec{D}$;

e) the probability $p_{0,C2}$ of having a bit at zero in the bitstream coding $\vec{D}$. Deduce the probability $p_{1,C2}$ of having a bit at one.

From now on, we consider the transmission of motion information (code $C2$) using the 4-QAM (QPSK) modulation (Figure 3.13 below).

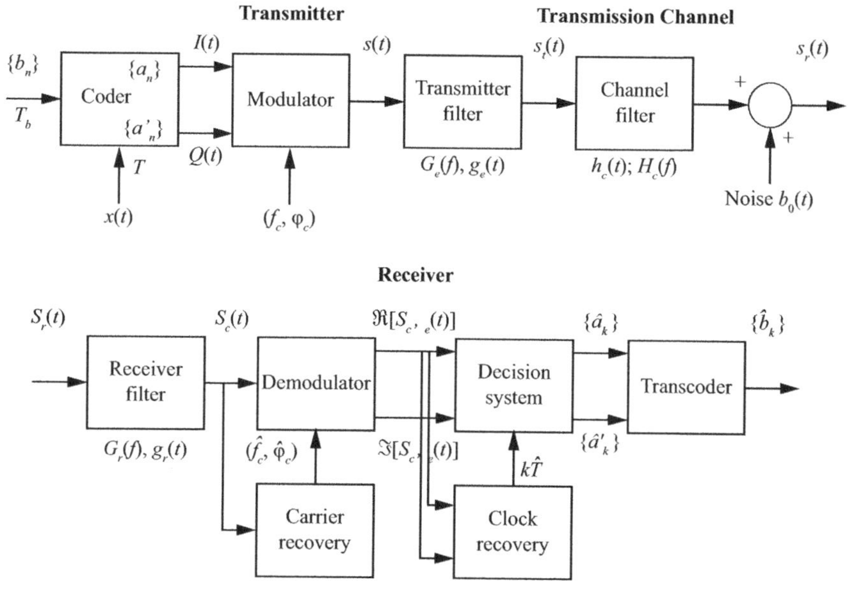

Figure 3.13. *Digital transmission system with quadrature amplitude modulation*

The symbols b_n are emitted every T_b seconds. The baseband encoder of the transmission system generates two baseband signals $I(t)$ and $Q(t)$ as follows:

– separation of the binary sequence $\{b_n\}$ into two binary sequences $\{b_{2n}\}$ and $\{b_{2n+1}\}$;

– transcoding the sequences $\{b_{2n}\}$ and $\{b_{2n+1}\}$ into two sequences of symbols $\{a_n\}$ and $\{a'_n\} \in \{1, -1\}$;

– pulse amplitude modulation using the basic pulse $x(t)$:

$$x(t) = \begin{cases} 1 & \text{for } t \in [-T/2, T/2[\\ 0 & \text{elsewhere} \end{cases}$$

The two baseband signals $I(t)$ and $Q(t)$ are expressed by:

$$I(t) = \sum_n a_n x(t - nT) \quad \text{and} \quad Q(t) = \sum_n a'_n x(t - nT)$$

and indicate that symbols $\{a_n\}$ and $\{a'_n\}$ are emitted every T seconds.

The modulator is defined by the carrier signal:

$$p_c(t) = V\exp[j(2\pi f_c t + \varphi_c)]$$

It constructs a real signal $s(t)$ by an adequate linear modulation of the digital signal to be modulated:

$$c(t) = I(t) + jQ(t)$$

We call $H_e(f)$ the frequency gain of the low-pass filter equivalent to the total filter of frequency gain $H(f)$ (which is a band-pass type around frequencies f_c and $-f_c$) and $y_e(t)$ the signal at the output of the equivalent low-pass filter (frequency gain $H_e(f)$) when the input signal is the previous signal $x(t)$ (noise excepted).

We note:

$$y_e(t) = p(t) + jq(t)$$

In the same way, by denoting $b(t)$ the noise $b_0(t)$ filtered by the receiving band-pass filter, we denote by $b_e(t)$ the complex envelope of the filtered noise $b(t)$:

$$b_e(t) = b_{e,p}(t) + jb_{e,q}(t)$$

In the following part of the problem, we assumed the two components (real and imaginary parts) of the complex envelope $b_e(t)$ to be Gaussian, with zero mean, of the same variance σ_e^2 and decorrelated.

5) a) Determine the complex envelope, noted $s_{c,e}(t)$, of the signal $s_c(t)$ at the output of the filter $H_e(f)$, and taking into account the noise $b_e(t)$. Particularize this later at the decision instants $t_k = kT$, that is $s_{c,e}(kT)$.

b) By separating the real and imaginary parts of the complex envelope $s_{c,e}(kT)$, determine for each part: the useful signal, the intersymbol interference (intra-channel, inter-channel) and the noise.

We consider that the amplitude spectrum $P(f)$ of signal $p(t)$ is a constant function, equal to T, on the frequency domain $[-1/2T, 1/2T]$, and zero elsewhere.

We consider also that the amplitude spectrum $Q(f)$ of signal $q(t)$ is 1/8 times that of $P(f)$.

6) a) Deduce precisely each of the 3 components of the real part of $s_{c,e}(kT)$. Do the same for the imaginary part of $s_{c,e}(kT)$.

Moreover, the signal-to-noise ratio after demodulation on each channel is:

$$\left[\frac{S}{b}\right]_{dB} = 20 \times \log_{10}\left[\frac{Vp(0)}{\sigma_e}\right] = 12.0412 \text{ dB}$$

b) Give the expression of the optimal threshold: $\mu_{0,e} = \mu_{0opt,e}$ and calculate its value as a function of the variance σ_e^2.

c) Calculate the various conditional probabilities of the following types:

$$Pr\{\hat{a}_k = a_j/a_k = a_i, m_l\}_{j\neq i}\,;\; a_j = \{-1,1\}\,;\; a_i = \{-1,1\}$$

necessary to calculate the probability of error $P_{e,a} = Pr\{\hat{a}_k \neq a_k\}$ on symbol a_k.

d) Deduce: $P_{e,a} = Pr\{\hat{a}_k \neq a_k\}$.

Similarly, deduce the probability of error $P_{e,a'} = Pr\{\hat{a}'_k \neq a'_k\}$ on symbol a'_k.

7) Finally, what is the probability P_d that a component d of the decoded motion vector is erroneous? Deduce the probability P_D that the decoded motion vector $\vec{D}$ is erroneous.

NOTE.– If X is a Gaussian random process, with zero mean ($m = 0$) and reduced standard deviation ($\sigma = 1$), we will assume that we have approximately:

$$Pr\{|X| > 3.57\} = 3.64 \times 10^{-4}\,;\;\; Pr\{|X| > 4.43\} = 4.4 \times 10^{-5}$$

$$Pr\{|X| > 4.57\} = 4 \times 10^{-6}\,;\;\; Pr\{|X| > 3.43\} = 6 \times 10^{-4}$$

Solution of problem 33

1) a) Entropy of a component:

$$H(d) = H(dx) = H(dy) = -\sum_{i=1}^{7} p(s_i)\, log_2\, p(s_i)$$

$$H(d) \cong -1.4427 \sum_{i=1}^{7} p(s_i) \log_e p(s_i)$$
$$\cong 2.41315 \text{ bits of information/component}$$

Recall:

$$\log_2 Z = \frac{\log_e Z}{\log_e 2} \cong 1.4427 \times \log_e Z$$

b) The components dx and dy have the same statistics and are coded separately, hence:

$$H(D) = 2 \times H(d) = 4.8263 \text{ bits of information/vector } \vec{D}$$

c) There are seven values per component dx and dy, then with a fixed-length code, 3 bits are required for coding dx and 3 bits for dy. Thus, $L_1 = 2 \times 3 = 6$ bits are required to encode the vector $\vec{D}$. The efficiency of the code is then:

$$\eta_1 = \frac{H(D)}{L_1} = \frac{4.8263}{6} \cong 80.44\%$$

d) The bitrate per second is:

$$D_{b_1} = L_1 \times K \times L = 6 \times 396 \times 25 = 59{,}400 \text{ bit/s}$$

2) Since dx and dy have the same statistics, one component is sufficient to determine the probability $p_{0,C1}$ of having a bit at zero.

s_i	s_1	s_2	s_3	s_4	s_5	s_6	s_7
$p(s_i)$	0.0625	0.125	0.25	0.35	0.125	0.0625	0.025
Code $C1$	000	001	010	011	100	101	110

Table 3.14. *Coding $C1$ of one component d of the displacement vector $\vec{D}$*

$$p_{0,C1} = \sum_{i=1}^{7} p(s_i) \times \frac{\text{Nb of zeros in } s_i}{l_i}$$

$$p_{0,C1} = \frac{1}{3}\{0.0625 \times 3 + 0.125 \times 2 + 0.25 \times 2 + 0.35 \times 1 + 0.125 \times 2$$
$$+0.0625 \times 1 + 0.025 \times 1\} = 0.54167$$

The probability of having a bit at 1 is then:

$$p_{1,C1} = 1 - p_{0,C1} = 0.45833$$

3) Huffman coding. Table 3.15 describes the Huffman coding process and the resulting code in the right-hand side column.

s_i	$p(s_i)_0$	$p(s_i)_1$	$p(s_i)_2$	$p(s_i)_3$	$p(s_i)_4$	$p(s_i)_5$	C_2
s_4	0.35	0.35	0.35	0.35	0.40	0.6 0	0 0
s_3	0.25	0.25	0.25	0.25	0.35 0	0.4 1	0 1
s_2	0.125	0.125	0.15	0.25 0	0.25 1		1 0 0
s_5	0.125	0.125	0.125 0	0.15 1			1 0 1
s_1	0.0625	0.0875 0	0.125 1				1 1 1
s_6	0.0625 0	0.0625 1					1 1 0 0
s_7	0.025 1						1 1 0 1

Table 3.15. *Huffman coding $C2$ of one component d of the displacement vector*

4) a) Average length of the codewords of code $C2$:

$$\overline{l_2} = \sum_{i=1}^{7} p(s_i) \times l_2(i)$$

$$\overline{l_2} = [(0.35 + 0.25) \times 2 + (0.125 + 0.125 + 0.0625) \times 3$$
$$+ (0.0625 + 0.025) \times 4] = 2.4875 \text{ bit/symbol}$$

b) Average length of the codewords in coding vector $\vec{D}$:

$$\overline{L_2} = 2 \times \overline{l_2} = 4.975 \text{ bit/vector } \vec{D}$$

c) The efficiency of code $C2$ is:

$$\eta_2 = \frac{H(D)}{\overline{L_2}} = \frac{4.8263}{4.975} \cong 97\%$$

d) The average bitrate D_{b_2} is:

$$D_{b_2} = \overline{L_2} \times K \times L = 4.975 \times 396 \times 25 = 49{,}252.5 \text{ bit/s}$$

e) Probability of having a bit at zero:

$$p_{0,C2} = \sum_{i=1}^{7} p(s_i) \times \frac{\text{Nb of zeros in } s_i}{l_i}$$

$$p_{0,C2} = \left[0.0625 \times \frac{0}{3} + 0.125 \times \frac{2}{3} + 0.25 \times \frac{1}{2} + 0.35 \times \frac{2}{2} + 0.125 \times \frac{1}{3}\right.$$

$$\left.+0.0625 \times \frac{2}{4} + 0.025 \times \frac{1}{4}\right] = 0.6375$$

The probability $p_{1,C2}$ of having a bit at 1 is then:

$$p_{1,C2} = 1 - p_{0,C2} = 0.3625$$

To summarize, the probability of issuing a bit at 0 (or at 1) is dependent on the code used, therefore the hypothesis that $Pr\{b = 0\} = Pr\{b = 1\} = 1/2$ is not realistic.

5) a) We have:

$$s_e(t) = V \sum_n [a_n + ja'_n]\, x(t - nT)$$

As the action of $x(t)$ at the input of the equivalent baseband system is $y_e(t) = p(t) + jq(t)$ at the output of filter of frequency gain $H_e(f)$ (noise-free), and taking into account the response of filtered noise turned into its equivalent baseband noise, i.e. $b_e(t)$, the response $s_{c,e}(t)$ is then:

$$s_{c,e}(t) = V \sum_n [a_n + ja'_n]\, [p(t - nT) + jq(t - nT)] + b_e(t)$$

For $t = t_k = kT$, $s_{c,e}(kT)$ is written:

$$s_{c,e}(kT) = V \sum_n [a_n + ja'_n]\, \{p[(k - n)T] + jq[(k - n)T]\} + b_e(kT)$$

b) Useful signal, intra-channel and inter-channel inter-symbol interferences.

We have successively:

$$\Re[s_{c,e}(kT)] = V\left\{a_k p(0) + \sum_{n\neq k} a_n p[(k-n)T] - \sum_{n} a'_n q[(k-n)T]\right\}$$
$$+b_{e,p}(kT)$$

$$\Im[s_{c,e}(kT)] = V\left\{a'_k p(0) + \sum_{n\neq k} a'_n p[(k-n)T] + \sum_{n} a_n q[(k-n)T]\right\}$$
$$+b_{e,q}(kT)$$

hence:

$$I_{m_l}(kT) = V\left[\underbrace{\sum_{n\neq k} a_n p[(k-n)T]}_{\text{ISI intra-channel}} - \underbrace{\sum_{n} a'_n q[(k-n)T]}_{\text{ISI inter-channel}}\right]$$

$$I'_{m'_l}(kT) = V\left[\underbrace{\sum_{n\neq k} a'_n p[(k-n)T]}_{\text{ISI intra-channel}} + \underbrace{\sum_{n} a_n q[(k-n)T]}_{\text{ISI inter-channel}}\right]$$

6) a) We have:

$$p(t) = F^{-1}\{P(f)\} = \frac{\sin[\pi\, t/T]}{\pi\, t/T}$$

$p(0) = 1$; $p(\pm iT) = 0\ \forall\, i \neq 0$; i: integer

$$q(t) = F^{-1}\{Q(f)\} = \frac{1}{8}\times p(t) = \frac{1}{8}\times\frac{\sin[\pi\, t/T]}{\pi\, t/T}$$

$q(0) = 1/8$; $q(\pm iT) = 0\ \forall\, i \neq 0$; i: integer

hence:

$$\Re[s_{c,e}(kT)] = \underbrace{Va_k}_{\substack{\text{Useful}\\ \text{signal}}} - \underbrace{V/8\times a'_k}_{\substack{\text{ISI}\\ \text{inter-channel}}} + \underbrace{b_{e,p}(kT)}_{\substack{\text{In-phase}\\ \text{noise}}}$$

$$\Im[s_{c,e}(kT)] = \underbrace{Va'_k}_{\substack{\text{Useful}\\ \text{signal}}} + \underbrace{V/8\times a_k}_{\substack{\text{ISI}\\ \text{inter-channel}}} + \underbrace{b_{e,q}(kT)}_{\substack{\text{In-quadrature}\\ \text{noise}}}$$

b) Expression of the optimal decision threshold $\mu_{0,e} = \mu_{0opt,e}$:

$$\mu_{0,e} = \frac{\sigma_e^2}{2Vp(0)} \log_e \left[\frac{p_{-1}}{p_1}\right] = \frac{\sigma_e^2}{2V} \log_e \left[\frac{p_{0,C2}}{p_{1,C2}}\right] = \frac{\sigma_e^2}{2V} \log_e \left[\frac{0.6375}{0.3625}\right] \cong 0.28 \frac{\sigma_e^2}{V}$$

c) Calculation of the conditional probabilities of the following types:

$Pr\{\hat{a}_k = a_j / a_k = a_i, m_l\}_{j \neq i}$

We have:

$$P_{e,a} = p_{-1} \left[\sum_{l=1}^{2} p_{m_l} \times P_{e_{-1}}\right] + p_1 \left[\sum_{l=1}^{2} p_{m_l} \times P_{e_1}\right]$$

with:

$$P_{e_{-1}} = P_{e(1/-1,m_l)} = \frac{1}{\sigma_e \sqrt{2\pi}} \int_{\mu_{0,e}+Vp(0)-I_{m_l}(kT)}^{\infty} \exp\left[-\frac{1}{2}\frac{b_{e,p}^2}{\sigma_e^2}\right] db_{e,p}$$

$$P_{e_1} = P_{e(-1/1,m_l)} = \frac{1}{\sigma_e \sqrt{2\pi}} \int_{-\infty}^{\mu_{0,e}-\left[Vp(0)+I_{m_l}(kT)\right]} \exp\left[-\frac{1}{2}\frac{b_{e,p}^2}{\sigma_e^2}\right] db_{e,p}$$

The signal-to-noise ratio is:

$$\left[\frac{S}{b}\right]_{dB} = 20 \times \log_{10}\left[\frac{Vp(0)}{\sigma_e}\right] = 20 \times \log_{10}\left[\frac{V}{\sigma_e}\right] = 12.0412 \text{ dB}$$

$$\rightarrow \frac{V}{\sigma_e} = 4 \rightarrow V = 4\,\sigma_e$$

Moreover, we have:

$m_l = a_k'$	$I_{m_l}(kT) = -V/8 \times a_k'$
-1	$V/8$
1	$-V/8$

Table 3.16. *Amplitudes of the inter-channel intersymbol interference*

Calculation of intervals and of conditional probabilities of error:

$m_l = a'_k$	$\mu_{0,e} + Vp(0) - I_{m_l}$	$P_{e_{-1}}$	$\mu_{0,e} - [Vp(0) + I_{m_l}]$	P_{e_1}
-1	$3.57\ \sigma_e$	1.82×10^{-4}	$-4.43\ \sigma_e$	2.2×10^{-5}
1	$4.57\ \sigma_e$	2×10^{-6}	$-3.43\ \sigma_e$	3×10^{-4}

Table 3.17. *Intervals and conditional probabilities of error*

d) We have:

$$P_{e,a} = p_{-1}\left[\sum_{l=1}^{2} p_{m_l} \times P_{e_{-1}}\right] + p_1\left[\sum_{l=1}^{2} p_{m_l} \times P_{e_1}\right]$$

or:

$$P_{e,a} = p_{-1}\left[p_{m_1} \times P_{e_{-1}} + p_{m_2} \times P_{e_{-1}}\right] + p_1\left[p_{m_1} \times P_{e_1} + p_{m_2} \times P_{e_1}\right]$$

with:

$$p_{-1} = p_{0,C2}\ ;\ p_1 = p_{1,C2}\ ;\ p_{m_1} = p_{-1}\ ;\ p_{m_2} = p_1$$

hence:

$$P_{e,a} = p_{0,C2}\left[p_{0,C2} \times P_{e_{-1}} + p_{1,C2} \times P_{e_{-1}}\right] + p_{1,C2}\left[p_{0,C2} \times P_{e_1} + p_{1,C2} \times P_{e_1}\right]$$

$$\rightarrow P_{e,a} = 0.6375[0.6375 \times 1.82 \times 10^{-4} + 0.3625 \times 2 \times 10^{-6}]$$

$$+0.3625[0.6375 \times 2.2 \times 10^{-5} + 0.3625 \times 3 \times 10^{-4}] \cong 1.1893 \times 10^{-4}$$

We evidently have: $P_{e,a'} = P_{e,a} = P_{e,b}$.

7) The probability P_d that a component d of displacement vector $\vec{D}$ be wrong decoded is:

$$P_d = \sum_{i=1}^{7} p(s_i) \times P_{e,d_i}$$

With: P_{e,d_i}, the probability that d_i represented by s_i be erroneously decoded. But the correct decoding of s_i is such that:

$$\left(1 - P_{e,d_i}\right) = \left(1 - P_{e,b}\right)^{l_i} \rightarrow P_{e,d_i} = 1 - \left(1 - P_{e,b}\right)^{l_i}$$

As we have:

$$P_{e,b} \ll 1 \rightarrow \left(1 - P_{e,b}\right)^{l_i} \cong 1 - l_i \times P_{e,b}$$

$$\rightarrow P_{e,d_i} \cong l_i \times P_{e,b}$$

then:

$$P_d \cong P_{e,b} \times \sum_{i=1}^{7} p(s_i) \times l_i = P_{e,b} \times \overline{l_2} \cong 1.1893 \times 10^{-4} \times 2.4875$$

$$= 2.9583 \times 10^{-4}$$

Thus, the probability P_D that the displacement vector $\vec{D}$ be erroneously decoded is:

$$(1 - P_D) = (1 - P_d)^2 \rightarrow P_D = 1 - (1 - P_d)^2 \cong 1 - (1 - 2P_d) \cong 2P_d$$

$$\cong 5.9167 \times 10^{-4}$$

PART 2

Practical Works

4

Study of the Transmission of Digital Information on Two-wire Cables

4.1. Introduction

The electrical transmission of information in digital form is not a recent phenomenon and it has even preceded the transmission of information in analog form, however it is only recently that digital transmissions have grown considerably. They used and still use, as a transmission medium, the important and classical infrastructure provided by the telephone and telegraph network of Telecommunications, infrastructure that has the advantage of existing.

However, the switched telephone network (the telex network being limited to very low modulation speeds) designed for the analog transmission of speech, only allocates a 300 to 3,400 Hz frequency band. This 3,100 Hz bandwidth allows, at the current time and under normal conditions, the transmission of digital information from 9,600 bit/s up to roughly 38 or even 56 Kbit/s. These limitations were due to amplitude and phase distortions and to the noise present in the transmission channel.

Developments in teleinformatics and the Internet (transmission of data, documents, still images, digital video-phone or television programs), the difficulties of performing correct equalization encountered in analog transmissions, while the needs of the telephone and its applications are also increasing, have led to the study and development of a high-speed universal network of digital transmissions of analog and digital information.

The study and development of such transmission systems is very complex. We will limit ourselves here to the study of the transmission medium that is the symmetrical twisted two-wire cable used in line equipment for digital transmissions at 2.048 Mbit/s, and also to the on-line codes necessary for the transmission.

This system is known as DT1 (first order digital transmissions in the European hierarchy of digital transmission systems, able to transmit 30 or 60 digital phone lines, each of them at 64 or 32 Kbit/s), the basic system for building digital multiplexes of higher orders:

- DT2: 8.448 Mbit/s, or 4 x DT1 or 120 or 240 phone lines;
- DT3: 34 Mbit/s, or 4 x DT2 or 480 or 960 phone lines;
- DT4: 140 Mbit/s, or 4 x DN3 or 1,920 or 3,840 phone lines;
- etc.

4.2. Recall of essential results on transmission line theory

When the dimensions of the cables (lengths) are comparable to the wavelength λ, one then deals with distributed constant circuits. The propagation time in this case is no longer negligible relative to the period of the wave that propagates on the structure. The theory of lines allows the study of these circuits.

It is shown that at any location on the line, along the axis of propagation z and at any time, the voltage $v(z)$ is:

$$v(z) = Ae^{-\gamma z} + Be^{\gamma z} = v_i(z) + v_r(z)$$

with:

- $\gamma = \alpha + j\beta$ the exponent of propagation (normalized per unit length);
- α, β the attenuation and the phase constants;
- A and B a pair of constants to be determined by the boundary conditions;
- $v_i(z)$ and $v_r(z)$ are the incident and the reflected voltages, respectively.

Moreover, the linear exponent of propagation γ and the characteristic impedance Z_c of the line are related to the distributed parameters of the line (r , l , c , g) with:

r: series resistance per unit length, for both conductors, in Ω/m

l: series inductance per unit length, for both conductors, in H/m

c: shunt capacitance per unit length, in F/m

g: shunt conductance per unit length, in S/m

by:

$$\gamma = \sqrt{z \times y} \text{ and } Z_c = \sqrt{z/y}$$

where:

$z = r + jlw$ is the impedance of the line per unit length

and:

$y = g + jcw$ is the admittance of the line per unit length

For a line of low losses (case in practice and neglecting the dielectric losses), we have:

$$\alpha \cong \frac{1}{2} r \sqrt{\frac{l}{c}} = k\sqrt{f}$$

due to the "skin effect", k is a constant and f is the working frequency.

$$\beta \cong w\sqrt{lc} \text{ ; } w = 2\pi f$$

is the pulsation;

$$Z_c \cong \sqrt{\frac{l}{c}} \text{ and } v_\varphi = \frac{w}{\beta} \cong \frac{1}{\sqrt{lc}} \cong \frac{v_l}{\sqrt{\varepsilon_r}}$$

is the phase velocity or propagation of the wave, where:

- v_l is the velocity of light;
- ε_r is the relative permittivity of the insulation used in the cable.

4.3. Practical study

There are two cables of different colors and with different propagation characteristics:

- gray cable: $v_\varphi = 2 \times 10^8$ m/s;
- black cable: $v_\varphi = 2.52 \times 10^8$ m/s.

4.4. Objectives

The most important parameters to know when you want to develop a cable transmission system include:

– the characteristic impedance $Z_C(\Omega)$;

– the attenuation α (dB/m ou dB/Km);

– the maximum possible bitrate.

In these practical works, you will try to determine by some adequate measurements the values of these important parameters on the cable sections that are at your disposal.

4.5. Measurement of the characteristic impedance Z_c by a reflectometry method (Time Domain Reflectometry: TDR)

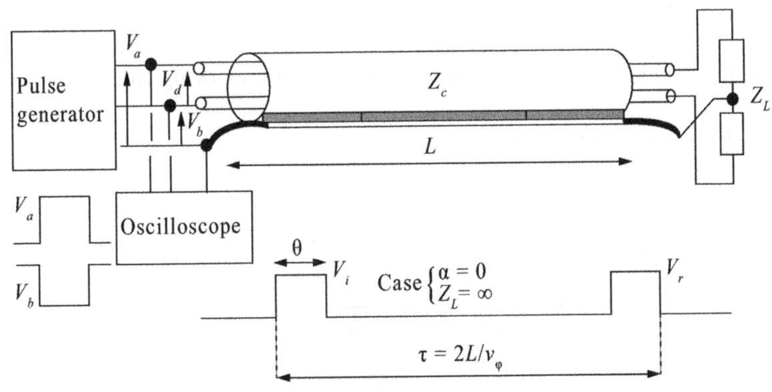

Figure 4.1. *Practical implementation of the TDR method*

To determine the characteristic impedance of the cable, the load Z_L is varied until the observed reflected pulse v_r is canceled (Figure 4.1), hence:

$$Z_C = Z_L.$$

Measure the characteristic impedance Z_C by the reflectometry method on the 10 m or 20 m cable section. In your opinion, will the measurement of Z_C be correct on a section of length >> 20 m? If not, why?

The same scheme is also used to measure the length L of the cable knowing the phase velocity v_φ. Indeed, measuring τ (for $Z_L = \infty$ preferently), we have:

$$2L = \tau \times v_\varphi$$

You have to measure the lengths of the three available cable sections.

4.6. Measurement of attenuation α as a function of frequency

When the cable is matched (the cable is loaded with its characteristic impedance), we then have:

– $v(z) = v_i(z) = A\exp(-\gamma z)$;

– $|v(z)| = A\exp(-\alpha z)$;

– $|v(0)| = A$: voltage delivered by the source generator at the input of the cable;

– $|v(L)| = A\exp(-\alpha L)$: voltage measured at the output of the cable (for $Z_L = Z_C$).

These two measurements make it possible to determine $\alpha(f)$:

$$\alpha(f) = \frac{1}{L} 20 \times \log_{10} \frac{|v(0)|}{|v(L)|} \quad (\text{dB/m})$$

Measure the variation of attenuation as a function of frequency.

Measurements will be made for the following values of the frequency f(KHz): 40; 60; 100; 200; 300; 500; 700; 1,000; 1,200; 1,500; 3,000 KHz; and for the cable section of median length.

Draw the curve $\alpha(f)$. Check that the measured law $\alpha(f)$ follows the theoretical law $\alpha(f)$.

4.7. Variation of the attenuation α as a function of length

For the frequency $f = 100$ KHz, determine the variation of the attenuation as a function of the length (for the three available sections). Deduce the 100 kHz attenuation produced by a cable length of 1,800 m (nominal length of a section for a DT1 two-wire cable).

4.8. Measurement of the bitrate D (bit/s)

The measurement is done in pulsed mode (see Figure 4.2), when the cable is matched (loaded with its characteristic impedance).

For each pulse duration $\theta(ns)$: 100; 200; 300; 400; 500 *ns* transmitted and using the pulse generator in dual pulse mode (the second pulse is transmitted with a delay τ adjustable with respect to the first), the time τ_{min} which must separate these double pulses will be determined so that the response due to the first pulse lowers to 50% of its maximum amplitude before the appearance of the response due to the second transmitted pulse.

Draw the curve: $\tau_{min} = f(\theta)$.

With this rather simplistic criterion, which does not take into account any signal processing (equalization) performed both in the transmitter and the receiver units, draw a line $\tau = 2\theta$ (case of the binary RZ code) for deducing the maximum bitrate (see Figure 4.2).

NOTE.– Measurements will be made on the medium length cable section (# 600 m).

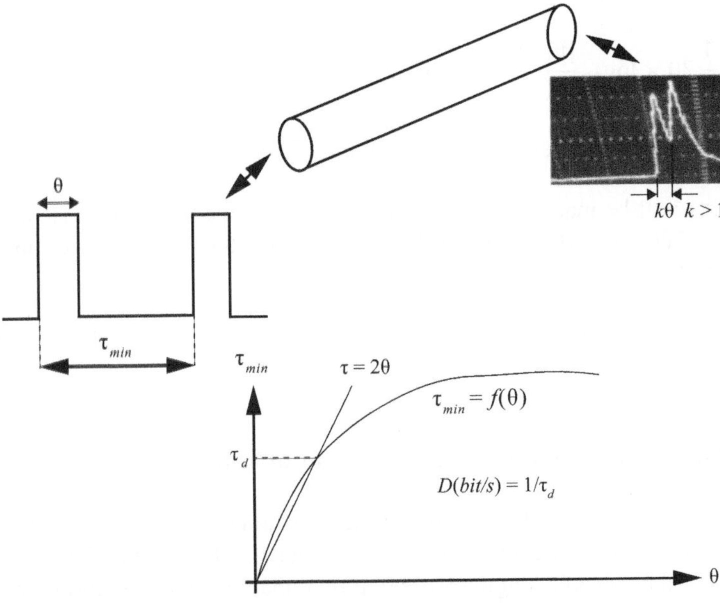

Figure 4.2. *Bitrate measurement scheme*

5

Study of Baseband Digital Transmission Systems for the Transmission of Analog Signals (Transmitter and Receiver)

5.1. Objectives

The goals of this lab work are to study the various functions necessary for serial transmission under digital form of analog signals. These are distributed in the transmitter on the one hand, and in the receiver on the other hand. Although most of them are present simultaneously in both of them, the receiver has some additional specific functions.

The study will largely be of a qualitative nature but will also include some measures allowing a quantitative assessment of the influence of certain parameters.

This study first deals with a pulse amplitude modulation transmission system with time division multiplexing. Then, the operation of a pulse code modulation (PCM) transmission system is analyzed. For these two systems, we will study in particular the two mechanisms of recovering the frame synchronization and the clock without additional links between transmitter and receiver other than that of transmission of the on-line code. In addition, for PCM transmission, codes for protection against transmission errors (error detection code, error detection and error correction code) are also used.

5.2. First part – Study of a pulse amplitude modulation and time division multiplex signal transmission system

A MODICOM2 educational electronic board from the company LJ Technical Systems (the reader must simply see the block diagram on the board) is at your disposal. It has two distinct parts: the transmitter and the receiver.

Its main features are as follows:

– time multiplexing of four channels, each associated with a signal on transmission and demultiplexing on reception;

– four sinusoidal signal generators, all synchronized, of frequencies $f_0 = 250$ Hz, 500 Hz, 1 KHz and 2 KHz respectively, and of variable amplitudes;

– a pure sampler with variable pulse duration (10 values $\theta_i = iT/10$ where T is the time unit);

– three operating modes corresponding to the interconnection signals between transmitter and receiver;

– a system for recovering the clock and a frame locking system in reception;

– four interpolation filters on reception.

Power is to be supplied to the electronic board $(+5\,V, 1\,A;\ \pm 12\,V, 1\,A)$.

The sampling frequency f_e of each of the four channels is fixed at 16 KHz ($T_e = \text{TX CH.0} = 62.5\ \mu\text{s}$) and therefore a unit of time T allocated to the transmission of a channel is equal to: $T = T_e/4 = \text{TX CLOCK}$.

The three operating modes correspond to three compromises between the number of connections between the transmitter and the receiver, the complexity of the receiver and the time required for the transmission of the signals:

– *mode 1*: three connections between transmitter and receiver; four analog channels; minimum receiver complexity:

TX CH.0 → RX CH.0

TX CLOCK → RX CLOCK

TX OUTPUT → RX INPUT

– *mode 2*: two connections between transmitter and receiver; four analog channels; regeneration of the clock by phase-locked loop on reception:

TX CH.0 → PLL I/P

TX OUTPUT → RX INPUT

– *mode 3*: a single connection between transmitter and receiver; three analog channels and a synchronization channel; regeneration of the clock by phase-locked loop and for the synchronization of the channels on reception:

TX OUTPUT → RX INPUT

5.2.1. *Experimental study*

IMPORTANT NOTE.– In all experiments, you will only power the boards (± 12 volts; 5 volts) when the wiring has been fully completed and checked.

In this lab work, you will study qualitatively and quantitatively the operation of this system of transmission of several signals by time multiplexing and pulse amplitude modulation according to its three modes of operation.

5.2.1.1. *Mode 1 of operation*

The parameter θ (pulse duration) is fixed at $T/2$ and the amplitudes of the four input signals are set to their maximum value, each of them being sent to one of the four inputs of the analog multiplexer of the transmitter.

Display the different signals available and the TX OUTPUT signal at the output of the transmitter by changing the amplitude of the analog signals. Simultaneously display the input and output of the interpolation filter corresponding to the channel receiving the signal at frequency: $f_0 = 250$ Hz.

By setting the maximum amplitude of the signal s_0 ($f_0 = 250$ Hz) on transmission, measure the output amplitudes of the interpolation filter for the nine possible values of the parameter θ (duty cycle control). Repeat the same measurements for the other three channels:

s_1 ($f_1 = 500$ Hz); s_2 ($f_2 = 1\,000$ Hz); s_3 ($f_3 = 2\,000$ Hz)

5.2.1.2. *Mode 2 of operation (without separate transmission of the clock)*

Make the following connections:

TX CH.0 → PLL I/P

SYNC → RX CH.0

CLK → RX CLOCK

Set the phase-lock loop switch to the PLL I / P position. Under these conditions, the phase-lock loop locks on the signal coming from the channel 0: TX CH.0 and generates two output signals:

SYNC → channel synchronization signal

CLK → regenerated clock signal

Check that the four channels are correctly reassigned and that the four output signals are correctly restored.

In particular, you will modify the amplitude and frequency of the signal transmitted on channel 0. Explain your conclusions.

5.2.1.3. *Mode 3 of operation (one link only between the transmitter and the receiver)*

Channel 0 is assigned to frame synchronization. For this, the SYNC LEVEL is connected to the CH.0 input of the transmitter, the other three inputs receiving three of the four available signals.

The SYNC → RX CH.O and CLK → RX CLOCK connections are maintained, however the TX CH.0 → PLL I/P connection is removed. It is replaced internally by switching the switch of the phase-lock loop, by the RX INPUT input signal of the comparator connection.

– Display the TX OUTPUT signals and one of the input signals on channels 1 to 3. Check that the amplitude of the pulse on channel 0 of the TX OUTPUT signal varies with the amplitude of the SYNC LEVEL.

– Set the amplitude of the synchronization detection comparator to synchronize correctly in reception for a maximum amplitude of the SYNC LEVEL. Then reduce the amplitude of the SYNC LEVEL signal. What are you observing? Explain your conclusions.

– Conversely, for a maximum amplitude of the SYNC LEVEL, vary the comparator threshold (COMPARATOR THRESHOLD LEVEL). What are you observing? Explain the phenomenon observed.

– Determine the two constraints on the level of the synchronization channel and on the threshold of the comparator for this system to work correctly in mode 3.

IMPORTANT NOTE.– Depending on the mode chosen, the phase of the regenerated clock (or not) at the receiver varies compared to the signal received at the receiver. Also, and more particularly for pulse durations $\theta > 70\,\%$ of T, it will sometimes be necessary to set the potentiometer of the clock control circuit to a value allowing a good reception operation. For this, it suffices that the falling edges of the clock obtained at the output of the clock phase shifter (signal at point 33) are located at the center of the interval separating two pulses of the re-amplified received signal (signal at point 41).

5.3. Second part – Study of a pulse code modulation (PCM) signal transmission system and transmission error control (error detector code and error corrector code)

You have at your disposal an educational electronic board MODICOM 3 from the company LJ Technical Systems. It actually consists of two distinct boards: one concerns the transmitter (MODICOM 3/1), the other the receiver (MODICOM 3/2) (the block diagrams are displayed on each of the electronic boards).

The main features of these boards are:

– time multiplexing of two signal transmission channels;

– two transmission rates: one fast (FAST mode, the normal operation), the other slow (SLOW mode, for operation analysis with the LEDs). The transmission bitrates are 240 Kbit/s and 1 bit/s respectively.

The transmitter includes two sinusoidal signal generators of frequencies $f_0 = 1$ KHz and $f_1 = 2$ KHz, of variable amplitudes which are synchronized, usable for the fast operating mode. In addition, there are two continuous signals of variable amplitudes usable for both the fast and slow operation modes.

The transmission and reception system can work according to three modes of operation concerning the control of transmission errors:

– no error checking: digitization and coding of 7-bit useful signal samples;

– odd/even parity bit error checking: encoding of 6-bit useful signal samples, 1-bit error detection control at the receiver;

– Hamming code error control at the transmitter: coding of the 4-bit signal samples used and addition of 3 control bits, detection and correction of a single error at the receiver;

– multiple possibilities of introduction of errors at the sending or receiving phase for analyzing the performance of the error control modes (detection/detection and correction of a single error).

Moreover, as for the system studied in part 1, it also has three modes of operation corresponding to the number of interconnection links between transmitter and receiver, and therefore included:

– a clock recovery system on reception;

– a frame synchronization generation system in transmission and frame detection and locking in reception.

The system operates according to the following basic principle. Each input signal is sampled at 16 KHz (sample-and-hold) and digitized on 7 bits with 2-order sampling interleaving because there are two transmission channels. A binary information frame of length 15 is formed whose structure is as follows:

$$\underbrace{\overset{b_{13}}{C_6}C_5C_4C_3C_2\,C_1\,\overset{b_7}{C_0}}_{\text{Canal 0}}\;\underbrace{\overset{b_6}{C_6}C_5C_4C_3C_2\,C_1\,\overset{b_0}{C_0}}_{\text{Canal 1}}\;S\;\rightarrow$$

where:

– S is a frame synchronization bit, first transmitted;

– $C_6 \cdots\cdots C_0$ corresponds to the 7-bit word allocated to the transmission of the sample of channel 0 or channel 1. The transmission time of a frame is $1/16{,}000 = 62.5\ \mu s$ (in fast mode);

– according to the selected error control mode, we have:

- $C_6 \cdots\cdots C_0$ corresponds exactly to the 7 scan bits $D_6 \cdots\cdots D_0$ of the considered sample if no error check,

- $C_6 \cdots\cdots C_1$ corresponds to the 6 most significant bits of digitization $D_6 \cdots\cdots D_1$ of the sample and C_0 is the parity check bit added if parity control is chosen (even or odd),

- $C_6 \cdots\cdots C_3$ corresponds to the respectively 4 most significant bits $D_6 \cdots\cdots D_3$ of digitization of the sample and $C_2\ C_1\ C_0$ is the sub-word of control for the detection and correction of a single error by Hamming code if this one is chosen.

Regarding the frame synchronization bit, this bit can be generated at the transmitter so that, on reception, it can almost certainly be identified in the received binary data stream. In the mode of operation 2, the transmitter uses a pseudo-random generator generating the repetitive sequence "000 100 110 101 111" of which successively a bit will be used as synchronization bit S per transmitted frame. The receiver also has this pseudo-random sequence in this mode and thus locks on it.

Finally, mode 3 allows, by the use of a clock regeneration block in reception and simultaneously by the use of the frame synchronization mode of mode 2, to regenerate the transmitted signals only from the global signal of transmission (RX DATA OUTPUT).

Two error generation blocks separately allow the transmitter and the receiver to insert errors at different levels. Each of them has four possibilities associated with a SF switch.

In the transmitter, if:

– SF1 = 1 → the bit D_6 after digitization of each sample is forced to 0;

– SF2 = 1 → the bit C_6 after coding of each sample is forced to 1;

– SF3 = 1 → the bit C_5 after coding is correctly transmitted but the error protection bits (C_0 or $C_1\ C_2$) are generated considering that the bit D_5 was at 1;

– SF4 = 1 → the frame synchronization bit S is no longer generated by the pseudo-random generator but is modified deterministically.

In the receiver, if:

– SF1 = 1 → defect in the phase-lock loop of the clock regeneration block (between the output of the "EXCLUSIVE OR" and the low-pass filter);

– SF2 = 1 → defect in the frame synchronization detection block;

– SF3 = 1 → defect in the detection block and correction of transmission errors by Hamming code given by the following Table 5.1;

– SF4 = 1 → defect in reconstituting the channel 0 signal by interrupting the input of the sampling and hold amplifier.

Wrong bit	Error indicated	Corrected bit
None	C_1	None
C_0	C_3	D_3
C_1	C_1	None
C_2	C_5	D_5
C_3	C_3	D_3
C_4	C_6	D_6
C_5	C_5	D_5
C_6	C_6	D_6

Table 5.1. *Types of defects in the detection and correction block of transmission errors by Hamming code (SF3 = 1)*

5.3.1. *Experimental study*

The multiple possibilities of operating of this system, including the aspect of generation and control of transmission errors, imply a rather rich experimental study.

5.3.1.1. *Study without error of transmission and without protection code*

5.3.1.1.1. Mode 1 of operation (clock and frame synchronization are also provided to the receiver)

– In fast mode, display the signals to be transmitted, the digital signal transmitted and the signals restored in reception for different amplitudes. What do you observe for large amplitudes? Using continuous signals, determine the useful amplitude range avoiding the saturation of the converters and associated codes.

– In slow mode, examine the operation of the transmitter and the receiver in detail. In particular, the value of the frame synchronization bit, the times of modification of the transmitted signal and the transmission order of the bits associated with the representation of the samples will be specified.

5.3.1.1.2. Mode 2 of operation (only the clock is also provided to the receiver)

The frame synchronization generator is used on transmission and the frame synchronization detector on reception.

– In fast mode, check the correct operation of the entire system for different input signals. In particular, it will be verified that the indicator of the bit synchronization counter is on. What do you observe if you turn off the frame synchronization generator at the transmitter? Same question if you put it back into operation.

– In slow mode, examine in detail the operation of the frame synchronization part at the transmitter and at the receiver the frame synchronization detection.

5.3.1.1.3. Mode 3 of operation (only the transmitted signal is used as a link between transmitter and receiver)

Due to the mandatory use of the clock regeneration block in reception, only the fast mode is possible.

– Set the correct operation of the clock regeneration block by using a continuous signal on transmission and adjusting the "TRIM" of the voltage-controlled oscillator of the clock regeneration block so that the synchronization bit counter LED remains on for all possible amplitudes.

– Check for different amplitudes of the sinusoidal signals that are to be transmitted that they are correctly restored. Make the appropriate measures.

5.3.1.2. *Study with protection code against transmission errors*

5.3.1.2.1. Error-free

– In slow mode and in mode 2, analyze the operation with a detector code of a single error with even and then odd bit of parity.

– Similarly, analyze the operation with a Hamming code correcting a single error. In particular, for each of the words to be coded $D_6\ D_5\ D_4\ D_3$, determine the corresponding codeword $C_6 \cdots\cdots C_0$.

– In fast mode and mode 3, check that the complete system is working correctly with the Hamming code. Are the signals restored correctly? What precision do we get on the amplitudes?

5.3.1.2.2. With errors in the transmitter or in the transmission channel

– In fast mode and in mode 3, without error protection, what do you observe if you set SF2 to 1? Same question with an error detector code. Explain.

– In the context of the previous question, but with a Hamming code, what do you observe? Explain.

– Set SF1 at 1 and explain what you observe in the three possible situations: unprotected, detection of an error, detection and correction of a single error.

– Set SF3 at 1. Same question as before.

– Finally, by setting SF4 at 1 and in the three possible situations, what do you observe? What proposal do you make so that in this context, the receiver synchronizes again correctly?

5.3.1.2.3. With errors in the receiver

In the same way as in the previous part, you will test and explain the overall operation of the system with various error modes and depending on the modes of protection against errors.

– Set SF1 at 1 and in mode 3, what do you observe? Explain.

– Set SF2 at 1 and in mode 2 or 3, what do you observe? Explain.

– Set SF3 at 1 and with Hamming coding, what do you observe? Also set the SF2 transmitter to 1. Is the result normal?

– Finally, set SF4 at 1 (with SF2 set at 0 in the transmitter) and observe the analog outputs at the receiver.

6

Study of On-line Codes for Baseband Modulation and Carrier Modulation

6.1. Objectives

In this lab work, you will study the various on-line codes for representing the binary information for its transmission on a communication channel. Two sets of codes are studied. The first provides a baseband signal. This one can be sent as it is. It can also modulate a carrier. The various digital modulations with carrier are also available (second way of transmission) and will be studied in principle and technique, especially at the demodulation level.

6.2. Description of the electronic boards

The MODICOM 5 system consists of two electronic boards. The MODICOM 5/1 board includes the conditioning of data (on-line coding) into baseband signals and, if necessary, carrier modulation and thus corresponds to the transmission part. The MODICOM 5/2 board performs the reverse operations and therefore corresponds to the reception part (see the block diagrams on the boards concerned).

The data conditioning part in baseband signals (information-to-signal coding) allows for working with the following on-line codes:

– NRZ-L coding (level type);

– NRZ-M coding (Mark (differential) type);

– RZ coding;

– Biphase coding (Manchester code);

– Biphase-M coding (Mark);

– Bipolar coding (also called AMI coding: Alternation Mark Inversion);

– RB coding (Return-to-Bias);

– Quaternary differential coding.

The carrier modulation part performs the following modulation types:

– ASK modulation: Amplitude Shift Keying modulation;

– FSK modulation: Frequency Shift Keying modulation;

– PSK modulation: Phase Shift Keying modulation;

– QPSK modulation: Quaternary Phase Shift Keying modulation.

The boards MODICOM 5/1 and MODICOM 5/2 are inter-connectable upstream with the MODICOM 3/1 board for the transmitter part (MODICOM 5/1) and downstream with the MODICOM 3/2 board for the receiver part (MODICOM 5 / 2). The MODICOM 5/1 and 5/2 boards have signal generators with a carrier at 1.44 MHz for one, and a carrier at 960 KHz for the other. Moreover, the latter also has a version in quadrature (Q) relative to the other (I).

More precisely, the MODICOM 5/1 board comprises the following elements:

– two inputs: a TTL binary signal input (TX DATA INPUT) and associated clock input (TX clock INPUT);

– direct signal outputs associated with on-line codes and the associated clock;

– two unipolar-bipolar connection blocks that can be used to build bipolar codes;

– an inversion block.

In addition, it is possible to simultaneously output a group of two digits (usable for QPSK modulation). Two modulation blocks with carrier and a block of summation make it possible to carry out the various types of digital modulation. On each modulation block, the gain and offsets of the carrier and the modulating signal are adjustable.

Similarly, the MODICOM 5/2 board (Receiver) includes the following elements:

– a rectifier detector block for amplitude modulation (ASK);

– a frequency demodulation block (FSK) with phase-lock loop;

– a phase demodulation block (PSK) with quadratic phase lock loop;

– a quaternary phase demodulation block (QPSK) with phase-lock loop (with power 4);

– two low-pass filters;

– a system for regeneration of bipolar signals into unipolar signals with double thresholds (data squaring circuits);

– a differential decoder;

– a clock regeneration system for the biphase code.

6.3. First part – Study of on-line codes for baseband digital transmission

IMPORTANT NOTE.– In all experiments, you will only power the boards (±12 volts; 5 volts) when the wiring has been fully completed and checked.

When using boards MODICOM 3/1 and 3/2, these will be used in the following modes and configurations:

– selection of fast mode;

– selection of SYNC CODE DETECTOR mode;

– no detector and corrector error code mode (A = B = 0);

– no addition of transmission errors or defects.

6.3.1. *Experimental part*

From the MODICOM 3/1 board and continuous analog input for it, set the signal amplitude so that the binary word transmitted in serial form by the channel is the following:

$$D_6 \cdots\cdots D_0 = 0\,1\,0\,0\,0\,1\,1$$

for example. However, you can modify the information transmitted at your convenience.

– Study successively the NRZ-L, NRZ-M, RZ codes and represent their observed chronograms. How does the NRZ-M code compare to the classic NRZ-L code?

– Study the biphase code (MANCHESTER) and represent the observed chronogram. Explain how the Manchester code works. The next step is to decode the Manchester code. Why does a simple clock regeneration circuit not work?

The clock regeneration specific circuit will be used for the biphase code. What does it produce as a regenerated clock signal?

– Study the biphase-M code and represent the observed chronogram. Based on your observations, build the biphase-M decoder and explain how it works.

– Study the bipolar code (AMI code) and represent the observed chronogram. To build it, you will connect the NRZ-M output to the input of the unipolar-bipolar converter and the RZ output to the DISABLE input of the same converter. Study the decoding of the bipolar code and carry it out by using the "Data Squaring Circuits" system and by properly adjusting the two thresholds of the two comparators.

It will then be possible to transmit sinusoidal analog signals over the whole system (MODICOM 3/1; 5/1; 5/2; 3/2) using any of these codes.

6.4. Second part – Study of digital modulations with carrier

We will study successively the digital amplitude modulation and demodulation (ASK), the digital frequency modulation and demodulation (FSK), the digital phase modulation and demodulation (PSK).

In analysis mode, the information sent will be taken equal to "0 1 0 0 0 1 1" for example.

6.4.1. *Amplitude shift keying modulation (ASK)*

An amplitude modulation will be performed using a modulation block and a 1.44 MHz carrier. The mode of the MODICOM 5/1 board will be set to position 1 and the SYNC CODE GENERATOR of the MODICOM 3/1 board to "OFF". Perform amplitude demodulation as follows:

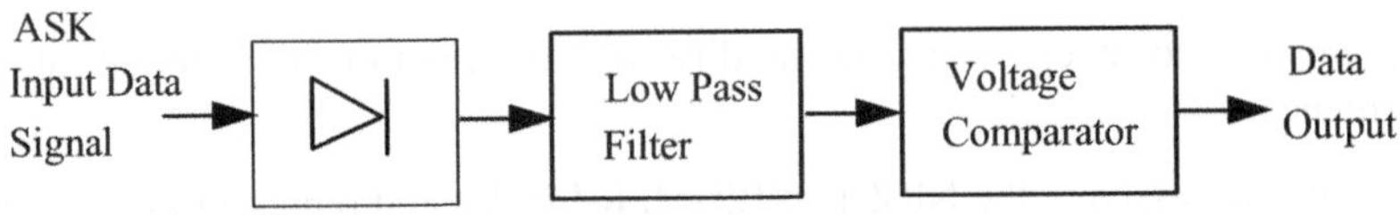

Figure 6.1. *Block diagram of the ASK demodulation*

The SYNC CODE GENERATOR of the MODICOM 3/2 board will be set to "ON" and the clock regeneration block will be used with an appropriate setting of "PULSE GENERATOR DELAY ADJUST".

You will adjust the gain and the offsets of the modulating signal and the carrier.

Examine the signals obtained at the output of each of the blocks of the demodulation system (the comparator threshold must be set correctly). Comments.

Change the amplitude of the transmitted signal (MODICOM 3/1) and check that the reconstructed signal is correct by setting the SYNC CODE GENERATOR of the MODICOM 3/1 board to “ON”.

6.4.2. *Digital frequency shift modulation (FSK)*

Perform a digital frequency modulation according to the block diagram of Figure 6.2.

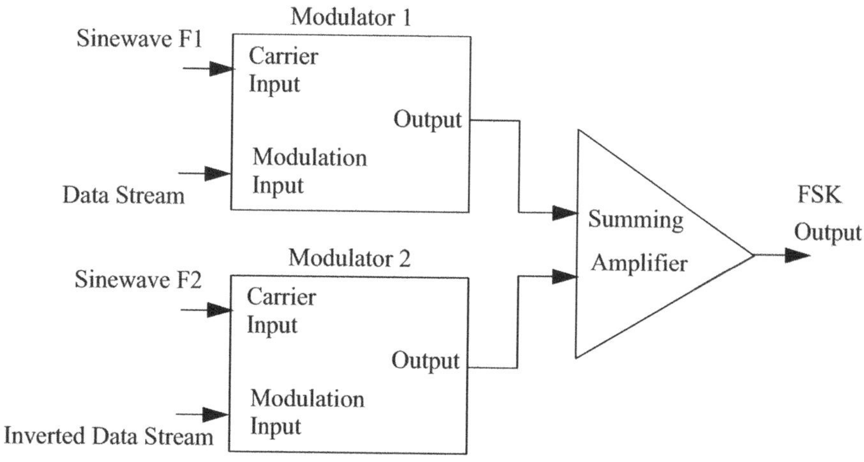

Figure 6.2. *Block diagram of FSK modulation*

In reception, the signal is decoded by means of a phase-locked loop, a low-pass filter and a comparator, according to the block diagram of Figure 6.3.

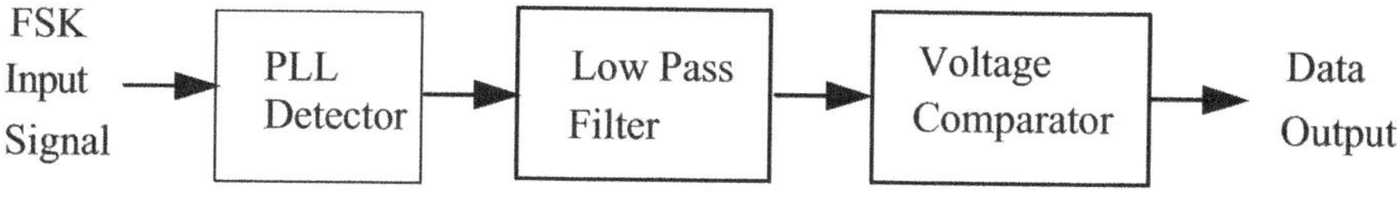

Figure 6.3. *Block diagram of digital FSK demodulation*

The two boards MODICOM 3/1 and MODICOM 3/2 will be used to generate the information to be transmitted with modulation and to reconstruct the signal on reception. Set the SYNC GENERATOR of the MODICOM 3/1 board to "OFF" and the SYNC CODE DETECTOR of the MODICOM 3/2 board to "ON".

– Analyze, for a configuration of information to be transmitted given by "0 1 0 0 0 1 1 1", the modulator and demodulator operations (you will adjust the two carriers used (960 KHz and 1.44 MHz) at the same amplitude. What do you observe before and after the low-pass filter? Adjust the threshold on the comparator so that the baseband signal is correctly reconstructed.

– By setting the SYNC CODE GENERATOR of the MODICOM 3/1 module to "ON", the system is fully operational for transmitting analog signals. Analyze the operation of the entire system.

6.4.3. *Phase shift keying modulation (PSK)*

Perform a digital phase modulation according to the block diagram of Figure 6.4.

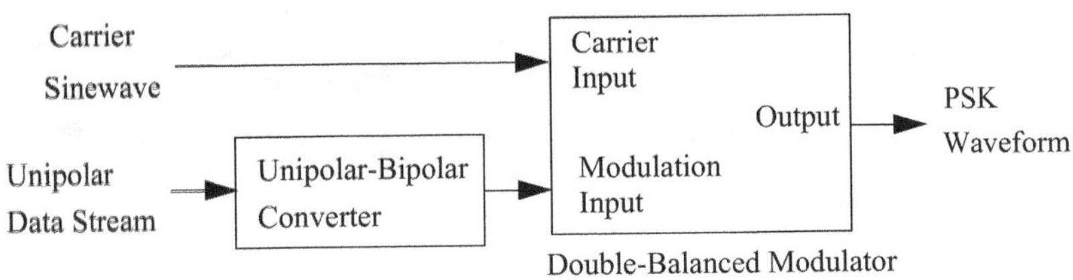

Figure 6.4. *Block diagram of a digital PSK modulation*

In reception, the signal is decoded in two steps.

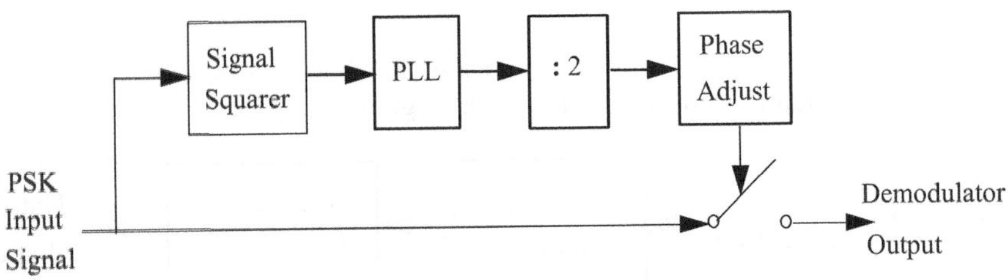

Figure 6.5. *Block diagram of a digital PSK demodulation*

A demodulator composed of four blocks performs a PSK demodulation (see Figure 6.5).

In a second step a low-pass filter and a comparator provide a logical signal.

– Show that if the code used at the input of the modulator is an NRZ-L code, there is ambiguity in the decoding.

– In order to remove this ambiguity, the code used will be an NRZ-M code and the complete scheme to be realized will be the following (Figure 6.6).

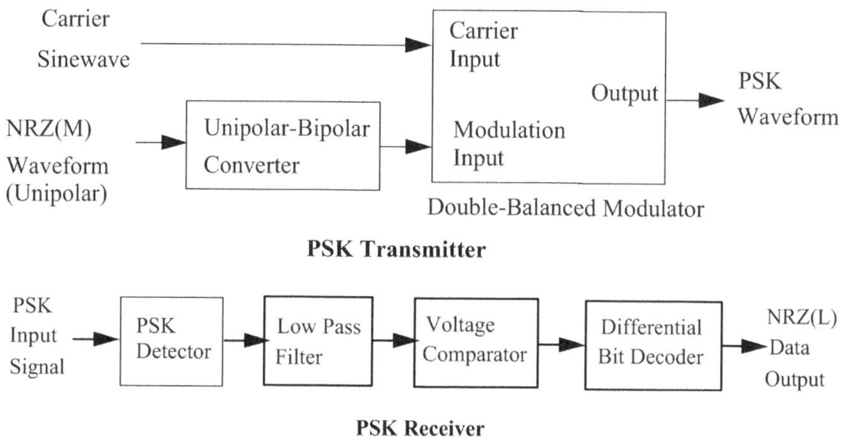

Figure 6.6. *Block diagrams of a digital PSK modulation and demodulation by using an NRZ-M code*

– A carrier at 960 kHz will be used. Using the word "0 1 0 0 0 1 1" to transmit, analyze the operation of the global system in detail. For that we will use an NRZ-L code at the beginning. Comments.

– To remove the ambiguity, use an NRZ-M code and analyze the transmission and reception.

Give your general conclusions on PSK modulation.

7

Study of a QPSK Modem Under MATLAB, Simulink, Communications and DSP

7.1. Objective

The objective of this lab work is the study, then the simulation under MATLAB, Simulink, Communications and DSP of a QPSK digital transmission modem (modulator and demodulator).

The complexity of telecommunications and signal processing systems has grown enormously in the last two decades.

The evaluation of performance on “hardware” prototypes is of course the best method to validate a concept, a structure or system. However, this approach is usually very time-consuming and expensive in terms of equipment. It is therefore understandable why the implementation phase intervenes only at the end of the development cycle.

The role of simulation is precisely to perform all these tests, at a lower cost both in terms of time and equipment.

The objective here is the study and the complete simulation of the proposed QPSK modem (see the block diagram in the Appendix).

The QPSK modem in question uses the following software modules, MATLAB, SIMULINK, COMMUNICATIONS and DSP:

– *MATLAB*:

- analysis, design, optimization;
- off-line data processing;

– *Simulink*:

- block diagram modeling;

- off-line simulation;

– *Communications Toolbox* (uses: SIMULINK and MATLAB): modeling and simulation of all the components of a digital or analog communication system "commlib";

– *DSP Blockset* (uses: MATLAB, SIMULINK, SIGNAL PROCESSING TOOLBOX): set of libraries specific to signal processing. They are used by SIMULINK "dsplib".

A MODEM (Modulator-Demodulator) is a system that modulates the baseband information at transmitter level and demodulates the received signal at receiver level to retrieve the information carried by a baseband signal.

The MODEM is decomposed into two essential functions:

– the transmission function (transmitter);

– the reception function (receiver).

The receiver consists of blocks that perform reverse functions to the transmitter blocks.

7.2. Required work

Under *Simulink*, open the "*mqpsk.mdl*" modem to view it, then execute the "*pmqpsk.m*" parameter file under MATLAB. The two files in question will be provided to you (see the different figures in the Appendix).

– Study very thoroughly the different blocks constituting the modem and simulate the whole for a time duration up to $1{,}000 \times T_b$. Observe and comment on the spatial diagram, the constellation diagram and the eye diagram. Then analyze the different intermediate signals.

– For different values of the signal-to-noise ratio (40 dB, 20 dB, 10 dB, 5 dB), qualitatively compare the results, the roll-off R parameter of the Nyquist filter being fixed at 0.5.

– Set the signal-to-noise ratio to 40 dB and vary the roll-off R parameter (R = 0.1, 0.3, 0.7 and 0.9). Analyze their influence on the demodulated signals (spatial diagram, constellation diagram and eye diagram).

– Replace the "QPSK map" block given in Figure 7.2 by that given in Figure 7.8 and redo the study done above.

Explain your conclusions.

7.3. Appendix: Diagrams of the QPSK modem and its different blocks

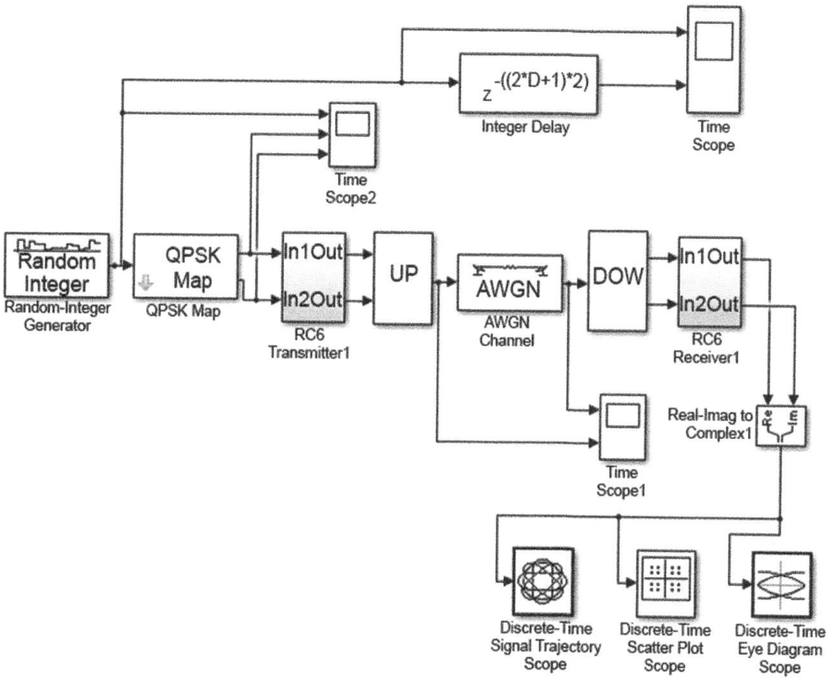

Figure 7.1. *Block diagram of a QPSK modem "mqpsk.mdl"*

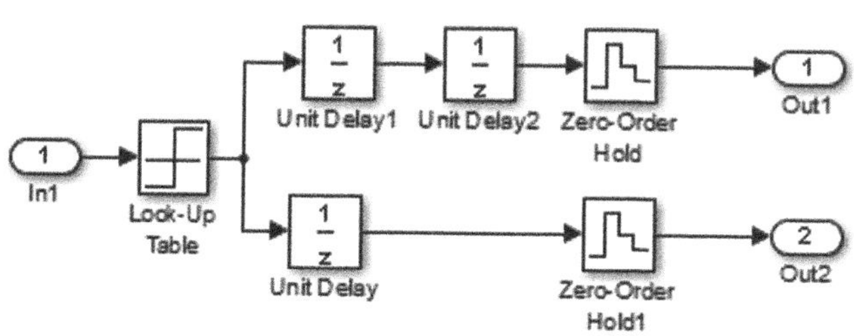

Figure 7.2. *Decomposition of function "Map QPSK"*

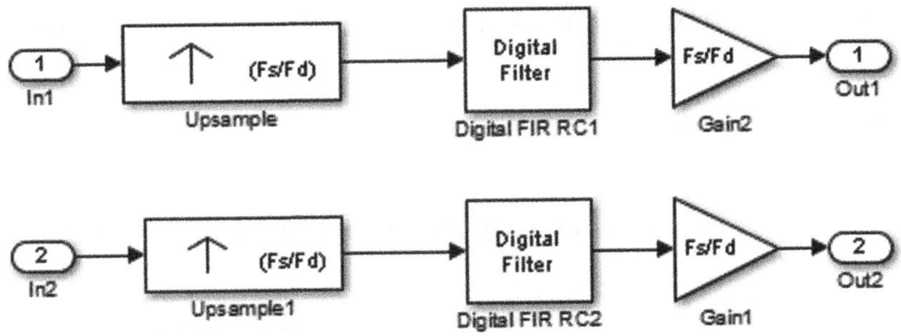

Figure 7.3. *Decomposition of function "transmitter1"*

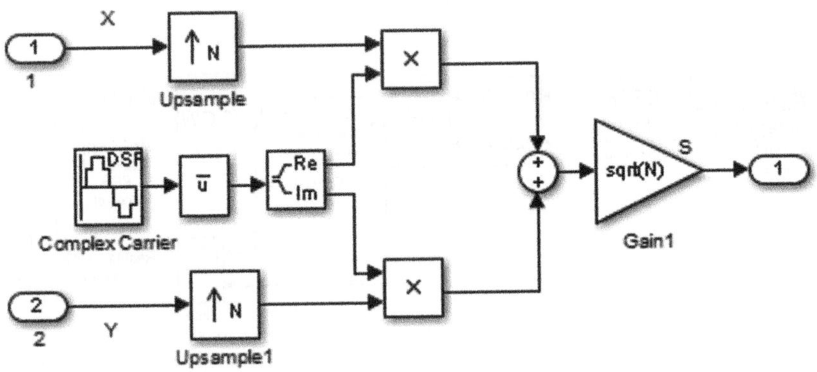

Figure 7.4. *Decomposition of function "up"*

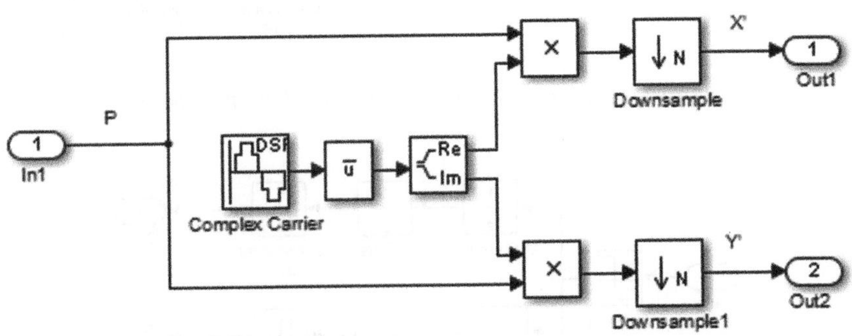

Figure 7.5. *Decomposition of function "down"*

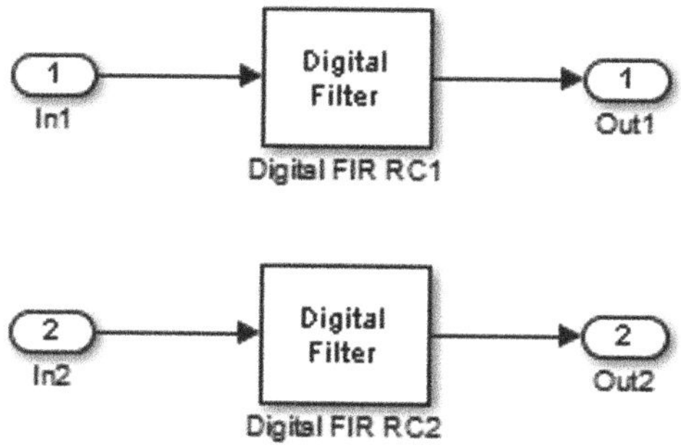

Figure 7.6. *Decomposition of function "receiver1"*

```
% pmqpsk.m
% Settings for mqpsk modem example
% Frequencies relations:
% Fs > Fc > Fd, with Fs >= 2Fc
% Fs: sampling frequency
% Fc: carrier frequency
% Fd: symbol frequency of input data
% Bit rate Fb=2*Fd (Hz) or bit/s
% Bit interval Tb =1/Fb s
% R: roll off factor
% D: delay of rcos FIR filter
% Ph: phase of the carrier
Fb = 50,000
Tb = 1/Fb
Td = 2*Tb
Fd = 1/Td
Fc = 2*Fd
Tc = 1/Fc
F s = 3*Fc
Ts = 1/Fs
Ph = 0
R = 0.5
D = 2
N = 1
```

Figure 7.7. *Parameter file "pmqpsk.m"*

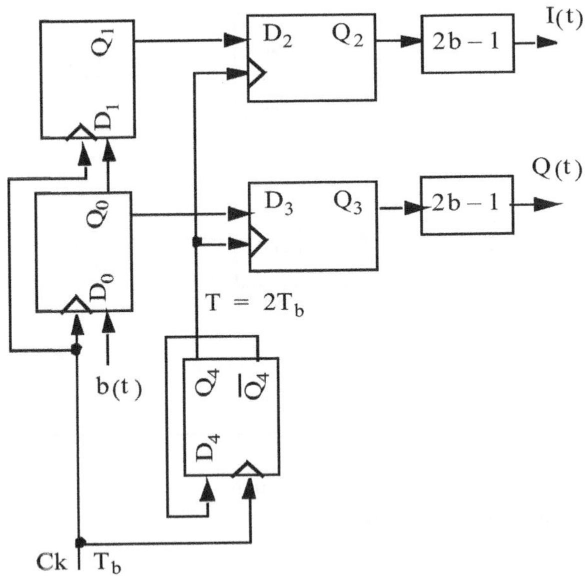

Figure 7.8. *QPSK map*

8

Study of a Coding and Decoding System by Cyclic Codes

8.1. Objective

The objective of this lab work is the study, then the complete simulation under MATLAB and Simulink, of a coding and decoding application using cyclic codes.

8.2. Recall of the principles of cyclic coding and decoding

Cyclic codes are block codes where the n symbols which constitute a word are considered as coefficient of a polynomial of degree $n-1$:

$$u^t = [u_{n-1}, u_{n-2}, \cdots, u_1, u_0] \Leftrightarrow u(x)$$
$$= u_{n-1}x^{n-1} + u_{n-2}x^{n-2} + \cdots + u_1 x + u_0$$

Any circular permutation on the symbols of a codeword gives a codeword:

$$[u_{n-1}, u_{n-2}, \cdots, u_1, u_0] \in C \Rightarrow [u_{n-2}, u_{n-3}, \cdots, u_0, u_{n-1}] \in C$$

The addition of two codewords is a codeword:

$$\forall\, u_i^t, u_j^t \in C \Rightarrow u_i^t + u_j^t \in C$$

The set of all the words of the code constitutes an *algebra*, while the set of words having a meaning constitutes an *ideal*.

The benefits of these codes are multiple:

– they are well suited to the detection of independent and packet errors;

– their implementation is easy, since the coding and decoding procedures can be made automatic by means of shift registers;

– their basic principle is based on the theory of polynomials and algebra.

8.3. Coding by division: systematic code

Let the information word represented by its polynomial:

$$i(x) = i_{m-1}x^{m-1} + \cdots + i_1x + i_0$$

The same word can be written:

$$\underbrace{0\,0\,\cdots 0}_{k}\,\underbrace{i_{m-1}\cdots i_1\,i_0}_{m}$$

Let us multiply $i(x)$ by x^k :

$$x^k i(x) = i_{m-1}x^{n-1} + \cdots + i_1x^{k+1} + i_0x^k$$

This shifts the word i of k positions to the left:

$$\underbrace{i_{m-1}\cdots i_1\,i_0}_{m}\,\underbrace{0\,0\,\cdots 0}_{k}$$

Now divide $x^k i(x)$ by $g(x)$ (whose degree is k):

$$x^k i(x) = g(x)q(x) + c(x) \quad \text{with} \quad d^\circ c(x) < k$$

$c(x)$ can be written:

$$\underbrace{0\,0\cdots 0}_{m}\,\underbrace{c_{k-1}\cdots c_1c_0}_{k}$$

Or again, the polynomial:

$$x^k i(x) + c(x) = g(x)q(x)$$

being a multiple of $g(x)$, is a polynomial of the code. The corresponding codeword consists of two disjoint sub-words:

$$\underbrace{i_{m-1} \cdots i_1\, i_0}_{m}\, \underbrace{c_{k-1} \cdots c_1 c_0}_{k}$$

This word has on the left the m bits of information. The following k bits are the control bits. The whole word of $m + k$ bits is a codeword.

In summary, we can code a word of information by:

1) multiplying it by x^k;

2) then dividing the result by $g(x)$. The remainder of the division provides the control bits.

8.4. Decoding by division: principle of calculating the syndrome

In general:

$$v(x) = u(x) + \varepsilon(x)$$

with:

– $v(x)$: received word;

– $u(x)$: codeword transmitted;

– $\varepsilon(x)$: possible error word.

The syndrome is:

$$s(x) = \text{Remainder}\left[\frac{v(x)}{g(x)}\right] = \text{Remainder}\left[\frac{\varepsilon(x)}{g(x)}\right]$$

If there have been errors, and if the erroneous received word $v(x)$ does not belong to the code, the division of the received word by $g(x)$ will give a non-zero remainder. For the error detection it is sufficient to add an "OR" gate whose inputs are the contents of the register.

8.5. Required work

Under Simulink, open the cyclic encoder-decoder: "*codec3f.mdl*" (provided to you, see block diagrams in the Appendix below) using a generator polynomial $g(x)$ of degree $k = 3$:

$$g(x) = x^3 + x + 1$$

– Study very thoroughly the different blocks constituting the encoder and the decoder.

– In the MATLAB workspace, set the following parameters:

$$n = 7, \qquad k = 3, \qquad T_b = 1$$

Then, in the Bernoulli generator of the transmission "channel" (Figure 8.4), set the "probability of a zero" parameter to 1.

Then run the simulation for up to $15 \times T_b$ (two codewords plus one bit). Analyze the results obtained for the encoder and the decoder.

– Calculate the codewords analytically and compare them with the result of the simulation.

– Now set in the Bernoulli generator of the transmission "channel", the "probability of a zero" parameter to 0.8 and start the simulation. Analyze the results obtained for the encoder and the decoder.

– Realize and test under Simulink the linear feedback shift register encoder using the preceding generator polynomial $g(x)$ given in Figure 8.7, and the associated decoder given in Figure 8.8.

– Realize and test under Simulink (see Volume 1, Chapter 4):

- the following generator polynomial: $g_1(x) = x^5 + x^2 + 1$ as a pseudo-random generator, starting from a non-zero initial state of the register.

- the following generator polynomial: $g_2(x) = x^5 + x^4 + x^3 + x^2 + 1$ as a pseudo-random generator, starting from a non-zero initial state of the register.

- the Gold generator polynomial $g_3(x)$ based on the two preceding generators $g_1(x)$ and $g_2(x)$ (which are preferred pairs).

What is the length n of the sequences generated?

What is the number of sequences generated?

Conclusions.

8.6. Appendix: Block diagrams

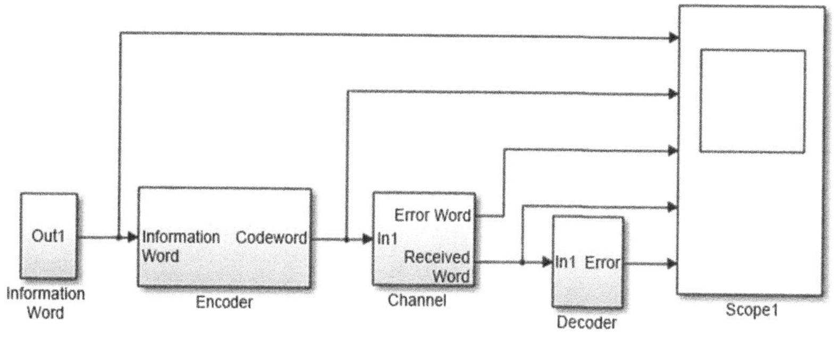

Figure 8.1. *Codec3f: Cyclic coding and decoding*

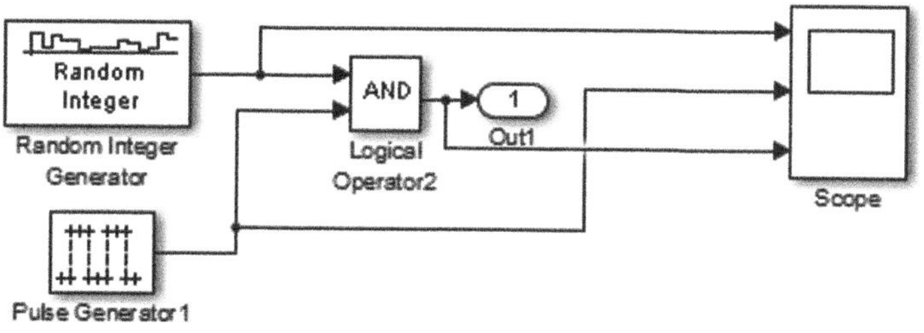

Figure 8.2. *Generation of the word information*

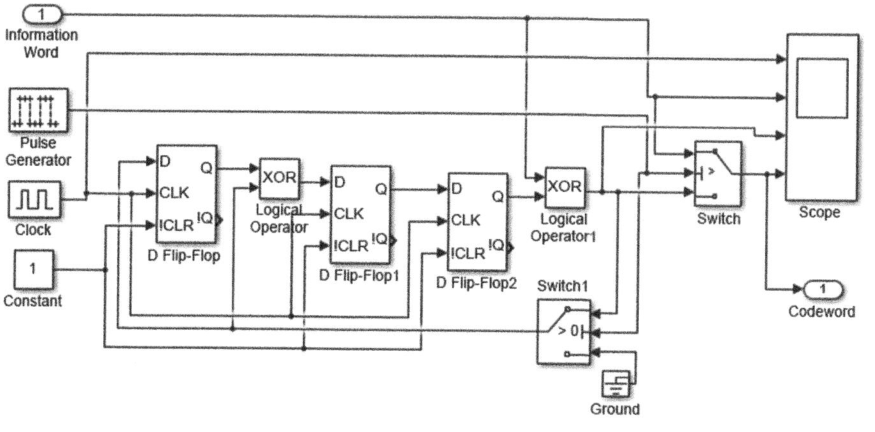

Figure 8.3. *Encoding diagram*

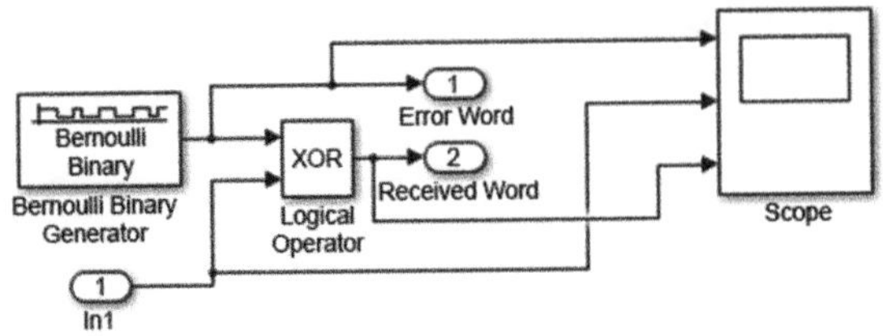

Figure 8.4. *Binary transmission channel generator*

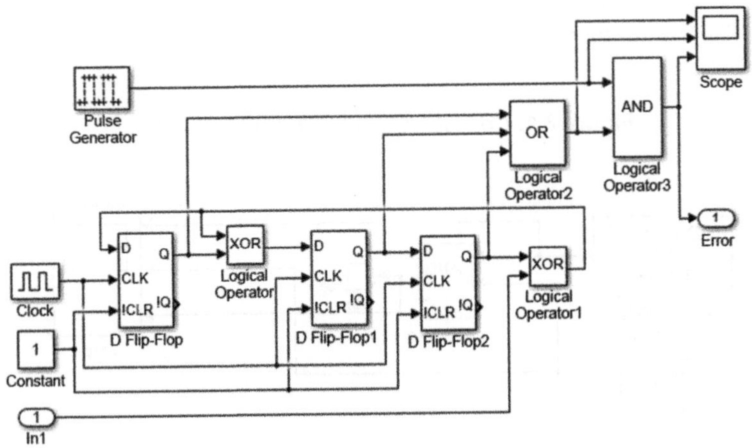

Figure 8.5. *Decoding diagram*

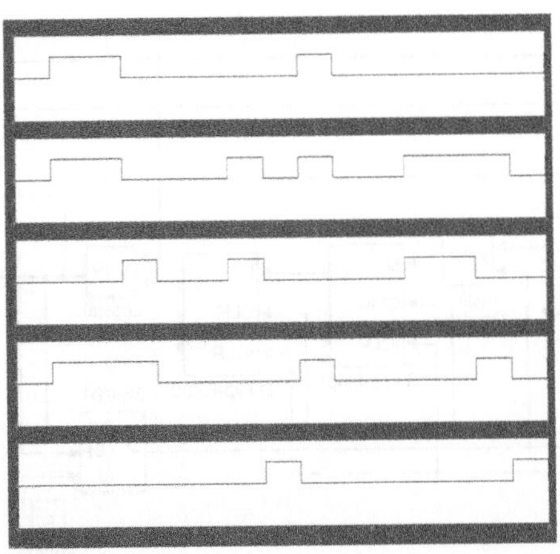

Figure 8.6. *Example of simulation results*

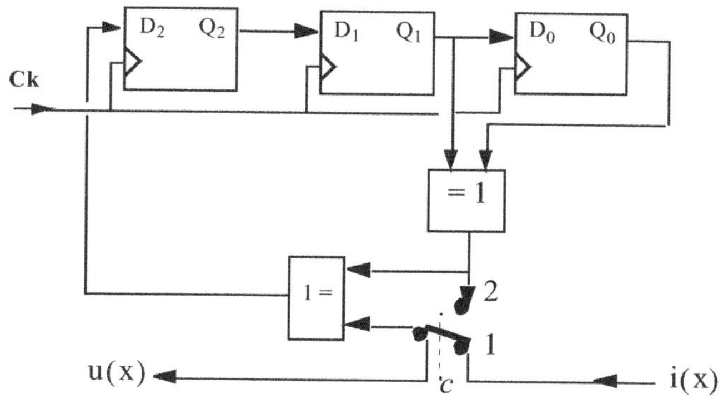

Figure 8.7. *Coder based on linear feedback shift register (LFSR)*

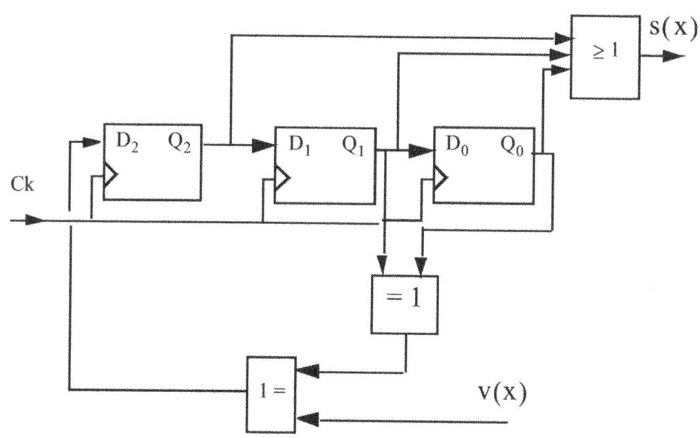

Figure 8.8. *Decoder based on linear feedback shift register (LFSR)*

References

Assaf, R., El Assad, S., Harkous, Y. (2010). *Simulation et égalisation adaptative des canaux de transmission*. Éditions universitaires européennes, Nantes.

Barba, D. (2019). Cours de théorie et transmission de l'information. Polytech, Nantes.

Baudoin, G. (2002). *Radiocommunications numériques 1*. Dunod, Paris.

Bic, J.C., Duponteil, D., Imbeaux, J.C. (1986). *Éléments de communications numériques*. Dunod, Paris.

Caragata, D., El Assad, S., Luduena, M. (2015). An improved fragile watermarking algorithm for JPEG images. *International Journal of Electronics and Communications*, 69, 1783–1794.

Charbit, M. (2003). *Systèmes de communications et théorie de l'information*. Hermes-Lavoisier, Cachan.

Combes, P.F. (1996). *Micro-ondes : lignes, guides et cavités, cours et exercices*. Dunod, Paris.

Csillag, P. (1990). *Introduction aux codes correcteurs*. Ellipses, Paris.

De Cannière, C., Preneel, B. (2006). Trivium specifications [Online]. Available at: http://www.ecrypt.eu.org/stream/p3ciphers/trivium/trivium_p3.pdf.

Dixon, R.C. (1994). *Spread Spectrum Systems with Commercial Applications*. John Wiley & Sons, Hoboken.

El Assad, S. (2001). Conception et simulation d'un modem MDP-8 cohérent et différentiel. *International Conference on Image and Signal Processing (ICISP 2001)*, Agadir, 845–852.

El Assad, S. (2019). Cours de communications numériques : bases et techniques. Polytech, Nantes.

El Assad, S., Barba, D. (2020). *Digital Communications 1: Fundamentals and Techniques*. ISTE Ltd, London, and John Wiley & Sons, New York.

El Assad, S. *et al.* (2003–2005). COBBALT : coopération ouest BAL en télécommunications. Internal report, Université virtuelle en Pays de la Loire.

El Assad, S. *et al.* (2006–2008). MASTER-FOAD, projet UVPL. Internal report, Université virtuelle en Pays de la Loire.

Fontolliet, P.G. (1983). *Systèmes de télécommunications*. Dunod, Paris.

Glavieux, A., Joindot, M. (1996). *Communications numériques*. Masson, Paris.

Guglielmi, M. (2004). *Signaux aléatoires : modélisation, estimation, détection*. Hermes-Lavoisier, Paris.

Haykin, S. (1988). *Digital Communications*. John Wiley & Sons, Hoboken.

MathWorks (2020). comm.KasamiSequence [Online]. Available at: https://fr.mathworks.com/help/comm/ref/comm.kasamisequence-system-object.html.

MacWilliams, F.J., Stoane, N.J.A. (1977). *The Theory of Error Correcting Codes*. Bell Laboratories, Murray Hill.

Maximov, A., Biryukov, A. (2007). Two trivial attacks on Trivium. In *Selected Areas in Cryptography*, Adams, C., Miri, A., Wiener, M. (eds). Springer-Verlag, Berlin, 36–55.

Paar, C., Oelzl, J. (2010). *Understanding Cryptography*. Springer-Verlag, Berlin.

Pätzold, M. (2002). *Mobile Fading Channels*. John Wiley & Sons, Hoboken.

Penaud, S. (2001). Étude des potentialités du chaos pour les systèmes de télécommunications : évaluation des performances de systèmes à accès multiples à répartition par les codes (CDMA) utilisant des séquences d'étalement chaotiques. PhD Thesis, University of Limoges.

Proakis, J.G. (1995). *Digital Communications*. McGraw-Hill, New York.

Roubine, E. (1970). *Introduction à la théorie de la communication, tome III*. Masson, Paris.

Sanchez, J., Thioune, M. (2008). *UMTS*, 3rd edition. Hermes-Lavoisier, Cachan.

Servin, C. (2009). *Réseaux & télécoms*. Dunod, Paris.

Sklar, B. (1988). *Digital Communications*. Prentice Hall, Upper Saddle River.

Spataru, A. (1987). *Fondements de la théorie de la transmission de l'information*. Presses polytechniques et universitaires romandes, Lausanne.

Stein, M. (1991). *Les modems pour transmission des données*. Masson, Paris.

Tomasi, W. (1998). *Electronic Communications Systems*. Prentice Hall, Upper Saddle River.

Webb, W., Hanzo, L. (1994). *Modern Quadrature Amplitude Modulation*. Pentech Press, London.

Index

D

E, F

G, H

Other titles from

iSTE

in

Information Systems, Web and Pervasive Computing

2020

CLIQUET Gérard, with the collaboration of BARAY Jérôme
Location-Based Marketing: Geomarketing and Geolocation

DE FRÉMINVILLE Marie
Cybersecurity and Decision Makers: Data Security and Digital Trust

GEORGE Éric
Digitalization of Society and Socio-political Issues 2: Digital, Information and Research

SEDKAOUI Soraya, KHELFAOUI Mounia
Sharing Economy and Big Data Analytics

2019

ALBAN Daniel, EYNAUD Philippe, MALAURENT Julien, RICHET Jean-Loup, VITARI Claudio
Information Systems Management: Governance, Urbanization and Alignment

AUGEY Dominique, with the collaboration of ALCARAZ Marina
Digital Information Ecosystems: Smart Press

BATTON-HUBERT Mireille, DESJARDIN Eric, PINET François
Geographic Data Imperfection 1: From Theory to Applications

BRIQUET-DUHAZÉ Sophie, TURCOTTE Catherine
From Reading-Writing Research to Practice

BROCHARD Luigi, KAMATH Vinod, CORBALAN Julita, HOLLAND Scott, MITTELBACH Walter, OTT Michael
Energy-Efficient Computing and Data Centers

CHAMOUX Jean-Pierre
The Digital Era 2: Political Economy Revisited

COCHARD Gérard-Michel
Introduction to Stochastic Processes and Simulation

DUONG Véronique
SEO Management: Methods and Techniques to Achieve Success

GAUCHEREL Cédric, GOUYON Pierre-Henri, DESSALLES Jean-Louis
Information, The Hidden Side of Life

GEORGE Éric
Digitalization of Society and Socio-political Issues 1: Digital, Communication and Culture

GHLALA Riadh
Analytic SQL in SQL Server 2014/2016

JANIER Mathilde, SAINT-DIZIER Patrick
Argument Mining: Linguistic Foundations

SOURIS Marc
Epidemiology and Geography: Principles, Methods and Tools of Spatial Analysis

TOUNSI Wiem
Cyber-Vigilance and Digital Trust: Cyber Security in the Era of Cloud Computing and IoT

2018

ARDUIN Pierre-Emmanuel
Insider Threats
(Advances in Information Systems Set – Volume 10)

CARMÈS Maryse
Digital Organizations Manufacturing: Scripts, Performativity and Semiopolitics
(Intellectual Technologies Set – Volume 5)

CARRÉ Dominique, VIDAL Geneviève
Hyperconnectivity: Economical, Social and Environmental Challenges
(Computing and Connected Society Set – Volume 3)

CHAMOUX Jean-Pierre
The Digital Era 1: Big Data Stakes

DOUAY Nicolas
Urban Planning in the Digital Age
(Intellectual Technologies Set – Volume 6)

FABRE Renaud, BENSOUSSAN Alain
The Digital Factory for Knowledge: Production and Validation of Scientific Results

GAUDIN Thierry, LACROIX Dominique, MAUREL Marie-Christine, POMEROL Jean-Charles
Life Sciences, Information Sciences

GAYARD Laurent
Darknet: Geopolitics and Uses
(Computing and Connected Society Set – Volume 2)

IAFRATE Fernando
Artificial Intelligence and Big Data: The Birth of a New Intelligence
(Advances in Information Systems Set – Volume 8)

LE DEUFF Olivier
Digital Humanities: History and Development
(Intellectual Technologies Set – Volume 4)

MANDRAN Nadine
Traceable Human Experiment Design Research: Theoretical Model and Practical Guide
(Advances in Information Systems Set – Volume 9)

PIVERT Olivier
NoSQL Data Models: Trends and Challenges

ROCHET Claude
Smart Cities: Reality or Fiction

SAUVAGNARGUES Sophie
Decision-making in Crisis Situations: Research and Innovation for Optimal Training

SEDKAOUI Soraya
Data Analytics and Big Data

SZONIECKY Samuel
Ecosystems Knowledge: Modeling and Analysis Method for Information and Communication
(Digital Tools and Uses Set – Volume 6)

2017

BOUHAÏ Nasreddine, SALEH Imad
Internet of Things: Evolutions and Innovations
(Digital Tools and Uses Set – Volume 4)

DUONG Véronique
Baidu SEO: Challenges and Intricacies of Marketing in China

LESAS Anne-Marie, MIRANDA Serge
The Art and Science of NFC Programming
(Intellectual Technologies Set – Volume 3)

LIEM André
Prospective Ergonomics
(Human-Machine Interaction Set – Volume 4)

MARSAULT Xavier
Eco-generative Design for Early Stages of Architecture
(Architecture and Computer Science Set – Volume 1)

REYES-GARCIA Everardo
The Image-Interface: Graphical Supports for Visual Information
(Digital Tools and Uses Set – Volume 3)

REYES-GARCIA Everardo, BOUHAÏ Nasreddine
Designing Interactive Hypermedia Systems
(Digital Tools and Uses Set – Volume 2)

SAÏD Karim, BAHRI KORBI Fadia
Asymmetric Alliances and Information Systems:Issues and Prospects
(Advances in Information Systems Set – Volume 7)

SZONIECKY Samuel, BOUHAÏ Nasreddine
Collective Intelligence and Digital Archives: Towards Knowledge Ecosystems
(Digital Tools and Uses Set – Volume 1)

2016

BEN CHOUIKHA Mona
Organizational Design for Knowledge Management

BERTOLO David
Interactions on Digital Tablets in the Context of 3D Geometry Learning
(Human-Machine Interaction Set – Volume 2)

BOUVARD Patricia, SUZANNE Hervé
Collective Intelligence Development in Business

EL FALLAH SEGHROUCHNI Amal, ISHIKAWA Fuyuki, HÉRAULT Laurent, TOKUDA Hideyuki
Enablers for Smart Cities

FABRE Renaud, in collaboration with MESSERSCHMIDT-MARIET Quentin, HOLVOET Margot
New Challenges for Knowledge

GAUDIELLO Ilaria, ZIBETTI Elisabetta
Learning Robotics, with Robotics, by Robotics
(Human-Machine Interaction Set – Volume 3)

HENROTIN Joseph
The Art of War in the Network Age
(Intellectual Technologies Set – Volume 1)

KITAJIMA Munéo
Memory and Action Selection in Human–Machine Interaction
(Human–Machine Interaction Set – Volume 1)

LAGRAÑA Fernando
E-mail and Behavioral Changes: Uses and Misuses of Electronic Communications

LEIGNEL Jean-Louis, UNGARO Thierry, STAAR Adrien
Digital Transformation
(Advances in Information Systems Set – Volume 6)

NOYER Jean-Max
Transformation of Collective Intelligences
(Intellectual Technologies Set – Volume 2)

VENTRE Daniel
Information Warfare – 2nd edition

VITALIS André
The Uncertain Digital Revolution
(Computing and Connected Society Set – Volume 1)

2015

ARDUIN Pierre-Emmanuel, GRUNDSTEIN Michel, ROSENTHAL-SABROUX Camille
Information and Knowledge System
(Advances in Information Systems Set – Volume 2)

BÉRANGER Jérôme
Medical Information Systems Ethics

BRONNER Gérald
Belief and Misbelief Asymmetry on the Internet

IAFRATE Fernando
From Big Data to Smart Data
(Advances in Information Systems Set – Volume 1)

KRICHEN Saoussen, BEN JOUIDA Sihem
Supply Chain Management and its Applications in Computer Science

NEGRE Elsa
Information and Recommender Systems
(Advances in Information Systems Set – Volume 4)

POMEROL Jean-Charles, EPELBOIN Yves, THOURY Claire
MOOCs

SALLES Maryse
Decision-Making and the Information System
(Advances in Information Systems Set – Volume 3)

SAMARA Tarek
ERP and Information Systems: Integration or Disintegration
(Advances in Information Systems Set – Volume 5)

2014

DINET Jérôme
Information Retrieval in Digital Environments

HÉNO Raphaële, CHANDELIER Laure
3D Modeling of Buildings: Outstanding Sites

KEMBELLEC Gérald, CHARTRON Ghislaine, SALEH Imad
Recommender Systems

MATHIAN Hélène, SANDERS Lena
Spatio-temporal Approaches: Geographic Objects and Change Process

PLANTIN Jean-Christophe
Participatory Mapping

VENTRE Daniel
Chinese Cybersecurity and Defense

2013

BERNIK Igor
Cybercrime and Cyberwarfare

CAPET Philippe, DELAVALLADE Thomas
Information Evaluation

LEBRATY Jean-Fabrice, LOBRE-LEBRATY Katia
Crowdsourcing: One Step Beyond

SALLABERRY Christian
Geographical Information Retrieval in Textual Corpora

2012

BUCHER Bénédicte, LE BER Florence
Innovative Software Development in GIS

GAUSSIER Eric, YVON François
Textual Information Access

STOCKINGER Peter
Audiovisual Archives: Digital Text and Discourse Analysis

VENTRE Daniel
Cyber Conflict

2011

BANOS Arnaud, THÉVENIN Thomas
Geographical Information and Urban Transport Systems

DAUPHINÉ André
Fractal Geography

LEMBERGER Pirmin, MOREL Mederic
Managing Complexity of Information Systems

STOCKINGER Peter
Introduction to Audiovisual Archives

STOCKINGER Peter
Digital Audiovisual Archives

VENTRE Daniel
Cyberwar and Information Warfare

2010

BONNET Pierre
Enterprise Data Governance

BRUNET Roger
Sustainable Geography

CARREGA Pierre
Geographical Information and Climatology

CAUVIN Colette, ESCOBAR Francisco, SERRADJ Aziz
Thematic Cartography – 3-volume series
Thematic Cartography and Transformations – Volume 1
Cartography and the Impact of the Quantitative Revolution – Volume 2
New Approaches in Thematic Cartography – Volume 3

LANGLOIS Patrice
Simulation of Complex Systems in GIS

MATHIS Philippe
Graphs and Networks – 2nd edition

THERIAULT Marius, DES ROSIERS François
Modeling Urban Dynamics

2009

BONNET Pierre, DETAVERNIER Jean-Michel, VAUQUIER Dominique
Sustainable IT Architecture: the Progressive Way of Overhauling Information Systems with SOA

PAPY Fabrice
Information Science

RIVARD François, ABOU HARB Georges, MERET Philippe
The Transverse Information System

ROCHE Stéphane, CARON Claude
Organizational Facets of GIS

2008

BRUGNOT Gérard
Spatial Management of Risks

FINKE Gerd
Operations Research and Networks

GUERMOND Yves
Modeling Process in Geography

KANEVSKI Michael
Advanced Mapping of Environmental Data

MANOUVRIER Bernard, LAURENT Ménard
Application Integration: EAI, B2B, BPM and SOA

PAPY Fabrice
Digital Libraries

2007

DOBESCH Hartwig, DUMOLARD Pierre, DYRAS Izabela
Spatial Interpolation for Climate Data

SANDERS Lena
Models in Spatial Analysis

2006

CLIQUET Gérard
Geomarketing

CORNIOU Jean-Pierre
Looking Back and Going Forward in IT

DEVILLERS Rodolphe, JEANSOULIN Robert
Fundamentals of Spatial Data Quality

Printed and bound by CPI Group (UK) Ltd, Croydon, CR0 4YY
16/09/2026
14963078-0006